Lecture Notes in Energy 18

For further volumes:
http://www.springer.com/series/8874

Lecture Notes in Energy

Lecture Notes in Energy (LNE) is a series that reports on new developments in the study of energy: from science and engineering to the analysis of energy policy. The series' scope includes but is not limited to, renewable and green energy, nuclear, fossil fuels and carbon capture, energy systems, energy storage and harvesting, batteries and fuel cells, power systems, energy efficiency, energy in buildings, energy policy, as well as energy-related topics in economics, management and transportation. Books published in LNE are original and timely and bridge between advanced textbooks and the forefront of research. Readers of LNE include postgraduate students and non-specialist researchers wishing to gain an accessible introduction to a field of research as well as professionals and researchers with a need for an up-to-date reference book on a well-defined topic. The series publishes single- and multi-authored volumes as well as advanced textbooks.

Brian Norton

Harnessing Solar Heat

Springer

Brian Norton
Office of the President
Dublin Institute of Technology
Dublin, Ireland

ISSN 2195-1284 ISSN 2195-1292 (electronic)
ISBN 978-94-007-7274-8 ISBN 978-94-007-7275-5 (eBook)
DOI 10.1007/978-94-007-7275-5
Springer Dordrecht Heidelberg New York London

Library of Congress Control Number: 2013949641

Printed on acid-free paper

Springer is part of Springer Science+Business Media (www.springer.com)

To Bahara, Pedram, Parisa and Arian

Preface

Systems engineered by man to harness solar heat in a controlled manner now include a diverse range of technologies each serving distinctive needs in particular climate contexts. This text covers the breadth of solar energy technologies for the conversion of solar energy to provide heat, either as the directly used output or as an intermediary to other uses such as power generation or cooling. It is a wholly updated, extended and revised version of *Solar Energy Thermal Technology* first published in 1992. The text draws on the author's own research and that of numerous colleagues and collaborators at Cranfield University, University of Ulster, Dublin Institute of Technology, Indian Institute of Technology Delhi and University of Nigeria.

The initial chapters deal with relevant fundamental aspects of solar energy meteorology, radiative heat transfer, material properties and energy storage. Solar energy collectors are discussed in detail before a set of chapters deal with each of the full range of applications. The early chapters consider: the solar energy resource; its distribution in geographical, spectral, skyward geometrical and temporal domains; the physics of solar energy absorption, transmission and loss at surfaces; and techniques for storing collected solar energy. Specific collector sub-systems are then discussed in Chaps. 7, 8 and 9. For each system, practical issues are discussed and a proven analytical procedure for predicting performance described. Similar analyses are presented in the concluding chapters on solar energy systems. These range from dryers to greenhouses to systems that render buildings solar energy systems in themselves and the associated design issues.

The context for any use of solar energy is the prevailing climate. This text, being global in scope, defines the most appropriate regions for particular technologies and applications. It is a research-orientated academic work citing publications on the peer-reviewed literature covering engineering and applied science topics intended for student use, as a reference tool for teaching solar energy as well as for those researching solar thermal applications in universities, industry or national/commercial laboratories. Insight into the challenges of implementation including practical constraints and operational considerations is provided to aid those undertaking feasibility studies, technical assistance, training assignments or operating testing facilities.

Harnessing Solar Heat

Brian Norton

Contents

Frequently-Used Nomenclature

A	Area, m^2
c	Specific heat capacity of the working fluid, $J\ kg^{-1}\ K^{-1}$
c	Velocity of light, $m\ s^{-1}$
C	Concentration ratio: ratio of aperture to absorber area of a concentrating collector
D	Layer thickness, m
D	Diameter, m
Dn	Dean number for the flow through a pipe bend
e	Electron charge
F_R	Heat-removal factor
g	Acceleration due to gravity, $m\ s^{-2}$
g	The fraction of the diffuse insolation exploitable by a concentrating collector
Gr	Grashof number
h	Heat transfer coefficient, $W\ m^{-2}\ K^{-1}$
h	Vertical distance m
I	Insolation Wm^{-2}
I_b	Direct beam component of total insolation, Wm^{-2}
I_d	Diffuse component of total insolation Wm^{-2}
K	Bailey number
1	Length, m
m	Fluid mass flow-rate, $kg\ s^{-1}$
K	Rate of heat transfer bv conduction
K	Thermal conductivity
L	Latent heat of vaporisation of water, $J\ kg^{-1}$
N	Refractive index
Nu	Nusselt number
n	Average number of reflection at the reflector, of solar radiation which reaches the absorber
Pr	Prandtl number
Q	Rate of energy output from a collector, $W\ m^{-2}$

r	Radius, m
R	Radius of the earth
Re	Reynolds number
R	Thermal resistance
T	Time
T	Temperature, K
T_a	Ambient air temperature, K
U	Overall heat-loss coefficient, W m^{-2} K^{-1}
W	Specific load ratio
X	Brooks number
Y	Heywood number
Z	Yellot number

About the Author

Professor Brian Norton is President of Dublin Institute of Technology. Previously, Professor Norton was Dean of Engineering and Built Environment at the University of Ulster and Professor of Built-Environmental Engineering. He has a B.Sc. (Hons) in Physics from the University of Nottingham; M.Sc. and Ph.D. degrees, in Engineering Experimentation and Applied Energy respectively, from Cranfield University; and a D.Sc. from the University of Nottingham. He is a Fellow of the Energy Institute, Engineers Ireland and Higher Education Academy, and a Chartered Engineer. Among his awards are the Napier Shaw Medal of the Chartered Institute of Building Services Engineers and the Roscoe Award of the Institute of Energy. He is an Honorary Fellow of the Chartered Institute of Building Services Engineers and a Fellow of the Irish Academy of Engineering. He is an Honorary Professor of both the University of Ulster and Harbin Institute of Technology, China and an Adjunct Professor at the University of Houston.

Chapter 1
Introduction

Here comes the Sun, it's all right

George Harrison (1969)

1.1 Key Contexts

The underlying imperatives driving use of solar heat include:

- Environmental sustainability: acceptance of an imperative to reduce both global greenhouse gas emissions and local air pollution with an appreciation that many energy reserves are finite
- Decentralised energy security: ameliorating reliance of centralised energy vulnerabile systems to geopolitics, price volatilities, natural disasters, disruption and attack.
- Commercial opportunities: there are gainful employment national and regional participation in solar heat manufacturing, services and provision.
- Economic rationality: in many contexts harnessing solar heat is the most viable option.

The applications of solar energy thermal systems include

- Heating systems, on a variety of scales, most often employing hot water or air.
- Large-scale high temperature solar power systems that generate electricity supplied to a grid.
- Solar heat driven systems for cooling, desalination, drying, cooking, water purification, refrigeration and industrial applications.

Systems that harness solar energy as heat should seek to provide an optimal life-cycle combination of:

- Have efficient solar energy conversion efficiency
- Match the temporal variation of the load diurnally and/or annually if necessary via the inclusion of energy storage

B. Norton, *Harnessing Solar Heat*, Lecture Notes in Energy 18,
DOI 10.1007/978-94-007-7275-5_1, © Springer Science+Business Media Dordrecht 2014

- Match the energy and exergy (i.e. temperature) of the load via, if required, energy storage or a heat pump
- Have low initial capital costs
- Have low running costs
- Incur low operational environmental impact
- Incur low embodied environmental impact
- Demonstrate high robustness, durability and safety

At a conceptual level, a solar thermal system enables solar energy collection, heated-fluid distribution, heat storage and control of heat withdrawal and/or circulation. Components can have single discrete functions or be inextricably coterminous with multiple roles. For example solar energy collection and storage are combined in applications as diverse as integral solar water heaters, non-convecting solar ponds and direct gain solar heating of buildings. Each system component's optimal specification will depend on the characteristics of the local solar energy resource, the ambient and load temperatures and temporal pattern of heat energy utilization.

At shorter time intervals (e.g., hourly or less), patterns of insolation, ambient temperature and the utilization of output energy (e.g., a pattern of hot water withdrawal from a solar energy water heater) are often stochastic, particularly in locations at high latitudes, in climates prone to tropical storms and in mountain areas. The variability of insolation patterns generally diminishes as the time-frames over which insolation is summed increase from hourly, daily, monthly, seasonal to annual. In certain sunnier climates, there are consistent solar energy patterns even over short intervals; though even in such locations attributes of the insolation such as the direct normal insolation on a solar tracking collector can change quickly. Load variations can be controlled to render them more predictable in particular individual applications.

Solar radiation has;

- Systematic variations; over familiar diurnal and annual cycles,
- Random variations; the major cause of which is the degree and character of cloud cover
- Low energy-density at higher latitudes; where it may only be just above 100 $W^{-2}m$ when averaged throughout the year, though about 900 Wm^{-2} at noon on a clear summer day;
- Often to be stored; to satisfy successfully many applications requirements for heat at night or during low-insolation periods.

Whilst solar energy and many heat loads have stochastic features, in contrast, collection, distribution and storage are more often deterministic systems acting as predictable intermediates between instantaneously less predictable inputs and outputs.

Solar thermal energy collectors fall into one of, at least, 15 generic types shown in Fig. 1.1. They are categorized by the extent to which they concentrate solar radiation and thus need to track the azimuthal path of the Sun across the sky.

The outcomes of component or system tests following internationally agreed rating procedures of solar thermal devices often inform component or system selection decisions. For small installations rated performance is often that of a

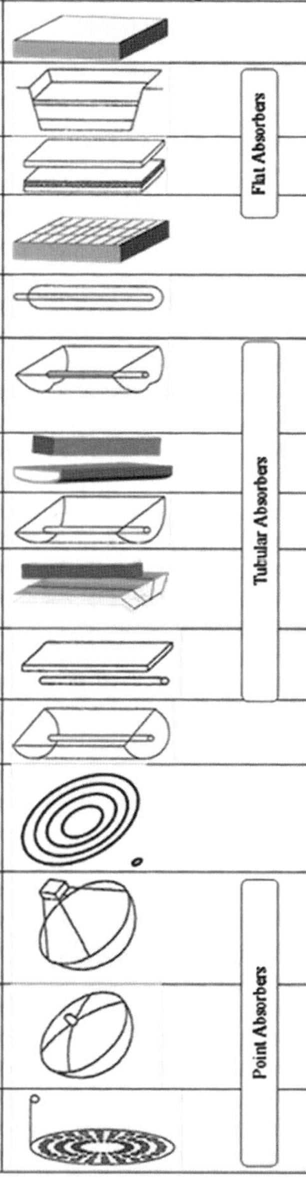

Motion	Solar Tracking	Collector Type: Name	Absorber type	Concentration Ratio, C_i for Direct Insolation	Indicative Temperatures T (K)
Stationary		Integral passive		$C \leq 1$	$300 < T < 320$
Stationary		Non-convecting solar pond	Flat Absorbers	$C \leq 1$	$300 < T < 360$
Stationary		Flat-plate absorber	Flat Absorbers	$C \leq 1$	$300 < T < 350$
Stationary		Flat plate photovoltaic/thermal	Flat Absorbers	$C \leq 1$	$300 < T < 360$
Stationary		Evacuated tube		$C \leq 1$	$320 < T < 460$
Stationary		Compound parabolic reflector	Tubular Absorbers	$1 \leq C < 5$	$340 < T < 510$
Motion	Single Axis	Compound parabolic reflector	Tubular Absorbers	$5 \leq C \leq 15$	$340 < T < 560$
Motion	Single Axis	Photovoltaic/thermal trough	Tubular Absorbers	$10 < C < 40$	$340 < T < 460$
Motion	Single Axis	Parabolic reflector	Tubular Absorbers	$15 < C < 40$	$340 < T < 560$
Motion	Single Axis	Linear Fresnel reflector	Tubular Absorbers	$10 < C < 40$	$340 < T < 560$
Motion	Single Axis	Linear Fresnel lens	Tubular Absorbers	$10 < C < 40$	$340 < T < 540$
Motion	Single Axis	Cylindrical reflector		$10 < C < 50$	$340 < T < 540$
Motion	Two Axis	Fresnel lens		$40 < C < 300$	$340 < T < 1000$
Motion	Two Axis	Parabolic dish reflector	Point Absorbers	$100 < C < 1000$	$340 < T < 1200$
Motion	Two Axis	Spherical bowl reflector	Point Absorbers	$100 < C < 300$	$340 < T < 1000$
Motion	Two Axis	Heliostat field	Point Absorbers	$100 < C < 1500$	$400 < T < 3000$

Fig. 1.1 Taxonomy of generic solar energy collector types with their salient features

single factory-built unit. For larger, more complex and/or specialised applications the use of computer-based design tools usually becomes necessary. The analytical models underlying design tools are also used for performance prediction. The relationships between these and other activities associated with solar thermal system development are summarised in Fig. 1.2.

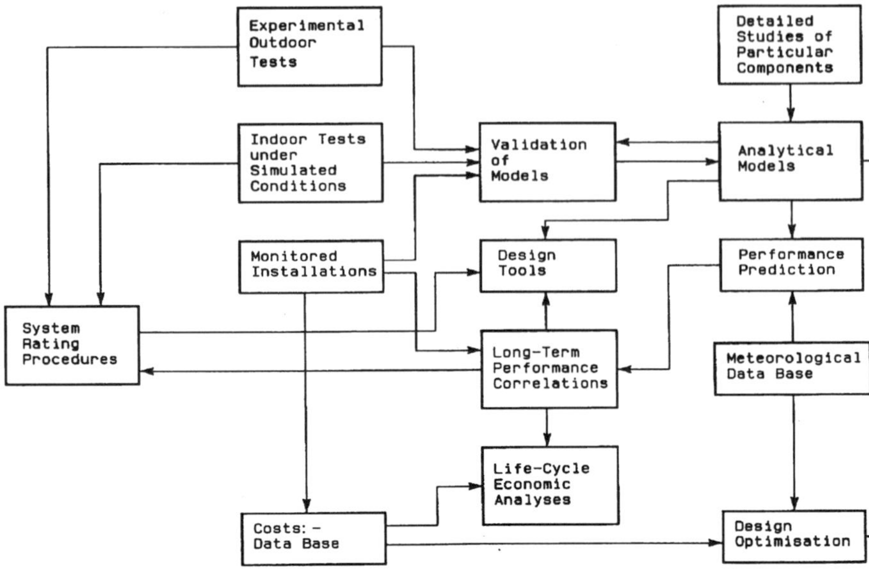

Fig. 1.2 Key aspects of and relationships between solar thermal systems design and innovation

1.2 Solar Energy Law, Politics and Economics

The consequences of prevailing market structures, public policy, tacit and overt political presumptions and social values regarding the environment combine to inform economic appraisals of the use of solar heating options. At the extremes these factors can lead inadvertently to either unnecessary barriers to the use of solar energy thermal systems from being required to either fulfill over-exacting economic criteria or to solar heating being favoured in inappropriate applications or locations. Balanced approaches to the comparative cost-benefit analysis of energy technologies (Holmeyer 1988) acknowledge that fossil fuel and nuclear energy supplies pass on to the general public a significant proportion of the costs of the adverse external consequences of their utilization. The latter include global and local environmental emissions, adverse health impacts or unwelcome changes in employment patterns. The most indirect and longer-term of these effects are impractical to include specifically in the energy bills paid by each and every energy customer; they are thus borne by society as a whole through some combination of taxation, lower gross domestic product, balance of payments deficits and/or poorer quality of life. By comparison fewer indirect costs generally arise from the minimal adverse effects of the utilization of solar heating technologies. Introducing subsidies to stimulate solar water heating use can, and has, developed particular markets. Those markets are however often difficult to sustain when such subsidies are removed. Market development is most successful when undertaken via regulatory requirements for solar heating (for example in new and refurbished buildings) so that it becomes a long-standing feature of a particular market.

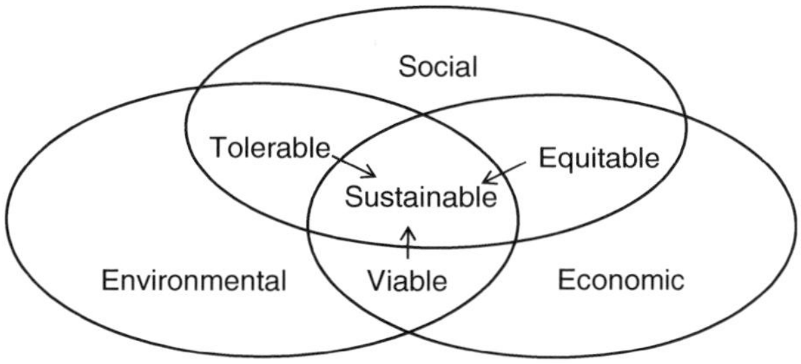

Fig. 1.3 Drivers for environmental sustainability

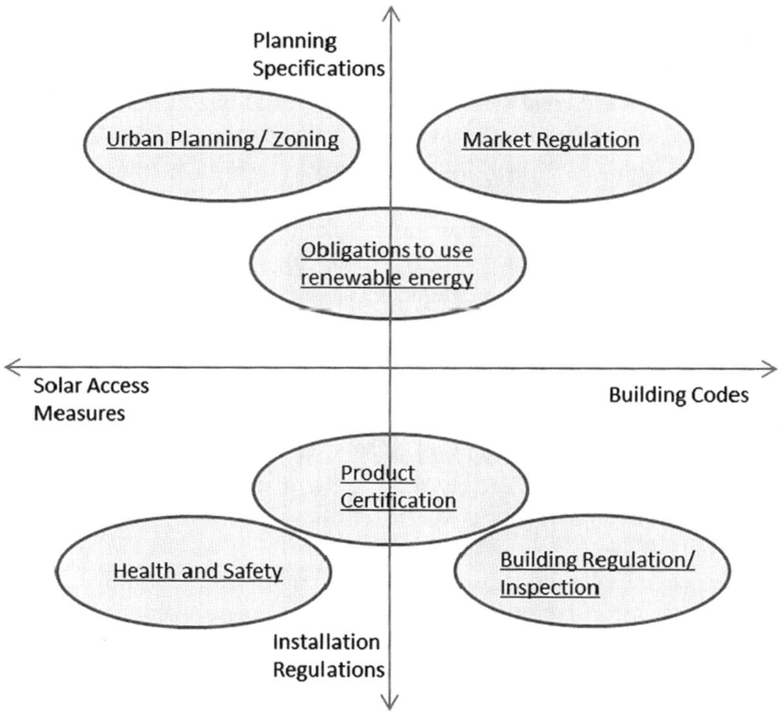

Fig. 1.4 Scope of legal oversight of solar heating

The key drivers to achieving environmental sustainability are shown in Fig. 1.3. Widespread use of an environmentally-benign option, such as solar heating, generally only ensues if solar heating is also sustainable from both social and economic perspectives. The scope of legal oversight summarised in Fig. 1.4 should seek ideally to ensure that solar heating is:

- Tolerable – via building codes and health and safety legislation
- Equitable – via urban planning/zoning processes
- Viable economically – via statutory obligations and equitable treatment of fossil and nuclear energy externalities.

Both local and/or national governments seek to control urban development to ensure (i) efficient use of community facilities, amenities and resources (ii) conservations of historic buildings or communities (iii) achieve segregation of activities such as transport corridors and heavy industry from residential areas and (iv) seek environmental sustainability. In addition the developers of new housing may oblige the subsequent homeowners to retain original design features via the use of restrictive covenants or similar measures. The latter can prevent the retrofitting of solar energy collectors to housing. In the USA many States have banned restrictive covenants that restrict the installation of solar collectors (Bronin 2009).

Many jurisdictions provide rights to unshaded access to solar energy arising from obligations to preserve pre-existing rights to daylight such as the UK rights based on the "Ancient Lights" law (Bickford-Smith and Francis 2007), or to secure sunlight for health (rather than specifically energy) in Japan (Takagi 1977). In contrast in the USA the legal system until recently did not confer property owners to rights to the solar energy incident on the property though some states notably Wisconsin, California and New Mexico have affirmed such rights (Paddick and Grinlinton 2010).

In virtually all jurisdictions worldwide, legislation exists to usually seek to ensure that permanent buildings are structurally sound, able to reasonably withstand possible local phenomena such as earthquakes, heavy storms, tornadoes, provide a sanitary and healthy interior environments and have low energy use. Such regulations are applied usually to newly-constructed buildings but in some countries have been extended to the refurbishment of existing buildings. Installation of solar water heaters is, in some jurisdictions, mandatory; though this may not be obviously apparent from published building codes. For example, in China many provinces have construction codes that merely request the inclusion of solar water heaters; in reality construction will not proceed unless solar water heaters are installed (Runqing et al. 2012).

Products accepted by the regulations of one country may be prevented entry to certain other markets. For example countries seek to prevent, what influential national interest groups see as, unfairly-low priced imports gaining market share at the expense of the products local firms. Import tariffs may be imposed by governments to remove such perceived competitive advantage. It is comparatively rare for there to be outright bans on imports, except in the case of politically-driven economic sanctions. However, the cost, to an exporter, of compliance with onerous national regulations and standards can be so insurmountable as to have a consequence distinguishable from an import ban. It is ironic that such actions may maintain the high cost of solar collectors in particular market when low initial installed costs are often necessary for solar water heaters to be viable economically.

Specific aspects of the design and installation of plumbing are specified by regulations in most jurisdictions. These regulations exist to ensure public and environmental health, consumer safety, water conservation and efficient use of energy. For example double-walled heat exchangers are often required in indirect systems with anti-freeze solutions. Pressurised systems and the community-use of large scale systems usually have additional requirements. In many jurisdictions only approved products may be installed that comply with specific standards. The

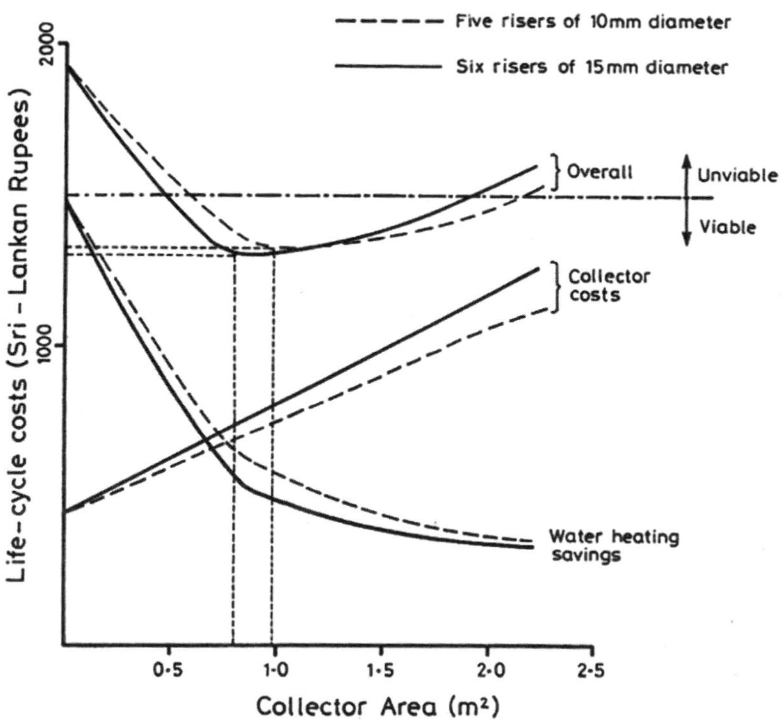

Fig. 1.5 Optimal areas of differing designs of flat plate collector for water heating in Sri Lanka

installation and operation of a solar water heater is governed usually by prevailing national legislation on health and safety that generally require some form of risk assessment prior to installation with procedures implemented to mitigate the most significant risks identified. The latter include providing temporary access to, and working on, roofs and procedures for welding of pipe joints. These risks together with the, often large, fraction of initial cost incurred in the installation stage has thus led to many manufacturers giving significant emphasis to systems being designed to simplify installation minimising the extent of specialist on-site work. If a system fails, the qualifications and training of the designers and installers with often have a bearing on where any negligence may be assigned. The liability to fines and other sanctions arising from a determination or admission of negligence is usually set by legislation. In some jurisdictions liabilities arise from case law can give rise to a requirement for installers to pay disproportionally high insurance premiums that increase installed cost.

The economic viability of solar energy use should be calculated on a life-cycle basis. For most applications there will be an optimal collector area. The latter can depend quite subtly on collector materials and design. For example as show in Fig. 1.5, differences in the number and diameters of risers in a flat plate collector altered significantly the optimal collector area for a particular system operating in the tropical climate of Sri Lanka (Norton and Perera 1991). The underlying reason for the existence of an optimal area is also illustrated in Fig. 1.5; as the collector area increases so do the cost of the collector, however at larger collector area the rate of savings of other fuels for auxiliary heating of water diminishes. The optimal size of a solar water heater can also be expressed in terms of solar savings fraction.

References

Bickford-Smith S, Francis A (2007) Rights of light. Jordan, London

Bronin SC (2009) Solar rights. Boston Univ Law Rev 89:1217–1232

Holmeyer OL (1988) The social costs of energy consumption – external effects of electricity generation in the Federal Republic of Germany. Springer, Berlin

Norton B, Perera KS (1991) Appropriate design of a solar energy water heater for Sri Lanka. ISES Solar World Congress, Denver, pp 3605–3610

Paddick L, Grinlinton D (2010) The role of feed-in tariffs in supporting the expansion of solar energy production. Univ Toledo Law Review 41:943–973

Runqing H, Peijun S, Zhongying W (2012) An overview of the development of solar water heater industry in China. Energy Policy 5:46–51

Takagi GF (1977) Designs on sunlight, solar access in the United States and Japan. Comp Law Rev 10:123–146

Chapter 2
The Solar Energy Resource

The sun shone, having no alternative, on the nothing new

Samuel Beckett, Routledge, London (1938)

2.1 Overview

The earth rotates at an axial tilt in an elliptical orbit around the sun producing the annual variation of intensity outside the earth's atmosphere (Lunde 1980) shown in Fig. 2.1. Beneath the atmosphere solar energy varies temporally and geographically in its

- Intensity
- The relative magnitudes of its direct and diffuse components and
- The skyward anisotropy of the diffuse component and
- In its spectral compositions.

Ninety nine percent of the thickness of the earth's atmosphere lies within a distance of about 30 km from the earth's surface. In passing through the atmosphere solar radiation is reflected, absorbed and scattered, diminishing total insolation and reducing its direct beam component. The intensity of solar energy received by a surface at ground level depends on the orientation of the surface in relation to the sun, the hour of day, the day of the year, the latitude and altitude at the point of observation and atmospheric conditions. The key factors that determine how much solar energy leaving the Sun is incident on a specific surface plane on earth are summarised in Fig. 2.2. Incident radiation from the sun arrives at the earth with a $1/2^\circ$ cone. When passing through a turbid atmosphere with large aerosol content of this angular cone broadens due to forward scattering. This broader cone is referred to as circumsolar radiation. In a clear atmosphere, direct solar radiation forms a large proportion of the solar radiation incident at the earth's surface. In a cloudy atmosphere, diffuse radiation becomes a significant fraction of the total solar radiation as a result of the scattering of the direct beam by clouds.

B. Norton, *Harnessing Solar Heat*, Lecture Notes in Energy 18,
DOI 10.1007/978-94-007-7275-5_2, © Springer Science+Business Media Dordrecht 2014

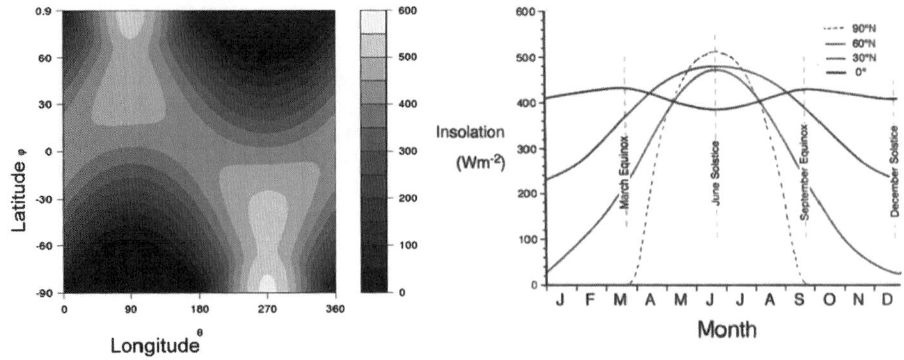

Fig. 2.1 Insolation on a horizontal plane just above the earth's atmosphere

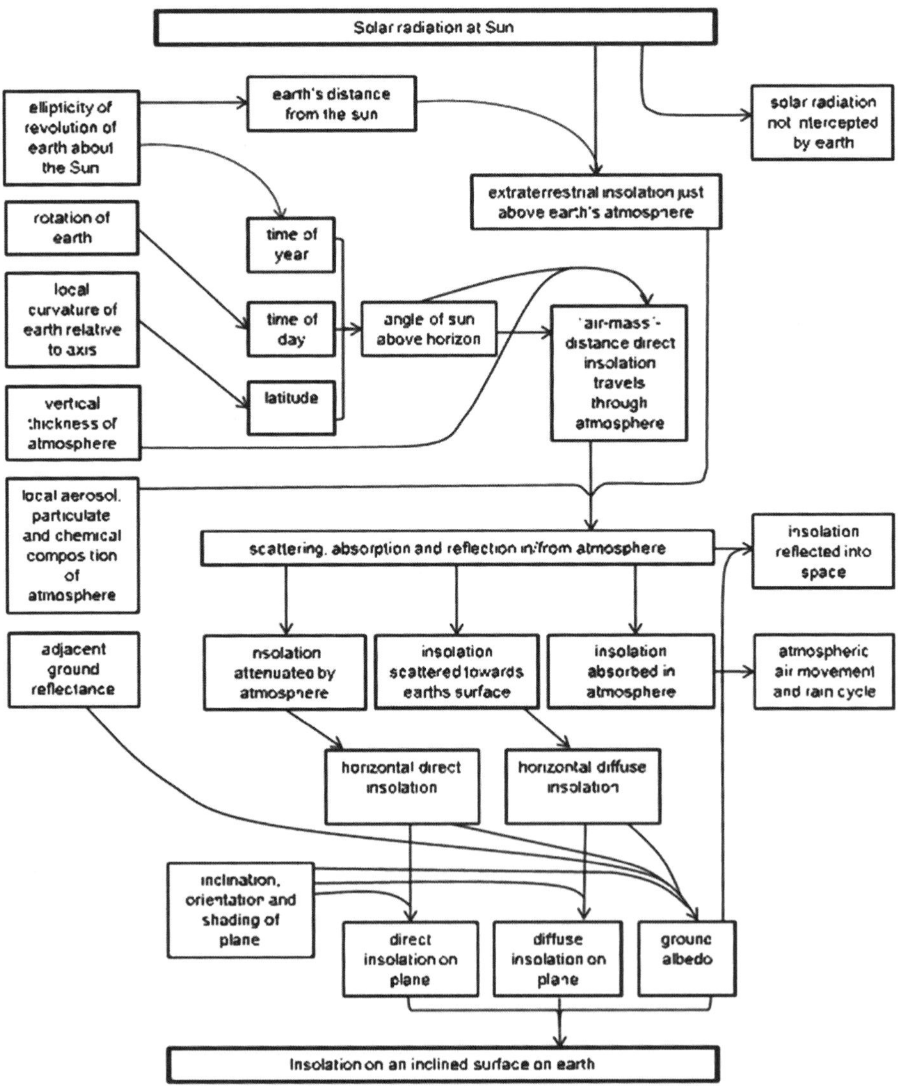

Fig. 2.2 Solar energy interception between emission from sun to a surface plane on earth

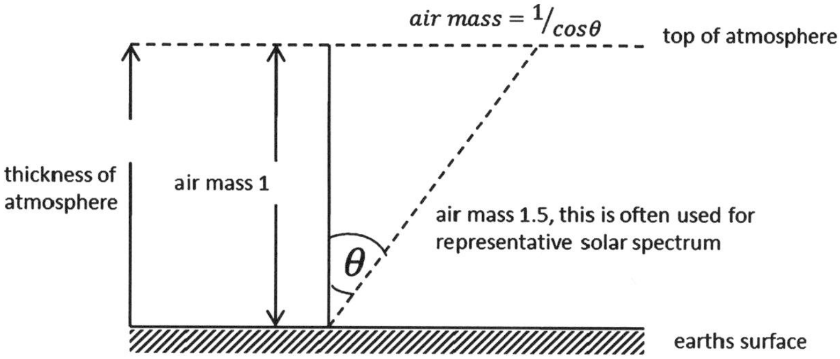

Fig. 2.3 Definition of air-mass

Scattered solar energy incident on the ground forms the diffuse insolation component. The extent of absorption and scattering of radiation by the atmosphere depends on the air-mass, i.e. length of the path traversed and the composition of the atmosphere. The traversed path for beam radiation is shortest when the sun is at zenith, the beam follows an inclined path in reaching the earth's surface. Air mass is defined as shown in Fig. 2.3.

Solar energy is radiation in a specific range the electromagnetic spectrum shown in Fig. 2.4. In passing through the atmosphere, ozone, water vapour dioxide, nitrogen, oxygen, aerosols, dust particles, and clouds all selectively attenuate particular solar radiation wavelengths by either absorption or scattering. Ozone, concentrated in a layer between 10 and 30 km above the earth's surface, with maximum concentration occurring between about 25 and 30 km, is a very strong absorber of solar radiation in the ultraviolet range between 0.2 and 0.29 μm, a relatively strong absorber in the range 0.29–0.34 μm and has a weak absorption in the range 0.44–0.7 μm. There are both geographic and seasonal variations in ozone concentration. Water vapour absorbs solar radiation strongly in wavelengths beyond about 2.3 μm with several absorption bands in the range of wavelengths between 0.7 and 2.3 μm. Oxygen absorbs solar radiation in a region of 0.762 μm. Carbon dioxide is a strong absorber of solar radiation in wavelengths beyond 2.2 μm. The effect of such absorptions on the solar spectrum received at the surface of the earth is summarised in Fig. 2.5.

Both the diurnal and annual patterns of insolation are stochastic in nature. In many climates it is not possible to predict exactly the insolation at any particular instant. However such instantaneous single values usually aggregate to form robust long-term statistical distributions. Thus it is possible to know that specific values for insolation *will* occur within a particular period of time but not precisely *when* within the period. Such correlations enable prediction of insolation in the absence of measured data. However they should be used with care as they cannot always be reliably extrapolated to different locations, even those with a similar climate (Van den Brink 1982).

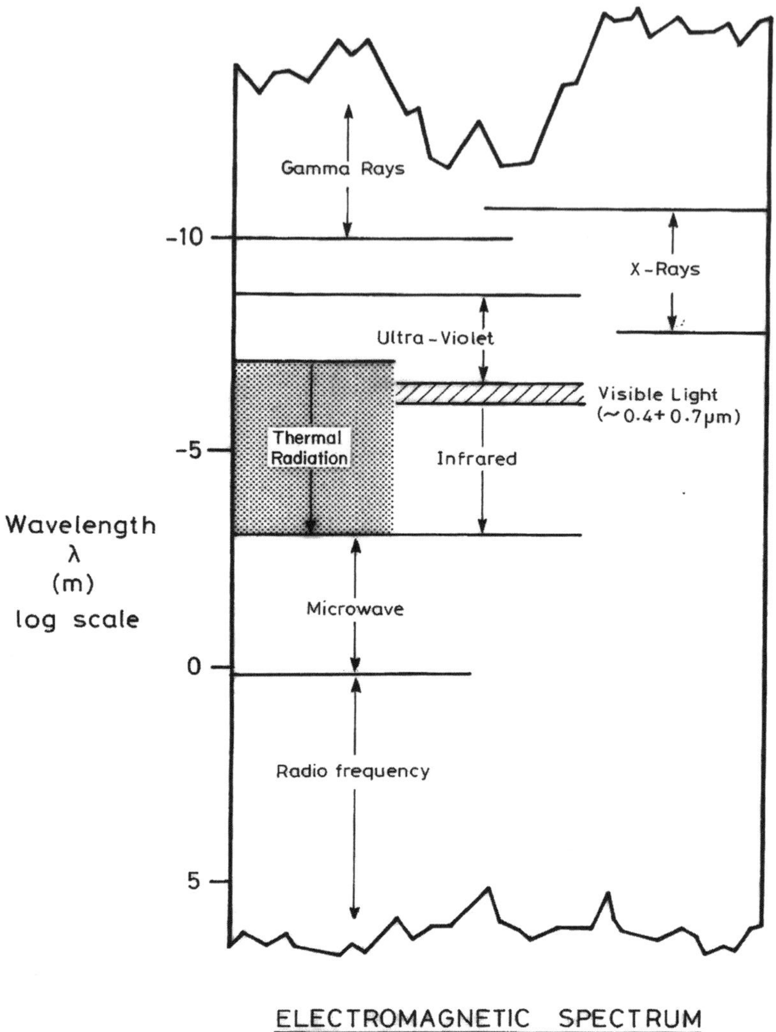

Fig. 2.4 Electromagnetic spectrum

2.2 Terrestrial Measurement of Solar Energy

Daily and hourly records of the amount of solar radiation received at any given location over the earth's surface have been essential for the design and optimisation of thermophysical systems utilising solar energy. Therefore, solar radiation measurements are made continuously at monitoring stations located at different parts of the world. Such measurements include the;

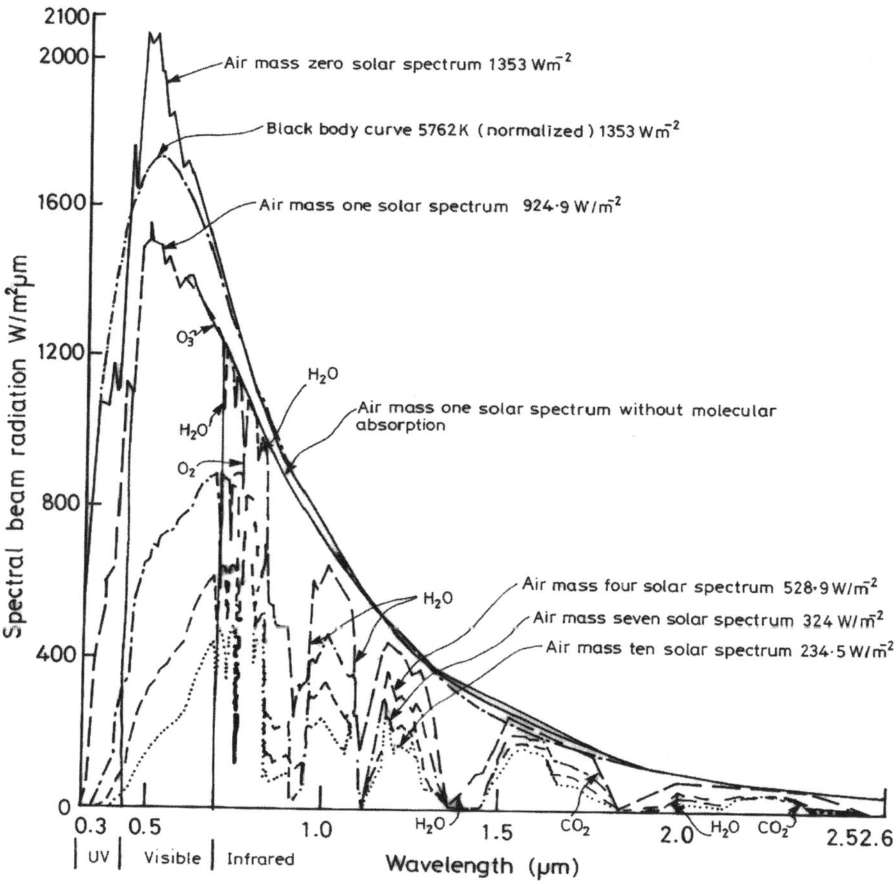

Fig. 2.5 Solar spectrum as attenuated by passage through different air masses

- Intensity of direct solar radiation at normal incidence
- Diffuse component on a horizontal surface
- Global solar radiation on a horizontal surface
- Total solar radiation on an inclined surface at a specified orientation
- Spectral distribution of over certain wavelength bands
- Solar radiation reflected from the ground

Instruments used for the measurement of solar radiation are either: pyranometers or phyrheliometers. A pyranometer measures the total solar radiation incident on a horizontal surface from the entire sky as shown in Fig. 2.6. A pyrheliometer measures the intensity of the direct solar radiation at normal incidence.

Diffuse solar radiation is measured most commonly by attaching a shadow-band to a pyranometer to obscure the direct rays of the sun from the sensing element. Commonly used constant-diameter shadow-bands are moved along an axis, parallel to the earth's axis, according to the solar declination angle. To determine shadow band

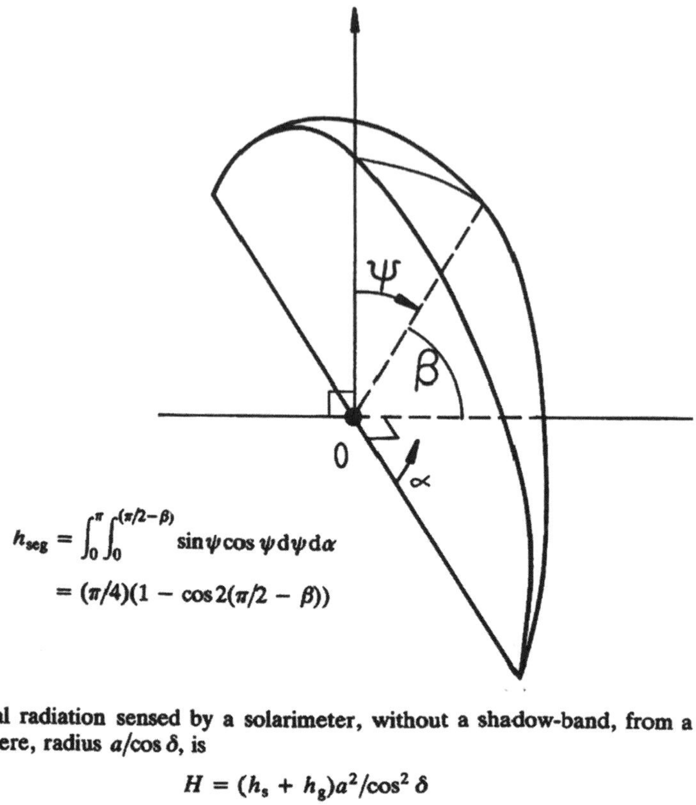

Fig. 2.6 Total radiation used by a solarimeter without a shadow band

correction factors on inclined planes under overcast conditions (Burek et al. 1988), a pair of pyranometers are mounted side-by-side on an adjustable-inclination south-facing plane; one measures global insolation, the other, fitted with a shadow band, measures diffuse. Both the shadow-band and the inclination of the solarimeters are adjusted periodically (e.g., every 2 weeks) to maintain the dome of the diffuse-measuring pyranometer in shadow and ensure that insolation is measured normal to the sun's rays at solar noon respectively. During uniformly cloudy conditions when the insolation is solely diffuse and isotropic, if no correction were necessary for the shadow-band then the measurements from the global-radiation pyranometer and the diffuse-radiation pyranometer with a shadow-band would be identical. In reality, the ratio of the measured total insolation to the measured diffuse insolation under such sky conditions is the correction factor required when using a shadow-band. This factor enables inclusion of the diffuse radiation from the direction of the sun obscured by the shadow band. The error is small, and for most applications, there is no need to distinguish between diffuse and direct radiation from the same direction. Because the shadow band need only be adjusted to correct for variations in the sun's declination angle only once every 7–14 days, simple methods (Burek et al. 1988) have practical

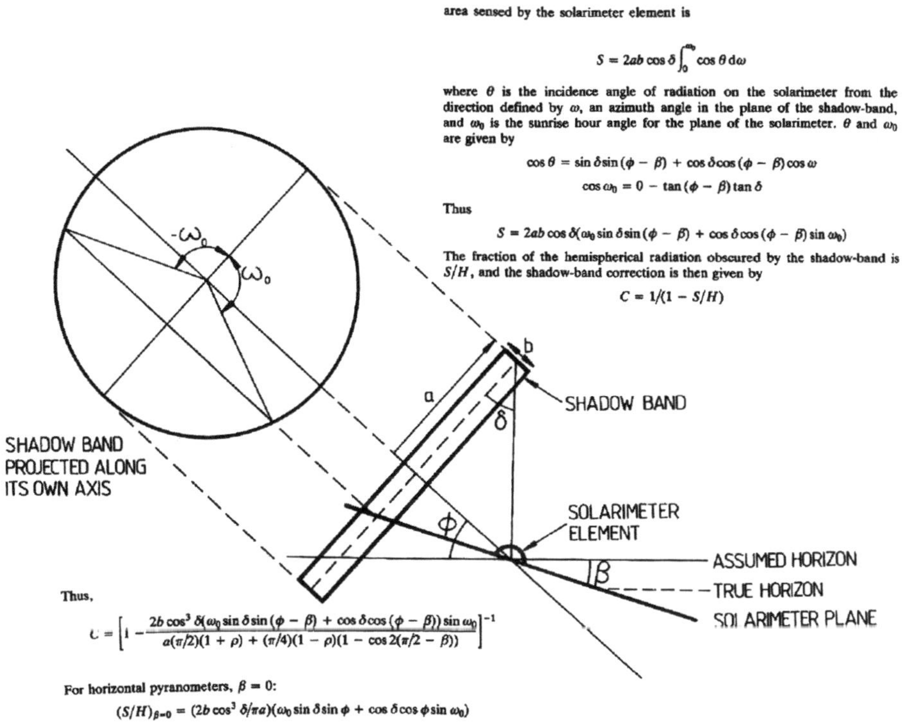

Fig. 2.7 Shadow band correction factors for inclined and a horizontal solarimeters

advantages when compared with the more accurate shade-disc method. The latter compares the 'true' value of diffuse radiation, as measured with a solarimeter fitted with a moving shade disc to obscure the direct radiation, with the reading from the pyranometer fitted with the shadow-band.

The geometry of the shadow-band in relation to the solarimeter has been analysed to determine the variation of the correction actor for horizontal solarimeters with the time of year, as a function of the solar declination and the sunrise and sunset hour angles (Drummond 1956; Robinson and Stoch 1964) and south-facing (in the northern hemisphere) inclined planes (Burek et al. 1988). Such a geometric correction as shown in Fig. 2.7 assumes implicitly an isotropic distribution of diffuse radiation. Anisotropic diffuse radiation introduces an error of only a few percent in an isotropic (i.e. geometric) correction (Ineichen et al. 1983). Only where such accuracy is essential should shadow-band corrections be used to account for anisotropic diffuse sky radiation. Either empirical or analytical distributions of diffuse radiation for clear skies can be used as neither approach is consistently better than the other (Rawlins and Readings 1986). Empirical correction methods have been based either on the time of year and the measured ratio of diffuse to global radiation (Painter 1981) or on global radiation, because the anisotropy of diffuse radiation can be related to this parameter (Mujahid and Turner 1980).

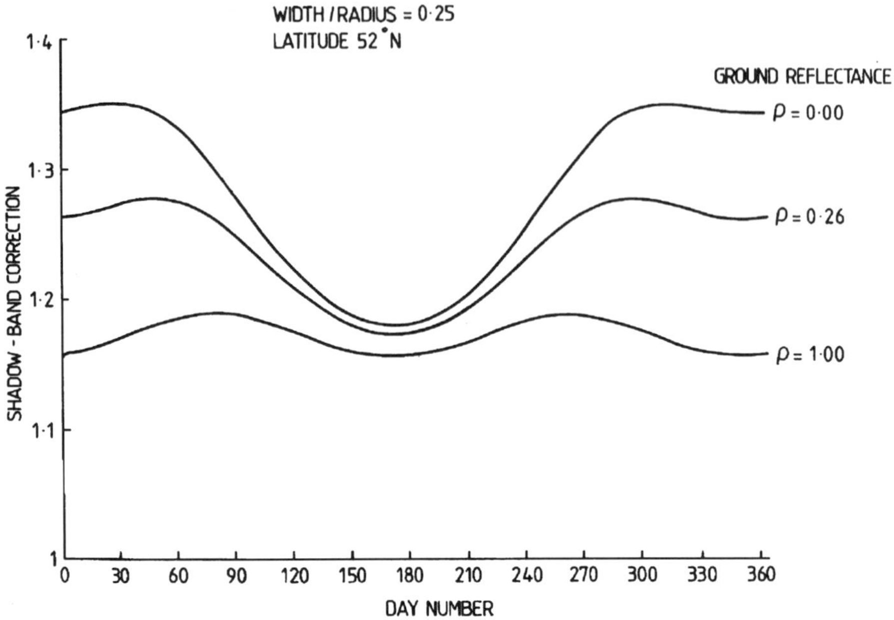

Fig. 2.8 Shadow band correction factors for a pyranometer perpendicular to solar incidence at solar noon

An analytical and experimental study of the effect of part of the inner surface of the shadow-band being illuminated by direct sunlight, especially at low solar altitudes showed that the effect of the width of the shadow-band on the anisotropy correction (considered separately from the geometric correction) was not significant (Steven and Unsworth 1980), except for very narrow bands, with width-to-radius ratio less than 0.15. It has been found that Drummond's (1956) correction function that implicitly assumed an isotropic radiation distribution was in reasonable agreement with daily-averaged data for cloudy days, whereas a correction with an anisotropic distribution fitted the data more closely for clearer days (Ineichen et al. 1983). Anisotropy correction of diffuse radiation measurements is thus, somewhat ironically, of some relevance for very accurate measurements of low diffuse insolation conditions.

The shadow band correction factors for inclined and horizontal solarimeter elements subject to a uniform solar radiation flux, are shown in Fig. 2.8 in which the incidence angle of radiation on the solarimeter is related to an azimuth angle in the plane of the shadow-band, and the sunrise hour angle for the plane of the solarimeter.

When seen from an altitude, the horizon of a flat landscape is below the horizontal; for example at an altitude of 5 m, the horizon is at an angle 2.2° below horizontal. For a roof-mounted pyranometer the shadow-board correction factors should include an altitude amendment to avoid an incorrectly high value. Additional correction is also required for inclined surfaces; Fig. 2.7 shows the

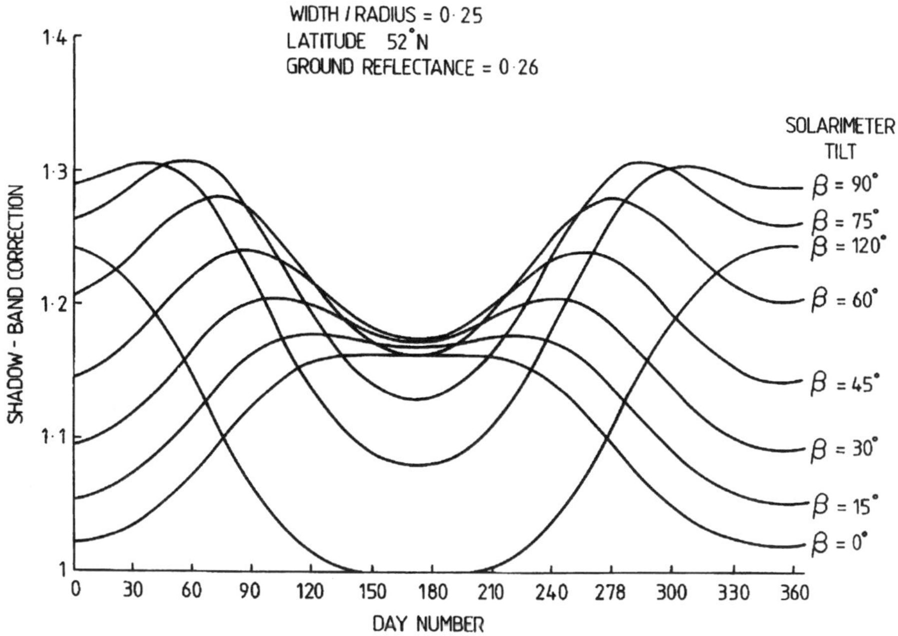

Fig. 2.9 Shadow-band correction factors for different pyronmeter inclinations

correction for a pyranometer tilted to be perpendicular to the sun at solar noon. The correction factor increases as decreases with angle of tilt because although less radiation reflected from the ground is obscured by the shadow-band, the sky radiation which is obscured is a greater proportion of the total available. The variation of the shadow band correction factor throughout the year at different pyranometer inclination angles and a ground reflectance of 0.26 (appropriate for green grass) is shown in Fig. 2.9.

At low tilt-angles, the curve in Fig. 2.9 has a minimum in winter and twin maxima in summer. At high tilt-angles, this is reversed, with a single minimum in summer, due simply to such a small section of the shadow-band being operative, at high summer tilt-angles. Indeed, if it were tilted further, there would be no need for a shadow-band during those parts of the year because the sensing element would be shaded completely from direct insolation. In deriving Fig. 2.8 it has been assumed that no light is reflected from the inner surface of the shadow-band, though it has been suggested that a value of 0.1 may be appropriate for the reflectance of a matt black surface (LeBaron et al. 1980). For the shadow-band geometry and ground reflectance indicated in Fig. 2.8, the shadow-based correction factor varies from 1.02 to 1.29 during winter between the horizontal and vertical pyranometer orientations. The variation is greater for lower values of ground reflectance and for shadow-bands with large width-to-radius ratios. Thus the potential error from such factors is relatively small but is at least as large as that an anisotropic diffuse-radiation distribution may introduce (Spencer et al. 1982) to shadow-band corrections that assume the diffuse sky to be isotropic.

2.3 Prediction of Solar Energy

The amount and type of solar radiation data available determines if it is appropriate to;

- Estimate insolation from nearby meteorological records of sunshine duration and degree of cloud cover (Iqbal 1983; Reddy 1987)
- Estimate the diffuse component from measurements of global insolation only (Liu and Jordan 1960)
- The use of design data based on semi-empirical relations, e.g. Hottel's clear day (Hottel 1976) established for cloudless conditions, the grey day derived from averaging over cloudy conditions (Stine and Harrigan 1985), and Standard Atmospheres and or correlation with ambient temperature (Norton and Abu-Ebeid 1989)
- Use measurements of global and diffuse insolation on horizontal surfaces from the nearest insolation measuring weather station
- Use satellite data (Perez et al. 2002)

Empirical formulas can be used to estimate insolation for locations at which no measurements are available. In many meteorological stations Campbell-Stokes and similar recorders record direct insolation above 200 Wm^{-2} up to 1981 and 120 Wm^{-2} since 1981, (Iqbal 1983). This is the basis of the "hours of sunshine" often used in daily weather reports. The threshold level of direct insolation is somewhat analogous to the concept of solar radiation utilisability (see Sect. 2.5). The duration of sunshine hours can be converted to insolation via various forms of the Angstrom correlation using extraterrestrial insolation and the solar geometric day length (and, in some versions, other factors) as normalising parameters. The Angstrom correlation is given in Fig. 2.10 (Angstrom 1924; Prescott 1940; Page 1961). Given the now ready availability of measured insolation data, the practical use of the Angstrom-type correlations – even the most refined (Reddy 1987) – has become very limited.

Various climatological parameters such as humidity, temperature, rainfall, number of sunshine hours, total amount of cloud coverage, have been used in developing empirical relations as substitutes for the direct measurement of insolation. However, as attenuation of solar radiation travelling through the atmosphere is a complicated stochastic process, the utility of such approaches is limited.

2.3.1 Insolation a Horizontal Plane

The ratio of the diffuse component to total insolation is dependent on the monthly mean clearness index (Liu and Jordan 1960). The clearness index is the ratio of the monthly mean daily insolation on the earth's surface to the extraterrestrial irradiation, both measured on horizontal plane at the same latitude as shown in Fig. 2.11. Subsequent developments have provided correlations that account of seasonal

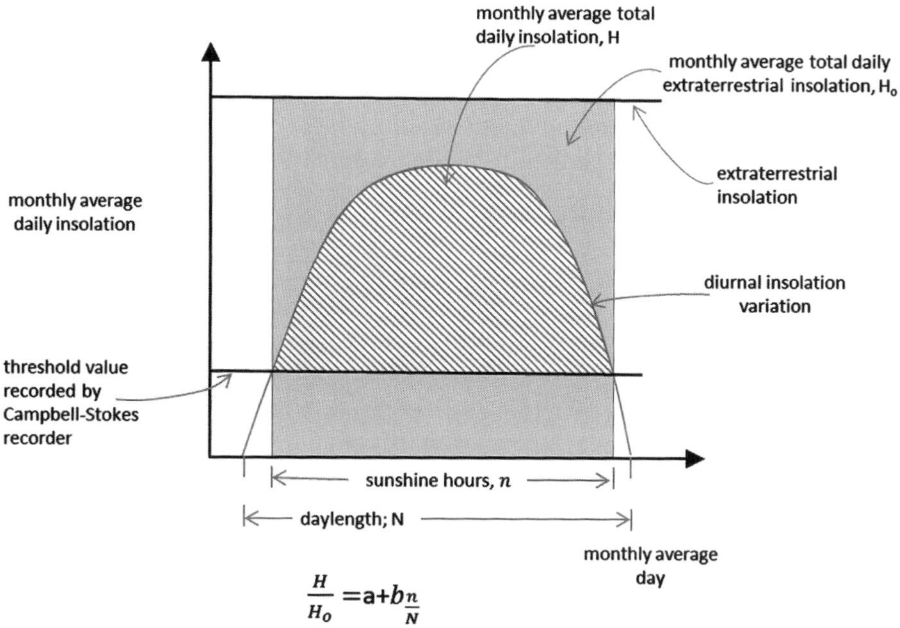

Fig. 2.10 Angstrom correlation of monthly-average daily insolation with apparent daylength

variations by including the sunset hour angle (Collares-Pereira and Rabl 1979; Erbs et al. 1982). It has been observed that for locations with similar climates there exist "generalised" cumulative frequency distribution curves for the daily clearness index as shown in Fig. 2.12 (Liu and Jordan 1963; Bendt et al. 1981; Theilacker and Klein 1980; Reddy et al. 1985; Hollands and Huget 1983). When the minimum and maximum average monthly clearness index are known, then the probability density function for clearness indices (Reddy 1987) can be produced as shown in Fig. 2.11.

Correlations that employ the clearness index are used to predict the daily diffuse component from the daily global incident insolation (Liu and Jordan 1960; Collares-Pereira and Rabl 1979) and can include seasonal variations as indicated by the sunset hour angle (Erbs et al. 1982) as shown in Fig. 2.13. Over the *long term*, in many climates, the diurnal variation of global insolation is symmetrical about solar noon (Liu and Jordan 1960). This observation has been used to develop a correlation between hourly and hourly and monthly mean daily horizontal global insulations (Collares-Pereira and Rabl 1979). Where the hour angle corresponds to the midpoint of the hour, expressions for hourly and monthly mean-daily horizontal diffuse components of insolation are shown in Fig. 2.13 (Liu and Jordan 1960). Hourly correlations of diffuse to global insolation have also been developed (Gordon and Hochman 1984; Erbs et al. 1982).

Monthly average clearness index K_T; average atmospheric insolation attenuation at a specific location over a particular month.

$$K_T = \frac{I_G}{I_{B_o}}$$

Bo extraterrestrial irradiance varies with the eccentricity of the earth's orbit;

$$I_{B_o} = \frac{24}{\pi} S\left\{1 + 0.33\cos\left(\frac{2\pi d_n}{365}\right)\right\}(\cos\varphi\cos\delta\sin\omega_s + \omega_s\sin\varphi\sin\delta)$$

S: Solar constant = 1367 W/m^2

d_n number of day in year

ϕ latitude

δ declination

ω sunrise hour angle

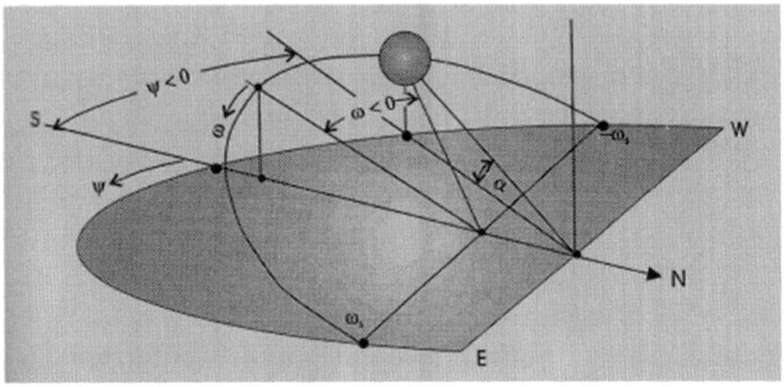

Fig. 2.11 Definition of clearness index

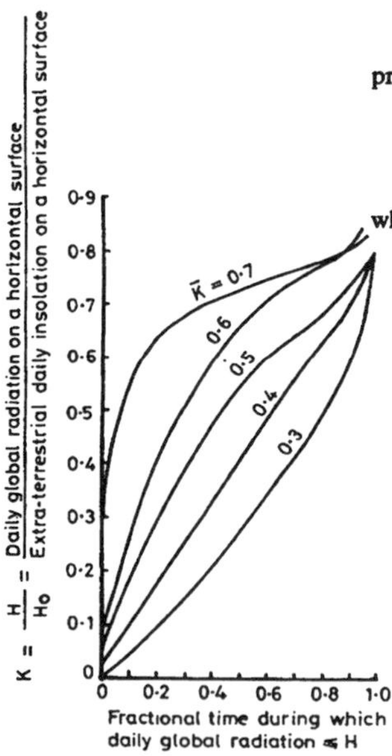

probability density function for clearness indices

$$P(K_T) = C\left[\frac{K_{T,max} - K_T}{K_{T,max}}\right]^n e^{\gamma K_T} \text{ for } K_{T,min} \leq K_T \leq K_{T,max},$$

where C is given by

$$C = \left[\int_{K_{T,min}}^{K_{T,max}} \left[\frac{K_{T,max} - K_T}{K_{T,max}}\right]^n e^{\gamma K_T}\right]^{-1}$$

cumulative frequency distribution is given by

$$F(K_T) = \int_{K_{T,min}}^{K_{T,max}} P(K_T)dK_T = \frac{\exp(\gamma K_{T,min}) - \exp(\gamma K_T)}{\exp(\gamma K_{T,min}) - \exp(\gamma K_{T,max})}$$

Fig. 2.12 Cumulative frequency distributions of daily clearness indices

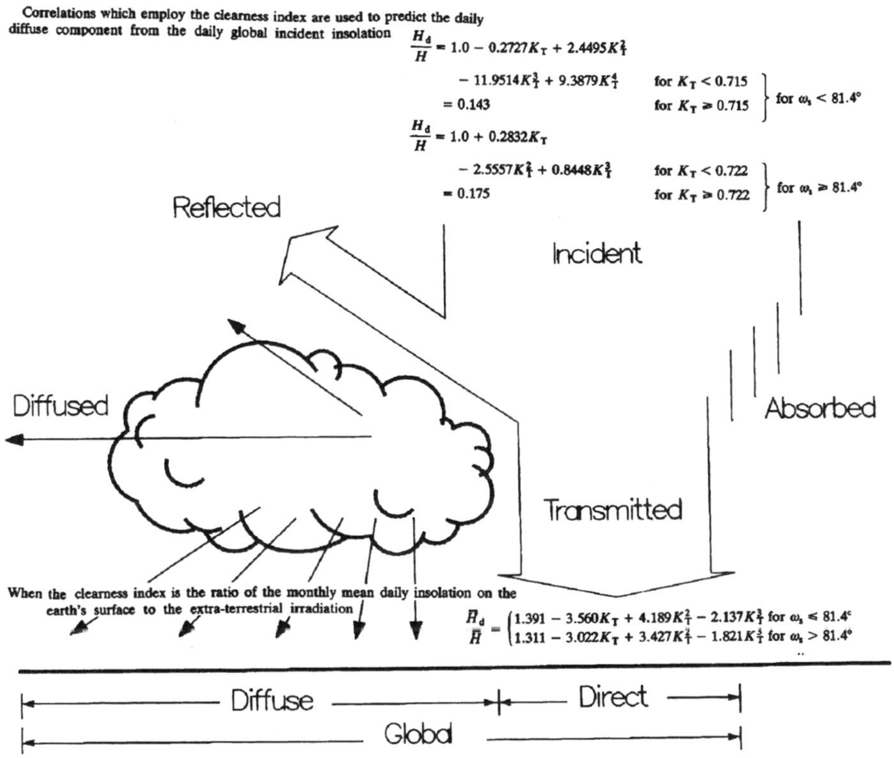

Fig. 2.13 Daily and monthly diffuse to total insolation correlations

2.3.2 Insolation on a Tilted Plane

Measured solar radiation data are now becoming available for an increasing range of locations in the form of hourly global and diffuse insolation on horizontal surfaces, and as monthly-averaged daily global insolation on horizontal surfaces. For a limited set of locations data is available for global insolation on tilted surfaces. However, it is not practicable to make measurements of global radiation on all the possible orientations of surfaces that may be needed, so calculations are used to derive radiation incident on such surfaces from horizontal surface measurements. Given a record of hourly solar radiation incident on horizontal surfaces it is possible to calculate the hourly incident energy on surfaces of any slope and orientation within the vicinity of the data-collection station. It is also possible to calculate the daily, monthly and annual incident energy for the year in question.

In calculating the insolation on an inclined plane most models treat the diffuse component like the beam component. This assumes that most of the diffuse

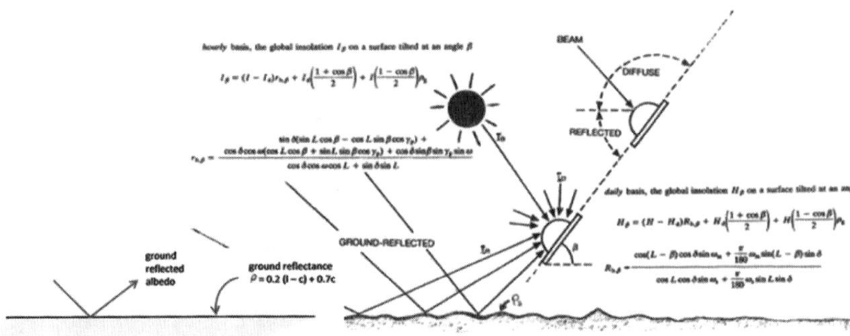

Fig. 2.14 Hourly and daily global insolation on a tilted plane

radiation comes from the circumsolar region of the sky, the implication being that most diffuse insolation arises from forward scattering. This approximation applies most closely to, and can be used for design purposes on cloudless days. The angular correction to be applied to the diffuse component is then the same as that for the beam component as shown in Fig. 2.14 which taken together with ignoring ground-reflected radiation, gives the total incident insolation on the inclined surface.

Figure 2.14 represents the "Liu and Jordan" (1962) model of insolation on inclined planes. This considers: beam radiation, diffuse solar radiation, and solar radiation diffusely reflected off the ground. It was assumed that, like isotropic diffuse solar radiation, ground-reflected radiation gives rise to an irradiance that is proportional to the appropriate view factor. A surface tilted at slope β from the horizontal has a view factor to the sky of $(1 + \cos(\beta))/2$, and that this is also the correction factor for an isotropic distribution of diffuse radiation. The view factor to the ground, assumed level and infinite in extent, must then be $(1 - \cos(\beta))/2$. If the ground is a perfectly diffuse reflector of constant diffuse reflectance (albedo) ρ for the global insolation, the radiation reflected off the ground onto the surface is shown in Fig. 2.14. The albedo ρ varies considerably between about 0.05 (for smooth water) and about 0.8 (for fresh snow), but most land surfaces have albedos around 0.2. A fractional time "C" (usually on a monthly basis) to represent duration of snow cover may be used, as shown in Fig. 2.14, to modify ground reflectance changes due to snow.

Liu and Jordan's (1962) model assumes that the ground is a Lambertian reflector reflecting all wavelengths equally both uniformly in all directions, and independently of solar elevation. This assumption is usual in treatments of ground-reflected solar radiation (Dave 1977), but, in practice, albedos vary significantly with the wavelength and angle of incidence of the radiation, and with the angle of view of the observer.

On an hourly basis the global insolation on a tilted surface (Liu and Jordan 1960) is also shown in Fig. 2.14.

2.4 Use of Satellite Information to Produce Solar Energy Data

Geostationary satellites orbiting the earth measure ground and cloud reflected radiation. This information is collated and analysed to produce solar radiation maps for monthly and annual insolation averages to a spatial resolution of 10 km by 10 km. Hourly radiance images from satellites are combined with daily snow cover data together with monthly averages of atmospheric water vapour, absorbing gases and atmospheric aerosols to determine hourly direct and diffuse insolation on a horizontal surface (Marion and Wilcox 1994; Maxwell et al. 1998; George and Maxwell 1999; Perez et al. 2002).

2.5 Solar Radiation Utilisability

For many solar thermal systems there exists a critical, or threshold value of insolation. For solar collectors, this is that insolation for which at a given ambient temperature, the heat gained equals heat losses as illustrated on the left hand side of Fig. 2.15 for an ambient temperature that varies over a day. Assuming that the monthly mean ambient temperature lies midway between the corresponding ambient morning and evening temperatures give the insolation thresholds shown on the right-handside of Fig. 2.15 for all days of a month a critical radiation ratio can be defined as shown in the right-hand side of Fig. 2.15.

When the ambient temperature remains constant, for a sinusoidal diurnal insolation pattern, the morning and evening thresholds of utilizable insolation are the same, a similar single value for the threshold of utilizable insolation can be calculated using a diurnal average ambient temperature. Solar radiation utilizability depends solely on insolation for a given location, month, the mean ambient temperature appropriate for the month and tilt of the system. The daily utilizability factor can be determined from the hourly utilizability fractions by weighting the

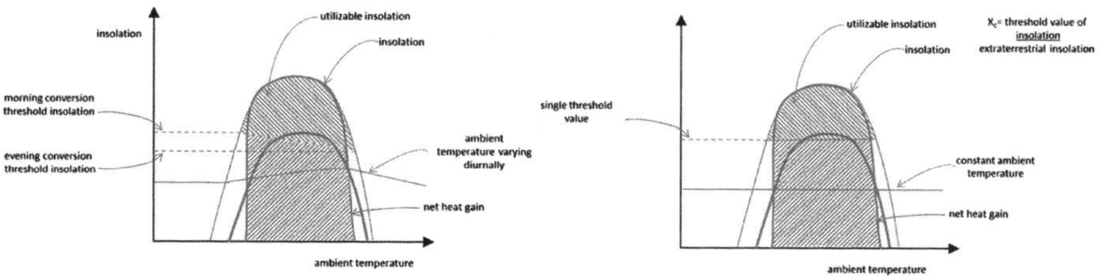

Fig. 2.15 Insolation utilisability threshold

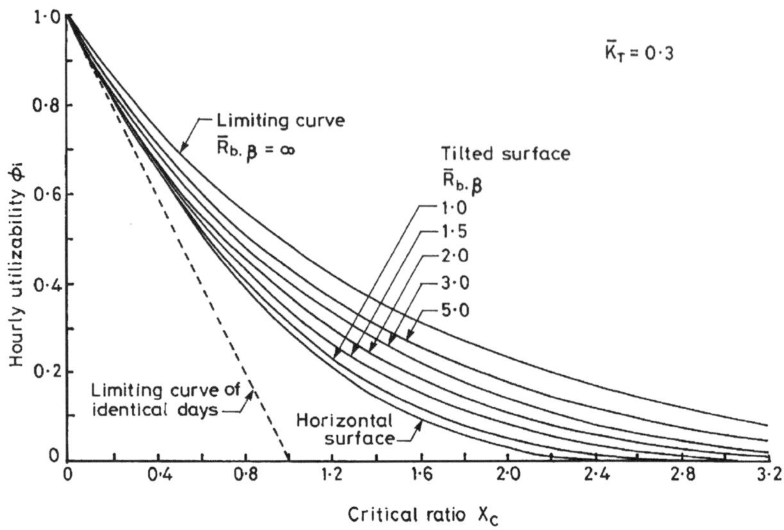

Fig. 2.16 Critical insolation ratio for different surface inclinations at a clearness index of 0.3

hourly insolation values (Clark et al. 1983). Generalizable hourly utilizability curves have been derived (Liu and Jordan 1965) from generalized clearness index curves. From those curves, the hourly utilizability on a monthly basis for equator-facing collectors can be deduced for different clearness indices, critical radiation ratios and different monthly average daily ratios of beam radiation on the tilted surface to that on a horizontal surface. These curves are illustrated in Figs. 2.16, 2.17 and 2.18. Similar forms of graph have been generated for generalised daily utilizability, (Klein 1978; Collares-Pereira and Rabl 1979; Theilacker and Klein 1980; Evans et al. 1982).

2.6 Daylight Data

Daylight is solar illumination in the interiors of buildings as perceived by the human eye. Solar radiation and the luminosity of daylight are related by the luminous efficacy of the radiation being considered. Luminous efficacy is defined as the quotient of luminous flux, as shown in Fig. 2.19. It is a function of the spectral distribution of the radiant energy and so varies with solar altitude, cloud cover, pollutant content of the atmosphere, and also the relative proportions of beam and diffuse radiation (Littlefair 1985, Robledo and Soler 2001).

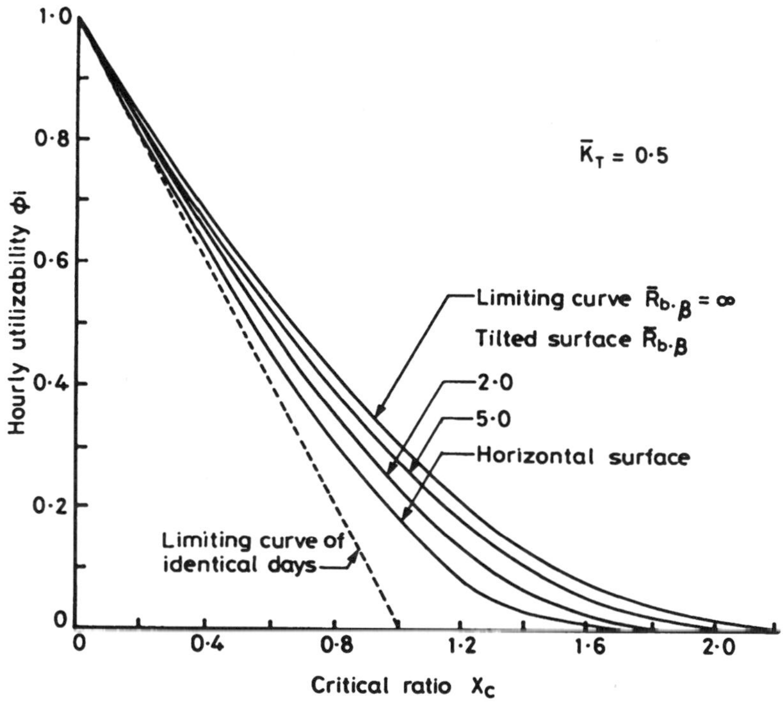

Fig. 2.17 Critical insolation ratio for different surface inclinations at a clearness index of 0.3

2.7 Geographical Availability of Solar Energy

A local climate is the consequence of a distinctive regional combination and interrelation of insolation, temperature, precipitation, humidity and wind speed and direction as driven by global air and ocean circulations, latitude above sea level and topography. The world can be divided into several distinctive climatic types (Trewarthu and Horn 1980) as shown in Fig. 2.20 and Table 2.1. The global distribution of insolation is shown in Fig. 2.21. Although latitudinal, variations of insolation may be seen for high latitudes, there are also appreciable longitudinal variations (particularly at equatorial latitudes) due to the alternation of continents and oceans over the earth's surface.

In addition to these macroscale factors, which determine climate type, mountainous and urban regions also have mesoclimatic effects. Urban areas often receive less insolation than their rural surroundings: the sunshine duration in industrial areas can be reduced by up to 20 % (Landsberg 1981). This reduction is largest at low solar elevations when the air mass is longest. Different impacts can arise seasonally in temperate climates, for example, if frequent low-level atmospheric inversions occur in winter and autumn, they contribute to the accumulation of pollutants and hence to solar radiation attenuation and a concomitant increased

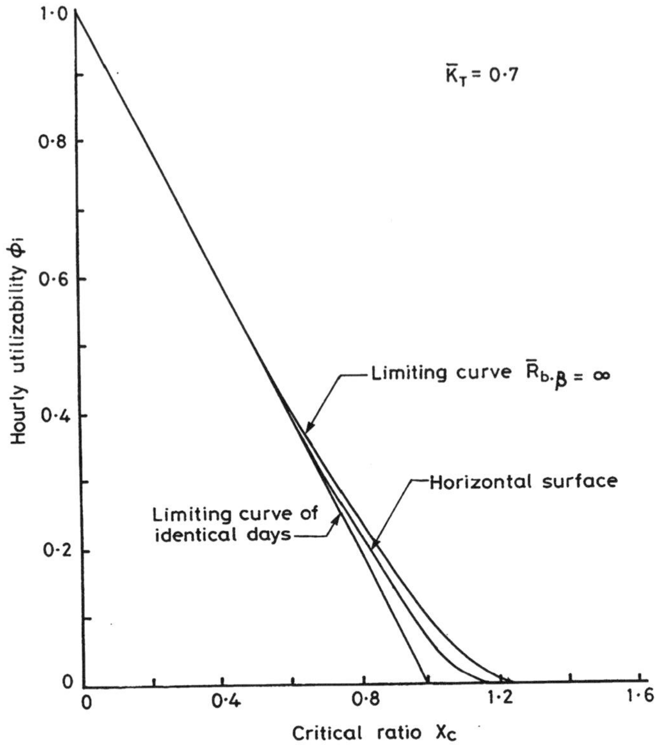

Fig. 2.18 Critical insolation ratio for a typical "clear" sky with a clearness index of 0.7 showing the limited change in solar energy utilizability in sunny climates

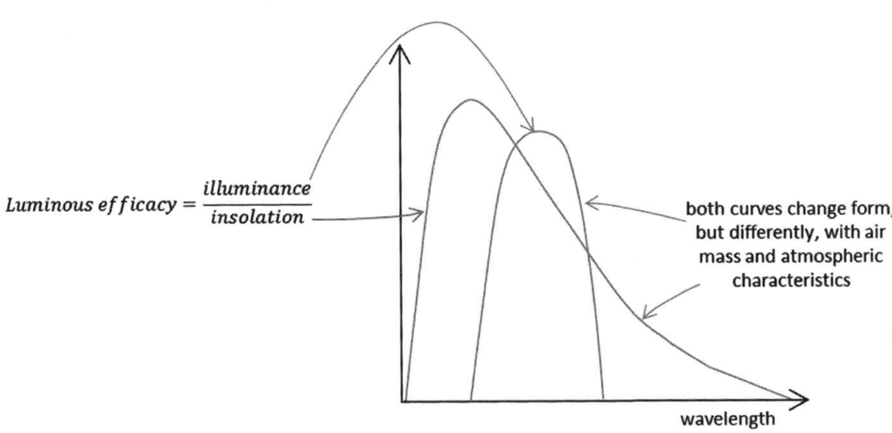

Fig. 2.19 Definition of luminous efficacy

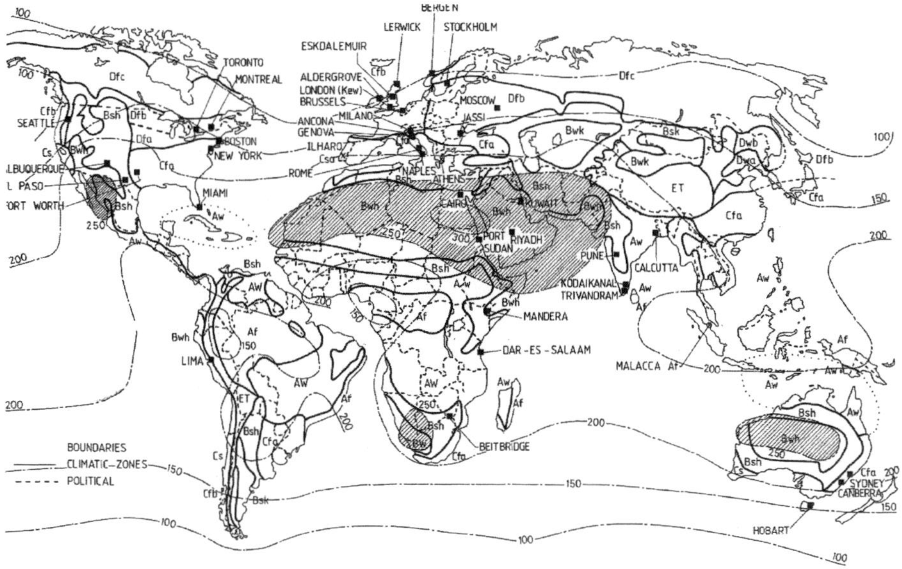

Fig. 2.20 Distribution of world climate types

Table 2.1 Climate types

Designation	Description
Aw	Tropical wet and dry climate: high temperature throughout the year, dry season in low insolation period or winter. At least 1 month had less than 6 cm of rain
Bsh	Semi-arid climate or steppe: average annual temperature over 18 °C. These locations are low-latitude, or tropical
Bwh	Arid climate or desert, average annual temperature over 18 °C
Cfa	No distinct dry season; the driest month of summer receives more than 3 cm of rain, hot summer average temperature of warmest month over 22 °C
Cfb	No distinct dry season; the driest month of summer receives more than 3 cm of rain, cool summer average temperature of warmest month under 22 °C
Csa	Summer dry: at least three times as much rainfall in the wettest month of winter as in the driest month of summer. Hot summer average temperature of warmest month over 22 °C
Dfb	Cold climate: humid winters and cool summer, average temperature of warmest month under 22 °C
Af	Tropical wet climate, rainfall of the driest month is at least 6 cm. Within this climate there is a minimum of seasonal variation in temperature and precipitation, both remaining high throughout the year
Bwk	Arid climate or desert, average temperature under 187 °C. These locations are middle-latitude, or cold desert and steppes
Cs	Summer dry, at least three times as much rain in the wettest month of winter as in the driest month of summer, and the driest month of summer receives less than 3 cm
Dfa	Cold climate with humid winters, hot summer, average temperature of warmest month over 22 °C
Dfc	Cold climate with humid winters, cool short summer, less than 4 months over 10 °C
ET	Tundra climate, average temperature of warmest month below 10 °C but above 0 °C

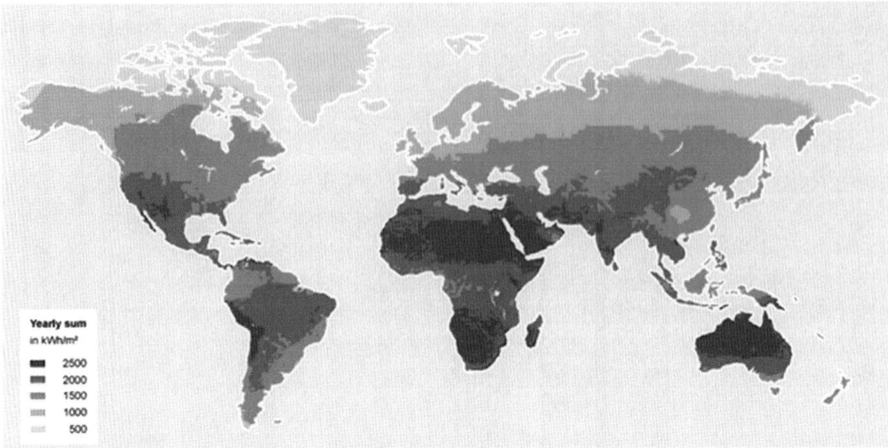

Fig. 2.21 Global distribution of insolation

diffuse component. However, again in temperate climates, in spring, generally-higher wind velocities and, in summer, greater atmospheric convection, both disperse pollutants with a relatively-smaller ensuent reduction in insolation. Maximum and minimum monthly values for ambient air temperatures lag very approximately a month behind those of the solar radiation: a manifestation of conduction of heat into, or out of, the ground. This lag is greater over the oceans at middle latitudes. The lag between temperature and insolation has been employed to develop an means of estimating monthly insolation from temperature data (Norton and Abu-Ebeid 1989).

For medium temperature applications, the ambient temperature can be a significant determinant of collector heat losses. The combination of different variations in cloud cover and ambient temperature can lead to solar thermal systems in ostensibly similar locations exhibiting very different behaviour. As an illustration of this, Fig. 2.22 shows that the numbers of days that a specified hot water demand can be met by an specific solar water heater across Europe. Insolation in winter determines the collector employed when very high solar savings fractions are sought. As can be seen from Fig. 2.22 in Europe, in locations such as Cyprus, solar water heater appropriately designed should be able to satisfy hot water demand all year round. In northern Europe as the same solar water heater is obviously unable to do this; as shown in Fig. 2.23, even if collector area is doubled from 5 to 10 m^2 in a northern European climate only a relatively small space heating solar fraction is achieved. The design of solar thermal system has to include interseasonal heat storage if a high solar function is required as shown in Fig. 2.24.

Though an interseasonal heat storage system may give a larger solar function, for water heating in particular this may be only in terms of energy not exergy, that is the desired end-use temperature may not be satisfied in winter and auxiliary energy will be used.

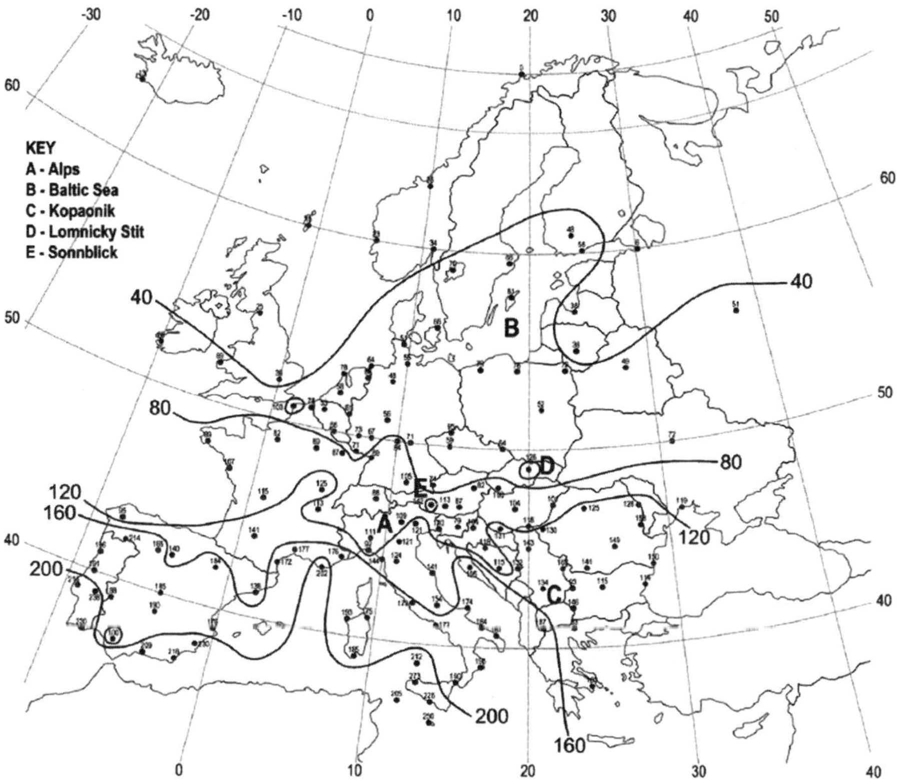

Fig. 2.22 Map of number of days of solely-solar hot water provision

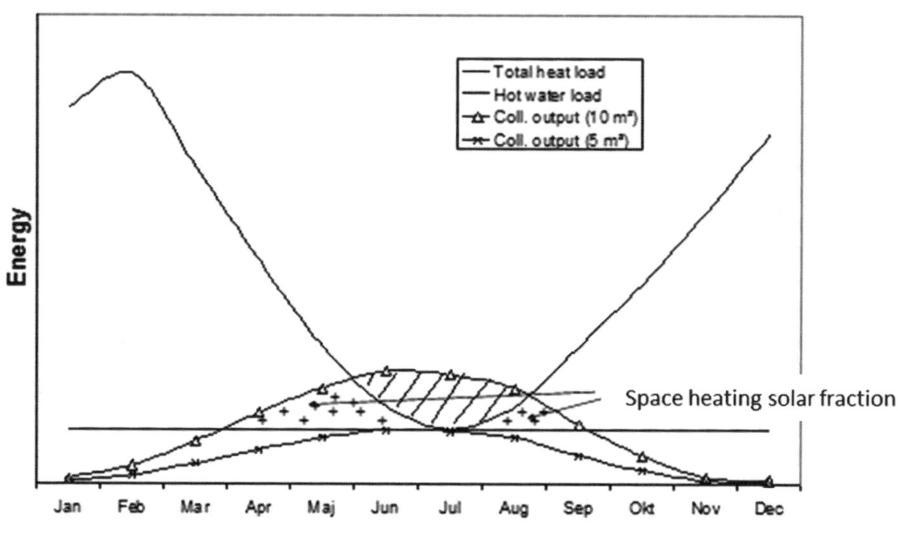

Fig. 2.23 Annual output for two specific collector areas

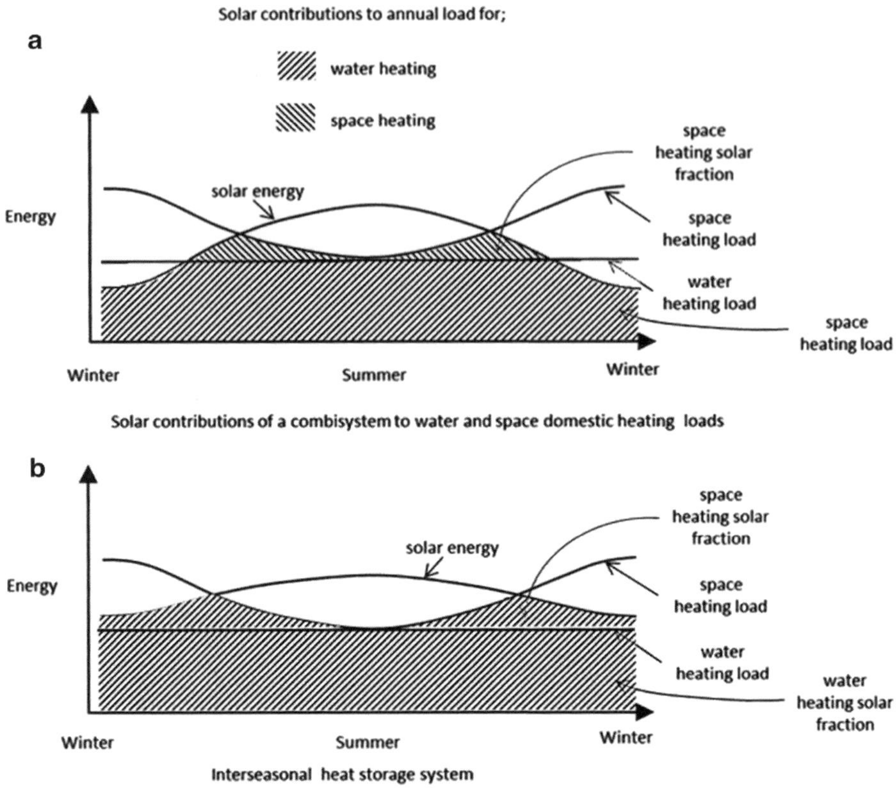

Fig. 2.24 Illustrative annual variations of average diurnal solar contributions of different solar energy systems for water and space heating loads in temperate climates

The geographic variation in the diurnal and annual variation of diffuse insolation is a critical factor limiting the use of concentrating collectors that collect solely direct insolation. The annual direct component at normal incidence is employed as a key selection factor for determining the most appropriate locations worldwide, as shown in Fig. 2.25, for concentrating solar thermal power generation. The example of Ghana, enlarged in Fig. 2.25, illustrates that there are very significant local variation in annual direct normal solar radiation, particularly in equatorial regions with monsoon or rainy seasons.

2.8 Solar Geometry

Solar energy originates in the interior regions of the Sun as a result of a hydrogen fusion reaction. Normal to the Sun's rays at the average sun-earth distance of 1.5×10^{11} m, the intensity of solar radiation incident per unit area measured

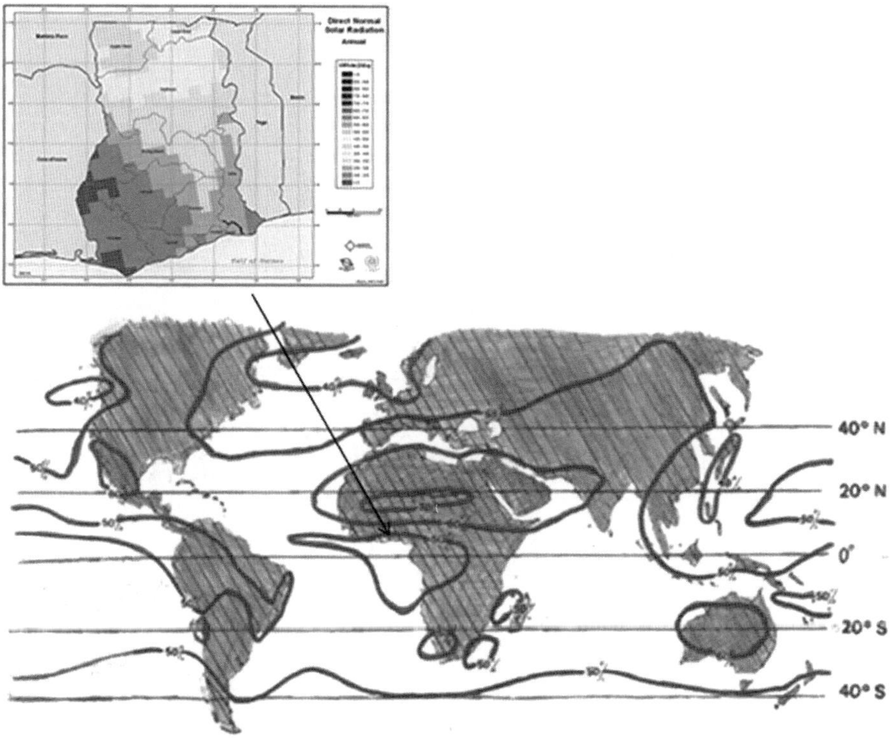

Fig. 2.25 Map indicating regions suitable for concentrating solar thermal power generation

outside the earth's atmosphere, is 1,367 W/m^2, this is called the solar constant. As the Earth's orbit is slightly elliptical, the intensity of radiation received outside the Earth's atmosphere varies ± 3.4 % over the year with the maximum intensity at the perihelion and the minimum at aphelion (Stine and Harrigan 1985).

Solar radiation varies over each day due to,

Geometrical factors:

- Influence of the slope and orientation of a surface on its interception of insolation;
- Obstruction of beam and diffuse solar radiation by neighbouring structures;
- Reflection of solar radiation from adjacent surfaces.

Physical factors:

- Total intensity and beam and diffuse insolation proportions due to changes in atmosphere conditions,
- The distribution of diffuse radiation over the sky and
- The variation of ground reflectance due to changes in ground conditions due to vegetation or snow.

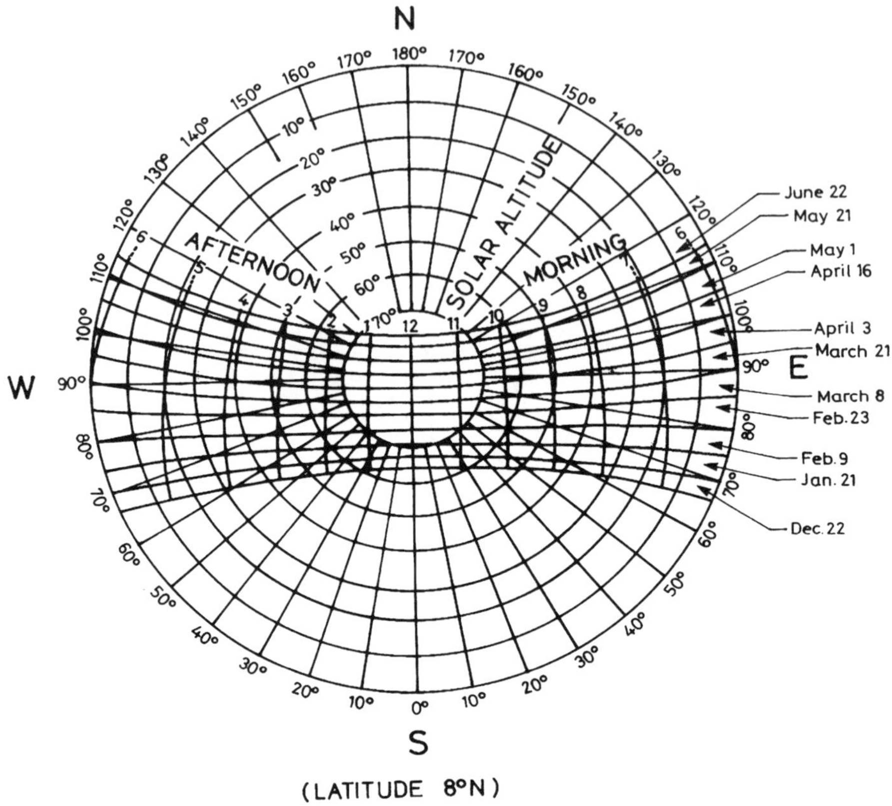

Fig. 2.26 A polar diagram

In the design of solar energy systems it is important to be able to predict the angle of incidence between the Sun's rays and a vector normal to the collection surface. The maximum amount of solar radiation reaching a surface is reduced by the cosine of the incident angle; referred to as the obliquity factor. Solar geometry is summarized in a polar diagram illustrated for 5 2° N and 8° W in Fig. 2.26.

For many solar thermal applications it is adequate to assume a uniform cloud distribution for the time of day. In an isotropic sky model, the diffuse component is assumed to be uniformly distributed over the sky; independent of direction. This is a reasonable approximation when there is a uniform cloud cover or when conditions are very hazy. Actual skyward diffuse radiance distributions can be both complex and transient depending on the form, height and movement of clouds and the air mass applicable. Concentrating solar energy systems with high concentrations do not collect diffuse insolation. Tracking systems are employed to align concentrators to be normal to incident solar radiation, no inclined plane solar incidences are involved. In concentrators the duration and intensity of direct normal incident insolation on the tracking place is a key design parameter.

2.9 Skyward Distribution of Diffuse Insolation

Air molecules (i.e. nitrogen, oxygen and other constituents) scatter radiation in very short wavelengths comparable to the size of molecules; this is called the Rayleigh scattering. Water droplets and aerosols scatter radiation whose wavelengths are comparable to the diameters of such particles. Therefore, an increase in the turbidity or dust content of the atmosphere or the cloud cover increases the scattering of solar radiation. As a result of scattering, part of the direct radiation is converted into diffuse radiation. Higher turbidity and cloud coverage increase the scattering of longer wavelength radiation which in turn causes the sky to be increasingly white. As a result of atmospheric scattering, some incident solar radiation is reflected back into outer space, while some of the scattered radiation reaches the earth's surface from all directions over the sky as diffuse radiation. Solar radiation that is neither scattered nor absorbed by the atmosphere, reaches the earth's surface as direct radiation. These processes are illustrated schematically in Figs. 2.12 and 2.13.

Diffuse solar radiation is usually anisotropic due to

- Overcast cloud
- Circumsolar brightening
- Horizon brightening and
- Broken cloud.

Broken cloud is a major cause of diffuse radiance anisotropy. An overcast sky of uniform cloud thickness, however, will be brighter near the zenith than near the horizon. This is opposite to the horizon brightening that occurs for a clear sky. Circumsolar brightening is caused by the strong tendency of dust matter suspended in the atmosphere, to scatter light by Mie scattering in the forward direction. Gas molecules scatter light by Rayleigh scattering which occurs equally in all directions. Thus, most of the beam radiation that is scattered by dust will deviate in a cone from the solar direction by only a few degrees from the direct beam. Consequently an isotropic sky around the Sun can have an insolation intensity up to ten times the mean skyward insolation depending on atmospheric dust content. Where air is very clear the circumsolar diffuse radiation is small, whereas in highly dusty or polluted air it produces a totally overcast sky. Atmospheric Rayleigh and Mie scattering give rise to spatially homogeneous turbidity that can be amenable to physical modelling (Kittler 1986). Clouds introduce spatial heterogeneity across a sky that can alter rapidly. Intermittent partial cloud cover is not modelled readily to produce generalizable predictions that can be compared with measurements. Indeed, conversely, it is correlation parameters obtained from measurements that enable the effects of partial cloud cover to be included in insolation models; for example the Perez tilted irradiance model (Perez et al. 1983, 1990a, b, c, 1993), includes two such parameters – a sky clearness parameter based on the ratio between the direct normal radiation and the diffuse horizontal radiation and a sky brightness parameter defined as the ratio of diffuse to extraterrestrial insolation. The sky brightness parameter is necessary to describe cloud opacity as the sky clearness parameter is constant under cloudy skies.

Even under cloudless skies the sky clearness parameter can inadequately represent turbid conditions.

Horizon brightening of clear skies arises because of greater the air mass traversed by insolation near the horizon which causes more insolation to be scattered toward the observer from near the horizon than from higher parts of the sky. This effect is due principally to Rayleigh scattering but will be enhanced by the presence of atmospheric dust that will be concentrated largely at low altitudes giving rise to low elevation Mie scattering.

To estimate the error introduced when assuming isotropic diffuse solar radiation, the diffuse radiance distribution was calculated from first principles for an atmosphere with no absorption or non-homegeity (Dave 1977). It was found that the assumption of isotropy would underestimate the diffuse insulation by a factor between 1 and 6. Liu and Jordan's (1962) model agrees well with observation at low irradiances (<300 Wm^{-2}), i.e. when the sky is overcast and the isotropic assumption is most valid but at higher irradiances >500Wm^{-2}, Liu and Jordan's model underestimated the irradiance on tilted surfaces by 3–20 % (Klucher 1979). Liu and Jordan's model is also deficient in predicting insolation on tilted surfaces for non-uniform clear but partly cloudy sky conditions (Klucher 1979).

Circumsolar diffuse insolation, when present in will cause measured diffuse insolation on an equator-facing surface to be greater than the diffuse insolation predicted from models that assume an isotropic sky (Lloyd 1984). Conversely, on a polar-facing surface when predicted assuming isotropy diffuse insolation will be overestimated. Horizon brightening renders insolation measured in any inclined plane greater than that predicted when assuming an isotropic sky. Predictions of global irradiance, assuming an isotropic diffuse insolation distribution, will be about 5 % too low for equator-facing surfaces and about 40 % too high for polar-facing surfaces (Lloyd 1984) but as insolation on a pole-facing surface is smaller (except for early and late parts of the day in summer at high latitudes), the absolute difference between measured and predoctral diffuse insolation will be similar for both these orientations.

The diffuse insolation distribution provided by Klucher's anisotropic 'all-sky' model (Klucher 1979) agreed with observation to within 5 % on average. It is based on the earlier 'clear-sky' model of Temps and Coulson (1977) who in turn applied two correction factors to the diffuse solar radiation term Liu and Jordan's approach. These factors were intended to represent the regions of anisotropy found in another more uniform overcast the diffuse radiance distribution Klucher (1979) extended Temps and Coulson's model by including a modulating function so that it could be used in all sky conditions, from clear to overcast.

Under overcast conditions, when the ratio of diffuse to global insolation, is unity, the all-sky anistropric model reduces to the Liu and Jordan isotropic model. For a clear sky, when the ratio of diffuse to global is small, the all-sky model approximates to the Temps and Coulson clear-sky model. Like Liu and Jordan, Hay (1979) resolved the total insolation on a tilted surface into three components: beam, diffuse and ground-reflected. Hay's anisotropic model differs from that of Liu and Jordan by assuming that all anisotropy in the diffuse insolation distribution on the tilted surface a circumsolar term is associated with circumsolar brightening as shown in Fig. 3.19. The two components are expressed in terms of a clearness index.

A comparison of models of diffuse solar radiation against three month's insolation data (Hogan and Loxsom 1981) concluded that neither Liu and Jordan's (1962) isotropic model, Hay's anisotropic model (Hay 1979) nor Klucher's all-sky model (Klucher 1979) provided accurate predictions for vertical surfaces. Klucher's all-sky model produced the most consistently correct results for their limited data base. Comparing the ground-reflection term used, Hogan and Loxom (1981) found that Klucher's model performed better than the other two, either with or without it. For example, the Klucher model correctly predicted the measured hourly global insolation 98.5 % of the time for a south-facing surface tilted at 20°. However, the same model showed significant errors in the predictions for vertical surfaces, facing either north, south, east or west, more than 50 % of the time. These three models were compared again (Ma and Iqbal 1983) using data collected at Woodbridge, Ontario. It was shown that Hay's and Klucher's models were equally accurate, and superior to Liu and Jordan's model, in predicting the insolation on inclined south-facing surfaces. The isotropic model underestimated the insolation consistently throughout the year; Hay's model also underestimated constantly but by a smaller amount; Klucher's model overestimated in summer and underestimated in winter. The maximum root mean square error for Klucher's and Hay's model was less than 15 % for slopes of up to 60°. All three models produced large errors at steep slopes (Ma and Iqbal 1983).

A model that compares favourably with those of Liu and Jordan and Hay and Klucher has been developed that includes (Perez et al. 1986):

- A geometrical representation of the sky dome incorporating independently variable circumsolar and horizon brightening;
- A parametric description of the insolation conditions,
- An experimentally-derived variation of circumsolar and horizon brightening with the insolation conditions.

Obstruction on the amount of solar radiation reaching a surface can cause

- The complete attenuation of beam radiation to form a shadow.
- Reduction of the diffuse component of the global insolation received.
- Less ground-reflected component to be incident
- Enhanced diffuse and specular reflection of insolation from its surfaces

The magnitude of these components will depend on the secularity and reflectance of the obstruction and incident angles and intensity of the insolation on its surfaces. Evaluating the solar energy lost and/or reflected to a surface by overshading requires numerical integration. Much work has been done on the analysis of shading geometries amenable to exact calculation as where analytical solutions can be found, they yield quicker results than those found by numerical integration (Sharp 1981, 1982).

References

Angstrom A (1924) Solar and terrestrial radiation. Q J Roy Meteorol Soc 150:121–126

Bendt P, Collares-Pereira M, Rabl A (1981) The frequency distribution of daily insolation values. Solar Energy 27:1–5

Burek SAM, Norton B, Probert SD (1988) Analytical and experimental methods for shadow-band correction factors for solarimeters on inclined planes under isotropically-diffuse and overcast skies. Solar Energy 40(2):151–160

Clark DR, Klein SA, Beckman WA (1983) Algorithm for evaluating the hourly radiation utilizability function ASME. J Solar Energy Eng 105:281–287

Collares-Pereira M, Rabl A (1979) Simple procedure for predicting long term average performance of non-concentrating and of concentrating solar collectors. Solar Energy 23:235–253

Dave JV (1977) Validity of the isotropic-distribution approximation in solar energy estimations. Solar Energy 19:331–333

Drummond AJ (1956) On the measurement of sky radiation. Arch Met Geophys Bioklim B7:413–436

Erbs DG, Klein SA, Duffie JA (1982) Estimation of the diffuse radiation fraction for hourly, daily and monthly – average global radiation. Solar Energy 28:293–302

Evans DL, Rule TT, Wood BD (1982) A new look at long term collector performance and utilizability. Solar Energy 28:13–23

George R, Maxwell E (1999) High-resolution ways of solar collector performance using a climatological solar radiation model. In: Proceedings of the annual conference of the American Solar Energy Society, Portland

Gordon JM, Hochman M (1984) On correlations between beam and global radiation. Solar Energy 32:329–336

Hay JE (1979) Study of shortwave radiation on non-horizontal surfaces. Canadian Climate Center. Report 79–12, AES, Downview

Hogan WD, Loxsom FM (1981) Preliminary validation of models predicting insolation on tilted surfaces. In: Proceedings of the annual meeting of the American section of the international solar energy society

Hollands KGT, Huget RG (1983) A probability density function for the clearness index, with applications. Solar Energy 30:195–209

Hottel HC (1976) A simple model for estimating the transmittance of direct solar radiation through clear solar atmospheres. Solar Energy 18

Ineichen P, Gremaud JM, Guisan O, Mermoud A (1983) Study of the corrective factor involved when measuring the diffuse solar radiation by use of the ring method. Solar Energy 31:113–117

Iqbal M (1983) An Introduction to solar radiation. Academic, Toronto

Kittler R (1986) Luminance model of homogeneous skies for design and energy performance predictions. In: Proceeding of the 2nd international daylighting conference, Long Beach

Klein SA (1978) Calculation of flat-plate collector utilizability. Solar Energy 21:393–402

Klucher TM (1979) Evaluation of models to predict insolation on tilted surfaces. Solar Energy 111–114

Landsberg HE (1981) The urban climate. Academic, New York

LeBaron BA, Peterson WA, Dirmhirn I (1980) Corrections for diffuse irradiance with shadowbands. Solar Energy 25:1–13

Littlefair PJ (1985) The luminous efficacy of daylight, a review. Light Res Technol 17:162–182

Liu BYH, Jordan RC (1960) The interrelationship and characteristic distribution of direct, diffuse and total solar radiation. Solar Energy 4:1–19

Liu BYH, Jordan RC (1962) Daily insolation on surfaces tilted towards the equator. Trans ASHRAE 526–541

Liu BYH, Jordan RC (1963) A rational procedure for predicting the long-term average performance of flat-plate solar energy collectors. Solar Energy 7:53–74

Liu BYH, Jordan RE (1965) Performance and evaluation of concentrating collectors for power generation trans. ASME Journal of Engineering for Power 87:1–7

Lloyd PB (1984) Solar energy for engineers. Helios 22: Solar energy unit, University College, Cardiff

Lunde PJ (1980) Solar thermal engineering. Wiley, New York

Ma CCY, Iqbal M (1983) Statistical comparison of models for estimating solar radiation on inclined surfaces. Solar Energy 31:31–317

Marion W, Wilcox S (1994) Solar radiation data manual for flat plate and concentrating collectors. Report NREL/TP-463-5607, National Renewable Energy Laboratory, Golden

Maxwell E, George R, Wilcox S (1998) A climatological solar radiation model. In: Proceedings of the annual conference of the American solar energy society, Albuquerque

Mujahid A, Turner WD (1980) Diffuse sky measurements and determination of corrected shadow band multiplication factors. ASME annual winter meeting, paper no 80-WA/Sol-26

Norton B, Abu-Ebeid M (1989) Estimation of mean monthly daily total insolation from mean monthly daily ambient temperature. Ambient Energy 10:151–162

Page JK (1961) The estimation of monthly mean values for daily total short wave radiation on vertical and inclined surfaces from sunshine hours for latitudes 40°N to 40°S. In: Proceedings of the UN conference on new sources of energy, Rome, pp 378–390

Painter HE (1981) The shade ring correction factor for diffuse irradiance measurements. Solar Energy 26:361–363

Perez R, Scott JT, Stewart R (1983) An anisotropic model for diffuse radiation incident of hopes of different orientations and possible applications to CPCs. Prog Solar Energy 6:883–888

Perez R, Stewart R, Arbogast C, Seals R, Scott J (1986) An anisotropic hourly diffuse radiation model for sloping surfaces: description, performance validation, site dependency evaluation. Solar Energy 36:481–497

Perez R, Ineichen P, Seals R, Zelenka A (1990a) Making full use of the clearness index for parameterising hourly insolation conditions. Solar Energy 45:111–114

Perez R, Ineichen P, Seals R, Michalsky J, Stewart R (1990b) Modeling daylight availability and irradiance components from direct and global irradiance. Solar Energy 44:271–289

Perez R, Seals R, Zelenka A, Ineichen P (1990c) Climatic evaluation of models that predict hourly direct irradiance from hourly global irradiance; prospects for performance improvements. Solar Energy 44:99–108

Perez R, Seals R, Michealsky J (1993) An all-weather model for sky luminance distribution – a preliminary configuration and validation. Solar Energy 50:235–245

Perez R, Ineichen P, Moore K, Kmiecik M, Chain C, George R, Vignola F (2002) A new operational satellite-to-irradiance model. Solar Energy 75:307–317

Prescott JA (1940) Evaporation from water surface in relation to solar radiation. Trans Roy Soc 54:114–118

Rawlins F, Readings CJ (1986) The shade ring correction for measurements of diffuse irradiance under clear skies. Solar Energy 37:407–416

Reddy TA, Kumar S, Saunier GY (1985) Review of solar radiation analysis techniques for predicting long-term thermal collector performance – applicability to Bangkok data. Renew Energy Review J 7:56–80

Reddy SJ (1987) The estimation of global solar radiation and evaporation through precipitation. Solar Energy 38:97–104

Robinson N, Stoch L (1964) Sky radiation measurements and correction. J Appl Meteorol 3:179–181

Robledo L, Soler A (2001) On the luminous efficiency of diffuse solar radiation. Energy Convers Manage 42:1181–1190

Sharp K (1981) Sun angles and shading analysis for surfaces at any tilt or azimuth. In: Proceedings of the 1981 annual meeting, AS/ISES

Sharp K (1982) Calculation of monthly average insolation on a shaded surface at any tilt and azimuth. Solar Energy 28:531–538

Spencer DW, Oettinger BS, Stewart R (1982) Diffuse band correction factors for short time intervals. Progress in solar energy. In: Proceedings annual general meeting of the American Solar Energy Society, pp 1253–1257
Steven MD, Unsworth MH (1980) Shade-ring corrections for pyranometer measurements of diffuse solar radiation from cloudless skies. Quart J Royal Meteorol Soc 106:865–872
Stine WB, Harrigan RW (1985) Solar energy fundamentals and design. Wiley, New York
Temps RC, Coulson KL (1977) Solar radiation incident upon slopes of different orientations. Solar Energy 19:179–184
Theilacker JC, Klein SA (1980) Improvements in the utilizability relationships. American Section of the international solar energy society. Proceedings, Phoenix, pp 271–275
Trewarthu GT, Horn LH (1980) An introduction to climate, 5th edn. McGrew-Hill, New York
Van den Brink GJ (1982) Climatology of solar irradiance on inclined surfaces IV- part II. Validation of calculation models, Royal Dutch Meteorological Institute (KNMI). Final report: EEC contract no ESF-006-80 NL (B)

Chapter 3
Optics and Heat Transfer in Solar Collectors

A couple of thousand years dropped in the ocean of time will completely exhaust the coal fields of Europe, unless, in the meantime, the heat of the sun be employed

John Ericsson (1868)

3.1 Overview

All substances emit electromagnetic radiation continuously in a particular range of the electromagnetic spectrum, the dominant form that such energy takes depends upon its nature and the form of the applied external excitation; electrical conductors emit radio waves when excited by an alternating current; certain elements emit X-rays if excited by atomic bombardment and if heated to a sufficiently high temperature all substances will emit visible light. Cosmic rays, X-rays, r-rays, visible light and radio waves are forms of electromagnetic radiation that, when absorbed by a substance, usually produces a very small heating effect. The wavelengths of the electromagnetic spectrum that interact with matter to produce significant radiative heating are confined to a band from approximately 0.1 to 100 μm; this includes a portion of the ultraviolet light together with all visible (0.40–0.7 μm) and infrared light bands. For many solar thermal systems the optical characteristics and geometries of aperture materials, reflectors and absorbers determines solar heat gains. How much of that heat is retained is determined largely by heat transfer (i) across air gaps, evacuated spaces, and insulation materials and (ii) provided by forced or buoyant removal of fluid from a collector.

B. Norton, *Harnessing Solar Heat*, Lecture Notes in Energy 18,
DOI 10.1007/978-94-007-7275-5_3, © Springer Science+Business Media Dordrecht 2014

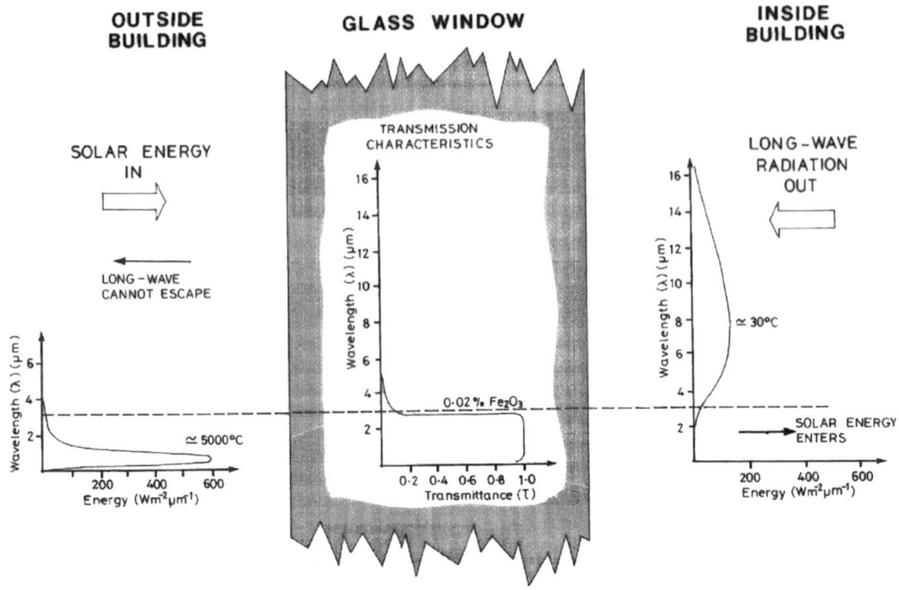

Fig. 3.1 Transmission through a glass pane

3.2 Glass

Low transmittance to infra-red radiation assists in the retention of solar gains, as illustrated for a glass window in Fig. 3.1. Glass is the oldest, and still used most commonly glazing material (Butti and Perlin 1980). The physical and optical properties of glass, having been investigated extensively, are well documented (Rawson 1982).

3.3 Plastic Materials

Most plastic materials have inferior light-transmission properties compared with glass (Oreski et al. 2010). Plastics tend to degrade when exposed to heat and the ultra-violet components of the solar radiation spectrum thereby reducing their useful life: Many plastics can only be used for a few years, compared with decades for glass. Despite these disadvantages, the use of plastic materials often can have a trade-off in terms of a lightweight structure, higher strength resulting in lower capital and/or running costs. Polyethylene is a widely used plastic film particularly in horticulture, where its short lifetime of about 1 year when exposed to typical weather conditions is appropriate for seasonally clad used for protective horticulture Polymers containing fluorine compounds have insolation transmission properties long-wave retention and/or resistances to ageing superior to polyethylene films. Polyamide, polystyrene, acrylics and polycarbonate, have been investigated for solar energy applications (e.g. as cover glazing for flat-plate collectors). Detailed mechanical properties of a wide range of plastics are available (Harper 1975), with some suitability for solar energy applications (Blaga 1978). Degradation processes for plastic films break up the long polymer chains, typically several thousand monomer units long. The strength and flexibility of a plastic film depend on the polymer chain length: the longer the chains, the less brittle the material. Degradation processes are;

- Thermal degradation
- Photo-degradation both involve the migration of hydrogen atoms and the formation of free radicals, commonly resulting in depolymerisation
- Oxidation also results in depolymerisation, especially at chain branches, and
- Mechanical breaking of chains by tearing, surface scratching and repeated flexing (Grassie 1972; Ranby and Rabek 1975).

The aging behavior of plastics is characterized in climates where consistently high levels of humidity, temperature and solar radiation accelerate the ageing process, even so a single test can still require several years. For accelerated aging, samples are subjected for several hundreds of hours to artificial accelerated weathering environments e.g. exposed to periods of intense light, to simulate solar irradiation, and sprayed with water, to simulate exposure to rain, in controlled cycles (Ranby and Rabek 1975). Lifetime modeling applied successfully to the ageing behaviour of polycarbonate (Kahlen et al. 2010a, b). In general, exposure of plastic films to weather reduces light transmittance. Veld and fracture strengths are reduced and plastic films become more brittle after ageing (Kimball and Munir 1978). Veld and fracture strengths decrease with ambient temperature but total elongation at fracture rise with ambient temperature. Consequently plastic films are weaker but more ductile at higher temperatures. Tearing of plastic greenhouse covers thus tends to take place during cold winter months, not only because of wind damage but also because plastic materials are more brittle at the lower environmental temperatures then prevailing. Thermal effects alone account for some 80 % of the decreases in light transmittance (Cheng and Bannerot 1983). Degradation due to solar radiation exposure has been observed to have only a small effect on particular 'uv - inhibited' samples, and wind, rain and dust have only slightly more significant effects (Cheng and Bannerot 1983). The degradation arising from simultaneous exposure to all three influences is little more than that due to exposure to thermal effects alone. Absorption and forward scattering of insolation are both affected similarly. Chemical degradation that weakens plastic material increases during prolonged periods of high insolation.

Thermotropic switchable plastic materials have been examined for use as aperture covers that become opaque before high temperatures are likely to damage an all polymeric collector (Wallner et al. 2008).

3.4 Transmittance of Solar Energy

To determine the angular variation of transmittance Fresnel's expression for the reflection of unpolarised radiation passing between media with different reflective indices as shown in Fig. 3.2 may be used. Various simplified expressions are shown in Fig. 3.3.

The variation in transmittance with incidence angle, as calculated by the approximate method of Duffie and Beckman (1974), of a aperture cover system comprising 1, 2 and 4 panes of float glass with an average refractive index of 1.526 for the solar spectrum and a thickness 4 mm is shown in Fig. 3.11.

The spectral transmittances of various types of glass and plastic suitable for use in collector systems, are available (Touloukian and Dewitt 1972; Meinel and

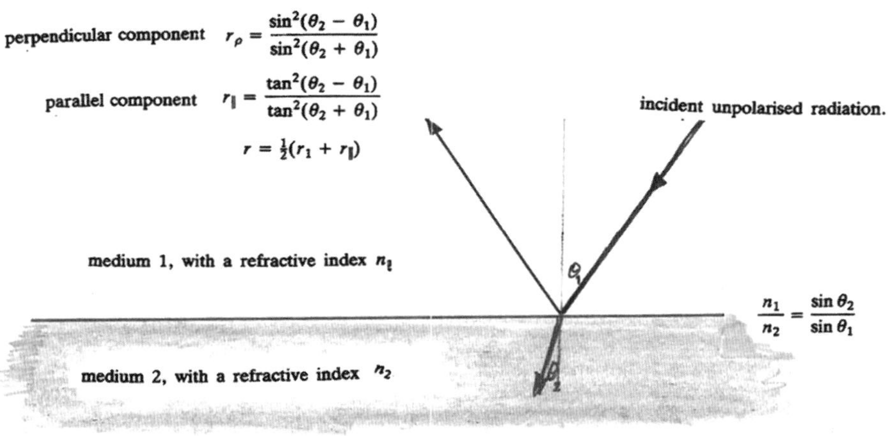

Fig. 3.2 Fresnel expression for radiation passing between media with different refractive indices

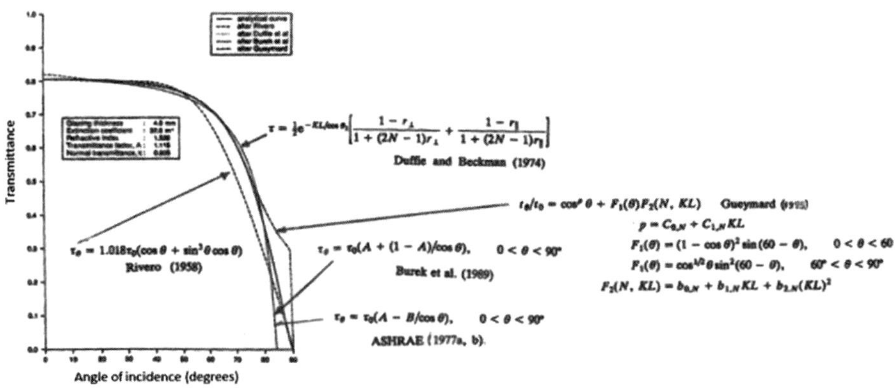

Fig. 3.3 Equation for the angular variation of transmittance of a single aperture glazing

Meinel 1976). Extinction coefficients are generally less well documented, although they can be calculated from available data. The manufacturers of these materials only usually provide spectral transmittance at normal incidence. This information is adequate only for general comparisons for detailed a hour-by-hour thermal performance simulations (variation in transmittance of glazing materials with incidence angle) is essential. When normal incidence transmittance is the only available data it may be extended to the full range of incident angles by curve fits such as shown in Fig. 3.3 (Rivero 1958). Such simple relationships can be adequate for the analysis of typical glazing materials. The measured variation in transmittance with angle for dirty window glass with a film of dirt also follows the form of a cosine function of the angle of incidence (Stephenson 1965).

Transmittance curves calculated using the expressions of Duffie and Beckman (1974); Riviero (1958); Burek et al. (1989); Gueymard (1989) to a transmittance curve for one sheet of float glass (4 mm thick) obtained by analytical means are in Fig. 3.3. The raised portion of the curve calculated using Riviero's (1958) method in Fig. 3.3, between 0° and 16° and that it diverges significantly from the analytically-derived curve at angles greater than 60° means that it is not a satisfactory approximation for general glazing applications. The approximation by Duffie and Beckman (1974) is essentially a slight modification of the rigorous analytical approach, and is so close to the true curve, that they are indistinguishable. The approximations by both Burek et al. (1989) and Gueymard (1989) exhibit good agreement for angles up to 60–70° but diverge markedly from the analytically derived curve at angles above 80°. This, however, would not be a problem in most practical situations as any contribution from transmitted radiation at such shallow angles would be small. The greater inherent simplicity of the method by Burek et al. (1989) with only one 'coefficient', the incidence transmittance factor, is its advantage over Gueymard's (1989) method that involves several equations and associated coefficients.

In many solar energy applications, a pane of transparent (or semi-transparent) material forms the aperture of the collector system. There are two parallel interfaces for each pane (e.g. air/glass and glass/air) with reflection at each interface. The radiation reflected at each interface however, is different for each component of polarization. The refractive index is also wavelength dependent, but this variation is usually negligible at solar energy wavelengths and for relevant materials is so average refractive indices for the visible spectrum are used usually. The effect of multiple reflections in an aperture pane, ignoring absorption losses is shown in Fig. 3.4.

Figure 3.4 shows the variation of transmittance with angle of incidence of a cover system comprising 1, 2, 3 and 4 sheets of non-absorbing glass with an average refractive index of 1.526 for the solar spectrum. In a real situation however, a glazing material exhibits some absorption due to the interaction of the incident radiation with the molecular structure of the medium. The absorbed energy is transformed into heat, which causes a rise in temperature of the medium. Bouguer's Law, that the absorbed radiation is proportional to the local intensity of the medium and the distance the radiation travels in the medium, is illustrated in Fig. 3.5.

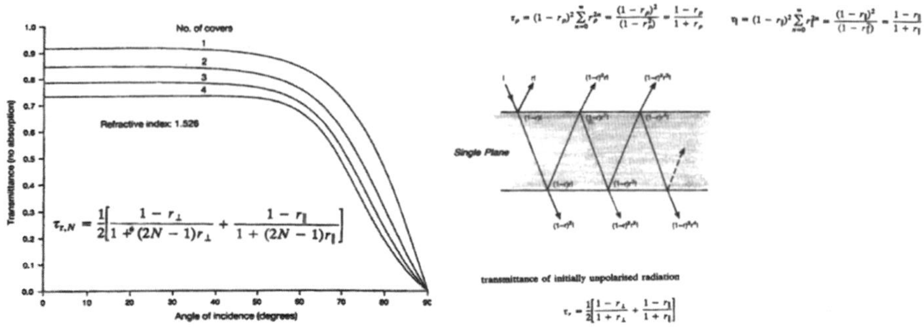

Fig. 3.4 Transmittance variation with incident angle for multiple aperture panes without absorption

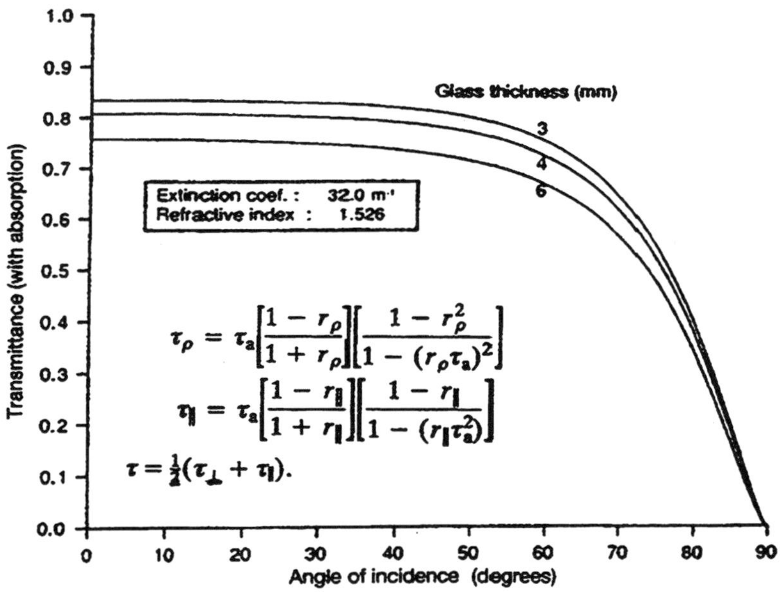

Fig. 3.5 Transmittance variation with incident angle for multiple panes with absorption

Refractive indeces and extinction coefficients are readily available, for aperture glazing materials suitable for solar energy applications (Harper 1975; Rawson 1982). Figure 3.5 shows the variation in transmittance with angle for a single glass cover of thicknesses 3, 4 and 6 mm.

For aperture glazings comprised of more than one pane of absorbing glazing material, both multiple reflections ensue within each pane and multiple reflections ensue between the individual panes. With detailed knowledge of the optical properties of the materials comprising each aperture pane ray-tracing (Whillier 1953a, b), or analytical (Siegel 1973; Shurcliff 1974; Sharafi and Mukminova 1975) methods can be used to generalise for any number of covers.

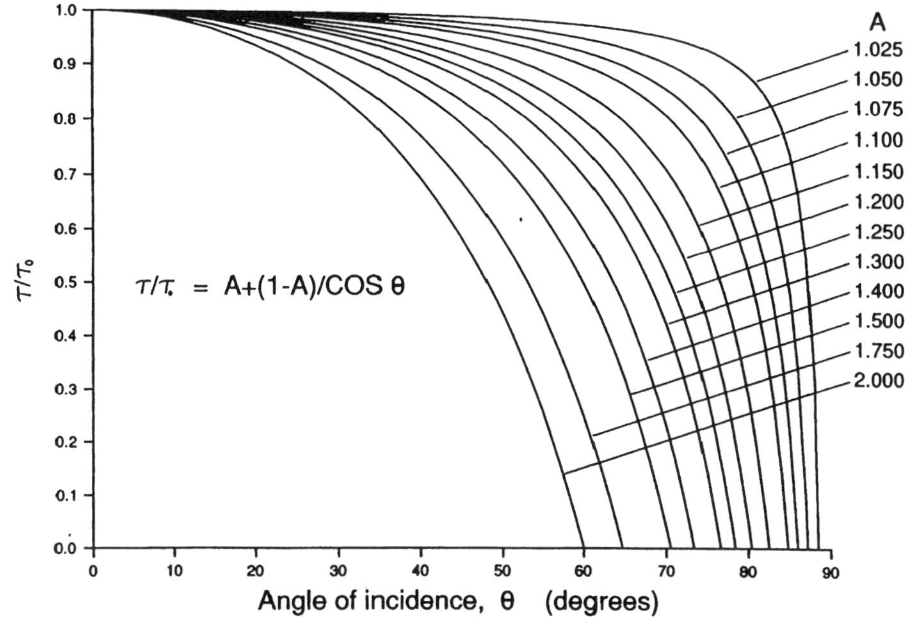

Fig. 3.6 Variations of transmittance with angle of incidence produced by the expression shown for different values of A

ASTM Standard E424 (Anon, ASTM 1985) describes a standard test to determine transmittance of global solar radiation on clear days close to the solar noon and with the plane of the test material orthogonal to the sun's rays. Modified versions of this method have been used to measure the variation of transmittance with incidence angle (Edlin 1959; Godbey et al. 1979; Robbins and Spillman 1980; Bhaduri and Nguyen 1983; Fintel and Jakubowski 1985). The standard test conditions specify a 'clear, sunny day', which, although implying a low proportion of diffuse insolation, are less than rigorous. In the ASHRAE standard testing procedure for the thermal performance of solar-energy collectors (Anon, ASHRAE 1977a), the much simpler expression was used for the incident angular variation of the transmittance of the beam radiation shown in Fig. 3.3. Burek et al. (1989) modified the ASHRAE expression is also shown in Fig. 3.3.

Angular dependence of transmittance can be described as a function of incidence angle, the normal transmittance and an incidence transmittance factor A as shown in Fig. 3.6. Simple techniques can be used to determine the value of the constant A for a wide variety of materials but should not be used at high incidence angles as τ tends to $-\infty$ as θ approaches 90°. Values for A have been provided for a range of plastic transparent and semi-transparent materials (Burek et al. 1989). A normalised transmittance is a particular angular transmittance divided by transmittance at normal incidence. An approximating function to the normalised transmittance curve is shown in Fig. 3.3 (Gueymard 1989).

The transmittance of photochromic and thermochromic devices are dependent on prevailing insolation and ambient temperature. They are thus useful for overheating prevention in solar energy collector aperture covers (Resch et al. 2009).

3.5 Solar Energy Reflector Materials

3.5.1 Flat Reflectors

An external flat reflector is a mirror built into the collector system but outside the sealed casing positioned to reflect additional insolation onto the aperture as shown in Fig. 3.7.

The reflector increases the effective collection area, by enabling greater incident angle acceptance and also, in integral passive solar water heaters, may also serve as moveable night-time insulation (Baer 1975). The first reported use of an external flat reflector utilised to enhance performance, was in 1911, when Shuman developed a water pumping system powered by a flat-plate/reflector assembly (Larson 1980). The instantaneous and integrated optical performance of various collector/reflector combinations has been studied (McDaniels et al. 1975; Baker et al. 1978; Grassie and Sheridan 1977; Wijeysundera 1978) with performance estimates and optimisation studies have been undertaken for both vertically and horizontally oriented systems (McDaniels et al. 1975; Seitel 1975; Kaehn et al. 1978). Theoretical calculations and experimental tests have considered specular, diffuse and combined specular/diffuse reflective surfaces (Grimmer et al. 1978). Such experimentally studies of reflector/flat-plate collector combinations have shown significant

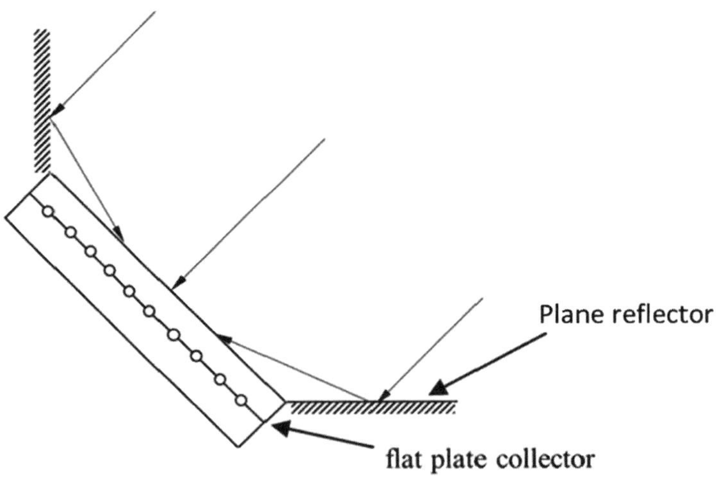

Fig. 3.7 A plane-mirror augmented flat plate collector

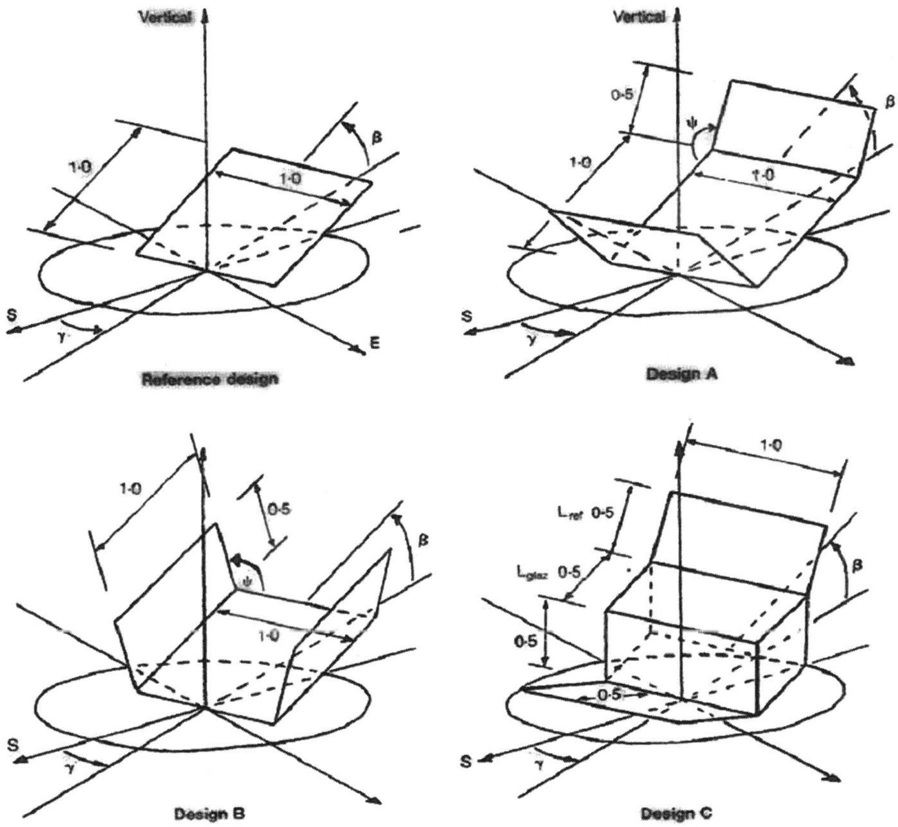

Fig. 3.8 Geometry of glazing and flat reflectors showing the glazing inclination and the surface azimuth angle

performance improvements over comparable unenhanced collector systems (McDaniels et al. 1975; Williams and Craig 1976; Grassie and Sheridan 1977; Weinstein et al. 1977; Kaehn et al. 1978).

Four designs shown in Fig. 3.8(i)–(iv), have been compared for use with integral passive solar water heaters at latitudes of 30°N and 45°N (Favard and Nawrocki 1981). The glazing/reflector combinations A to C were compared to the reference design, which comprised a simple inclined glazed aperture and no reflector.

The optimal angle between reflector and aperture glazing at 30°N latitude for each month of the year is shown in Fig. 3.9. Figure 3.10 shows the total energy transmitted through the glazing at 30°N, for the different designs. Adjusting the reflectors of design A monthly gave a 27 % improvement in daily total energy collection over the reference design, but only a 16 % improvement if the reflectors were fixed at the yearly optimum. Monthly adjustments to the reflectors of design C gave a 70 % improvement, and utilising the fixed yearly optimum a 50 % improvement over a C design without reflectors, but did not show significant improvement over the reference design. Design B offered little advantage over the reference design.

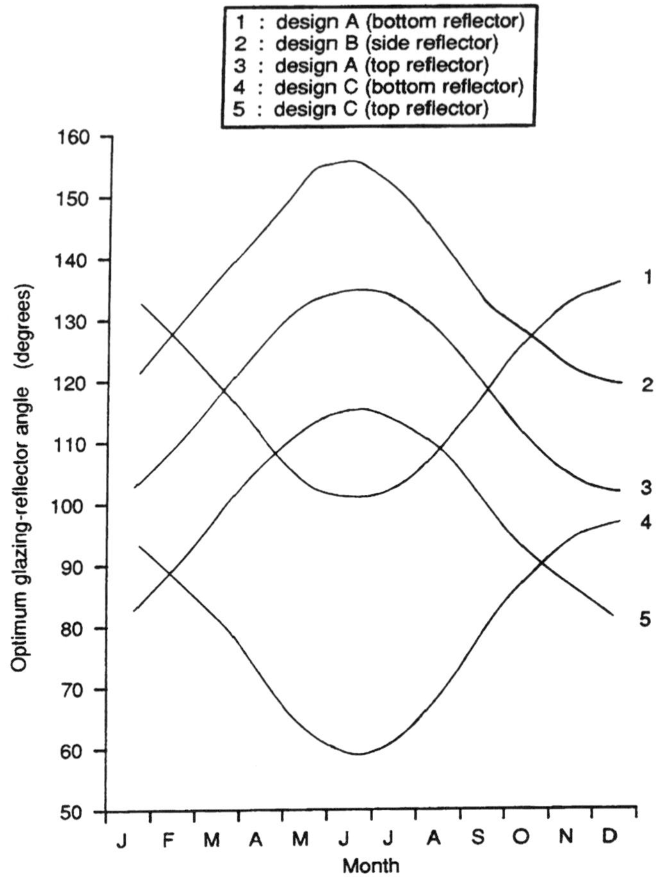

Fig. 3.9 Optimal flat reflector/glazing angles

Figure 3.11 shows the increase in can be seen from Fig. 3.20 that increasing the reflector length by more than 2.5 times the glazing length provided no additional improvement.

3.5.2 Parabolic Reflectors

The two types of line-axis parabolic solar-energy concentrators are in common use are compound parabolic concentrating (CPC); and parabolic-trough concentrating (PTC). The geometries of both these concentrators are illustrated in Fig. 3.12a, b respectively.

A line-focus CPC concentrator is characterised by an acceptance half-angle as shown in Fig. 3.12 that determines the maximum attainable concentration ratio, is

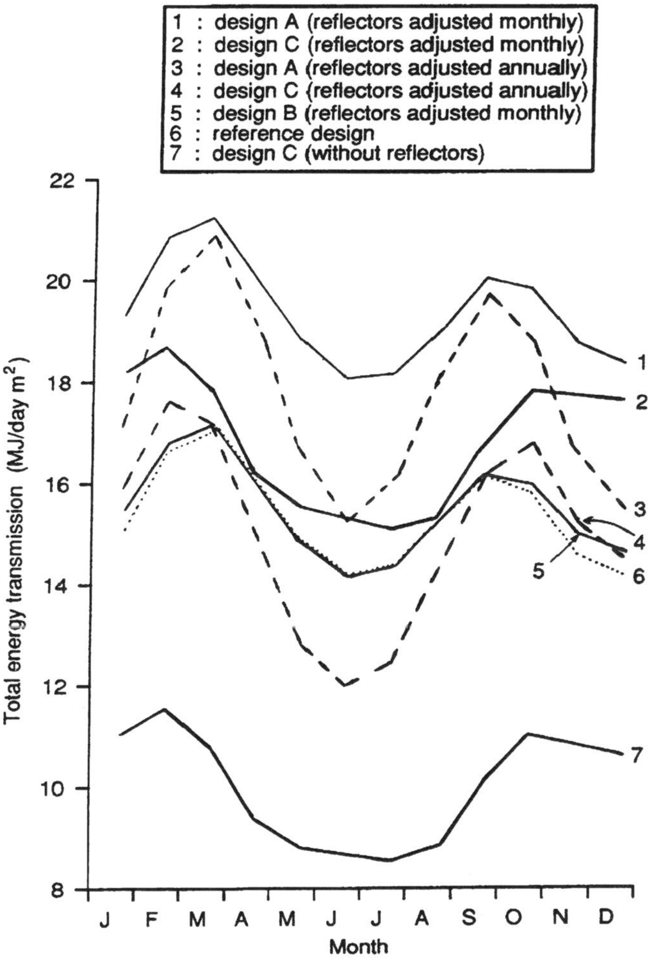

Fig. 3.10 Comparison of energy transmitted for monthly–adjusted flat different reflectors

given in Fig. 3.13 (Winston 1974). This maximum concentration ratio can be achieved only by a full-height CPC, i.e. no truncation is applied at the top of the reflectors, and if the absorber is of optically-correct area with an absorber area $1/C_{max}$ of the aperture area.

Definitions of concentration ratio and its limits for two and three dimensional concentrators are provided in Fig. 3.13 concentration ratio of real system is lower than given in Fig. 3.13 because truncation of the concentrator top, undertaken normally to reduce the capital cost (Carvalho et al. 1985); and oversizing of the absorber's diameter, to allow for optical scatter introduced by imperfections arising during manufacture and operation. Absorbers of non-circular cross-section may also be employed.

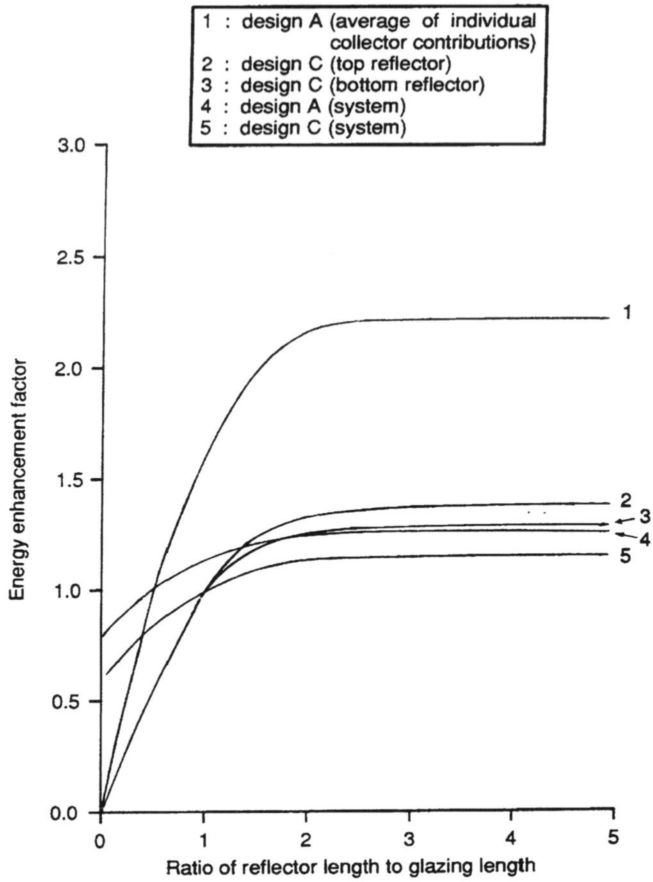

Fig. 3.11 Total energy transmitted through the glazing as a function of reflector length to glazing length ratio for the reflectors individually and for the complete systems

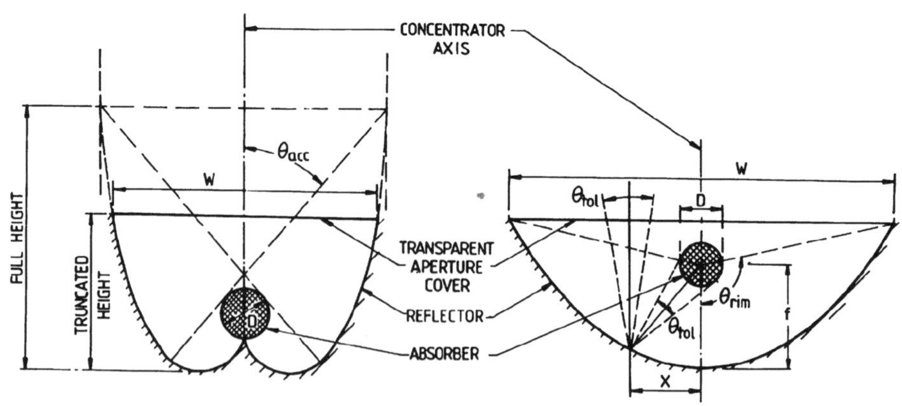

Fig. 3.12 Geometries of compound parabolic and parabolic trough reflectors

- **Concentration ratio (C) is defined as the ratio of the aperture area to the receiver/absorber area, i.e.:**

$$C = \frac{A_a}{A_r}$$

- **the maximum possible concentration is**
- **a single axis tracking collector given by:**

$$C_{max} = \frac{1}{\sin(\theta_m)}$$

where θ_m is the half acceptance angle

- **for two-axes tracking collector:**

$$C_{max} = \frac{1}{\sin^2(\theta_m)}$$

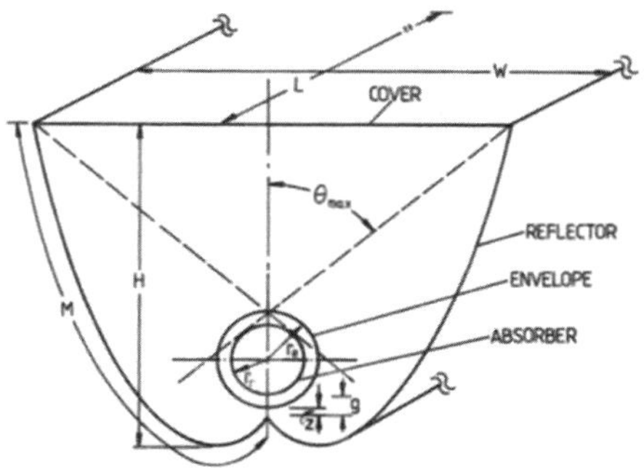

- **For a perfect collector and tracking system C_{max} depends on a width of the sun's disk 32'.**
- **For single axis tracking:**
  C_{max} = 1/sin(16') = 216
- **For full tracking:**
  C_{max} = 1/sin²(16') = 46,747

Fig. 3.13 Concentration ratio

A particular shape is require for a specularly reflecting surface to convert a collimated beam of light of any distribution into another specified distribution over an arbitrary absorber surface (Burkhard and Shealy 1975). In a PTC direct insolation is focused on a line-axis absorber, to return high optical efficiency periodic solar tracking is required (Look and Sundvold 1983; Cachorro and Casanova 1986; Evans 1977; Mills 1986). A PTC collector, shown schematically in Fig. 3.12, is assumed to track the Sun continuously, so any ray entering the concentrator parallel to its axis will, either after reflection or directly, intercept the tubular absorber. The finite diameter of the absorber allows some additional rays, not parallel to the concentrator axis, to reach the absorber. This can be expressed by a local tolerance angle θtol (see Fig. 3.12), whose value varies according to the position at which a particular light ray is incident on the reflector surface, as shown in Fig. 3.14. A mean tolerance angle, θ_{m}, is defined as the average of the local tolerance angles across half the aperture width, W/2 as shown in Fig. 3.15 where $g_{D,\ PTC}$ is the exploitable fraction of the diffuse insolation incident at the aperture.

In Fig. 3.15 depicted, by a single curve, is the exploitable fraction of the diffuse insolation, assuming an isotropic angular the diffuse insolation distribution. Unlike the mean tolerance angle, the fraction of diffuse insolation exploitable by a PTC is not dependent on the rim-angle of the PTC collector. The rim-angles less than 90° the fraction of diffuse insolation exploitable by a PTC decreases. When diffuse radiation is assumed to exist only in a range of −90 to 90° with respect to the concentrator axis (Rabl 1985). In reality, however, for rim <90° albedo and diffuse consolation originates from the ground and from a portion of the sky respectively; their relative contributions depend on the inclination of the reflector trough.

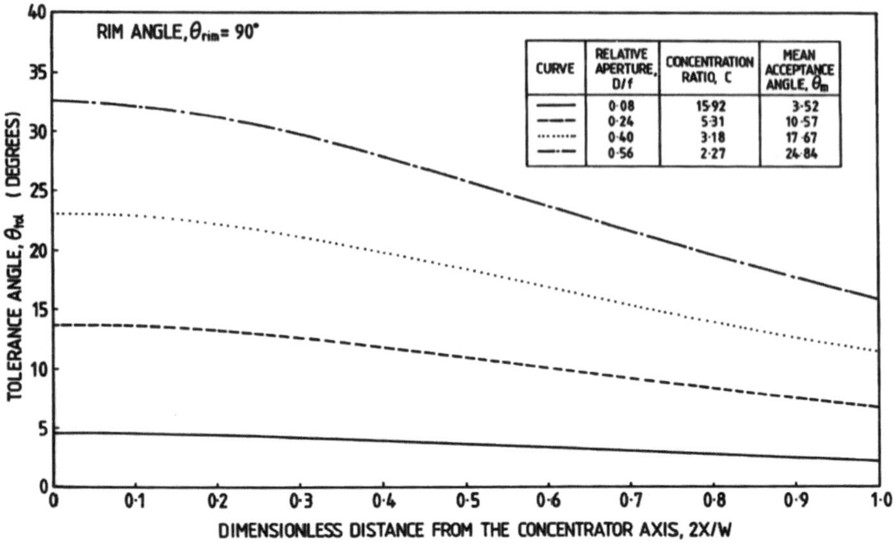

Fig. 3.14 Tolerance angle variation in a parabolic trough collector

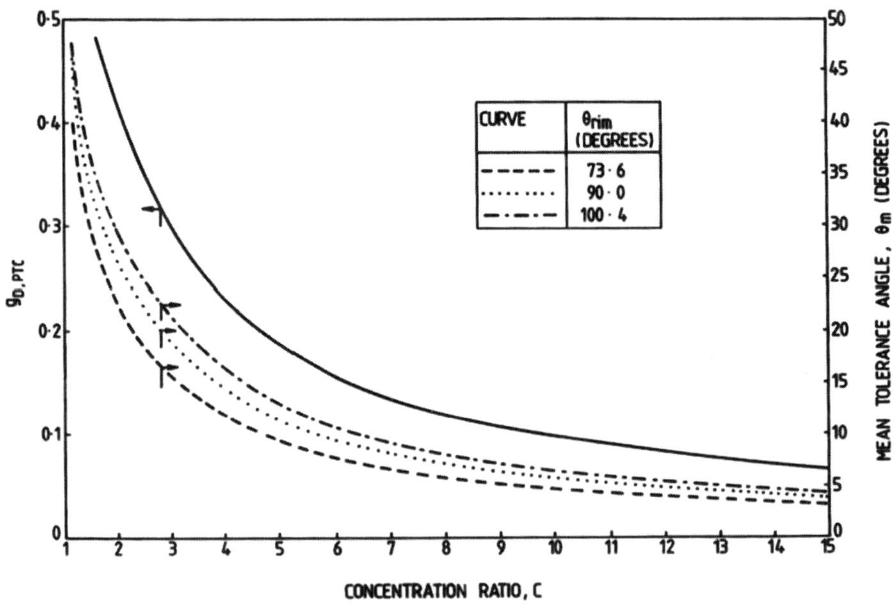

Fig. 3.15 Mean tolerance angles and exploitable diffuse insolation for parabolic tough collectors with concentration ratios in the range $1 < C < 15$

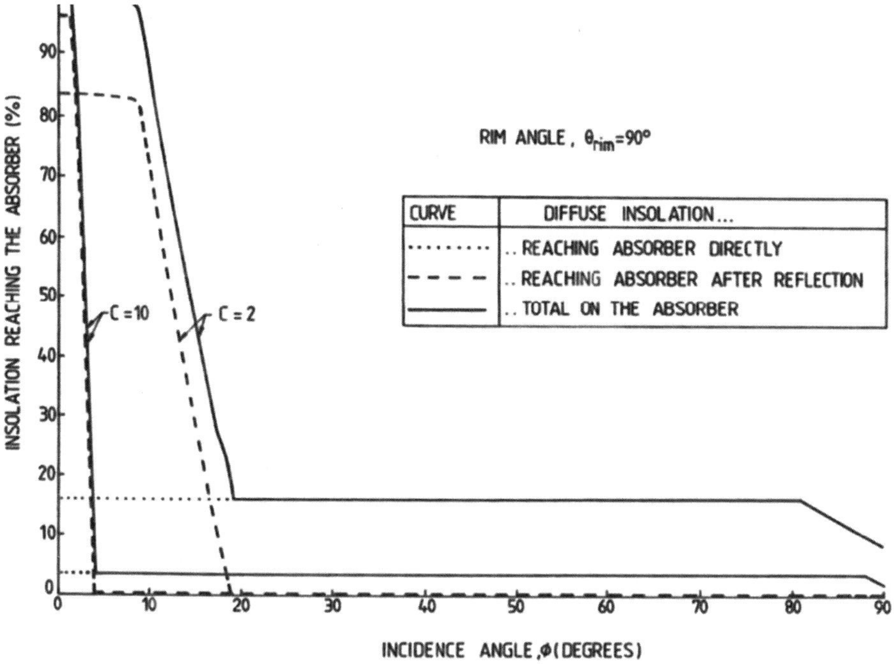

Fig. 3.16 Diffuse insolation paths in a parabolic trough collector

Unlike non-concentrating collectors, only a fraction of the diffuse insolation is exploitable by concentrating collectors. This can be shown by considering the factors for the radiation exchange between absorber-aperture and aperture-absorber respectively (Rabl 1976).

For a CPC collector the absorber-aperture exchange factor is unity, as any ray emitted from the absorber will either directly, or after one or more reflections, reach the aperture. Thus if an isotropic diffuse distribution is assumed for the diffuse radiation then the aperture-reflector radiation exchange factor also represents the exploitable part of the diffuse insolation of a CPC collector.

Unlike the acceptance of the CPC collector, the mean acceptance angle of a PTC collector is not an intrinsic optical property. $\theta acc_{\mathrm{PTC}}$ (Rabl 1985) is that tolerance angle that corresponds to the rim of the reflector cross-section. The latter definition quantifies a PTC collector average optical behavior and it is thus comparable with CPC collector acceptance angle. The direct and the diffuse components of the insolation can reach the PTC absorber via two different routes: either directly, i.e. without the participation of the reflector; or indirectly, i.e. via a single reflection at the reflector (the optical design of a PTC collector ensures that a light ray reaching the absorber with the participation of the reflector will experience no more than one reflection). The diffuse insolation delivered to the absorber is shown in Fig. 3.16.

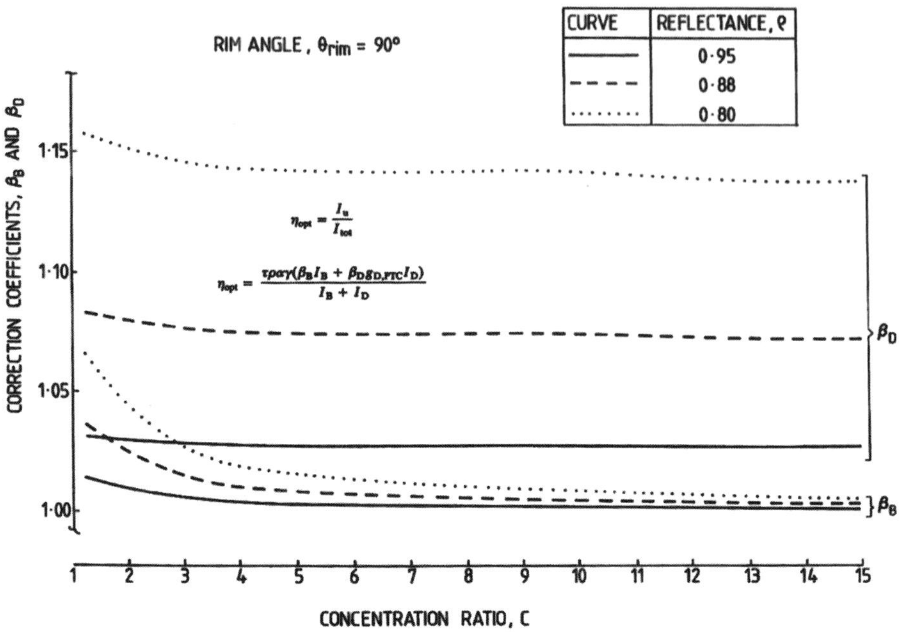

Fig. 3.17 Variations of correction coefficients β_B and β_D for parabolic trough collectors with concentration ratios in the range $1 < C < 1S$ together with the expression for overall optical efficiency.

For an ideal CPC the exploitable part of the diffuse insolation is $^1/c$. For a PTC collector, the reflector-absorber radiation exchange factor is less than unity as the absorber can "view" itself on the reflector. Thus, the exploitable part of the diffuse insolation of a PTC collector, is less than $^1/c$ and can be calculated by numerical integration.

At incidence angles near normal, all the diffuse insolation reaches a PTC absorber, most of it after reflection. However, at incidence angles greater than a critical angle, depending on the concentration ratio, the diffuse insolation can only reach the absorber directly. The curves shown in Fig. 3.16 detail quantitatively the contribution of the diffuse insolation to the collected solar energy and can thus be used to calculate the correction coefficients β_B and β_D that account for the direct and diffuse insolation respectively that reach the PTC absorber directly; that is unattenuated by reflector losses. The variation of these correction coefficients with the concentration ratio is shown in Fig. 3.17 for three reflector reflectances. It can be seen in Fig. 3.17 that both correction coefficient are slightly greater than unity. It was found that the correction coefficient exhibits negligible dependence on the rim angle of the concentrator; thus, values for a rim-angle of 90° can also be used for other rim-angles.

In Fig. 3.17 the intercept factor γ (Rabl and Bendt 1982) accounts for the optical losses occurring in a real PTC due to optical errors.

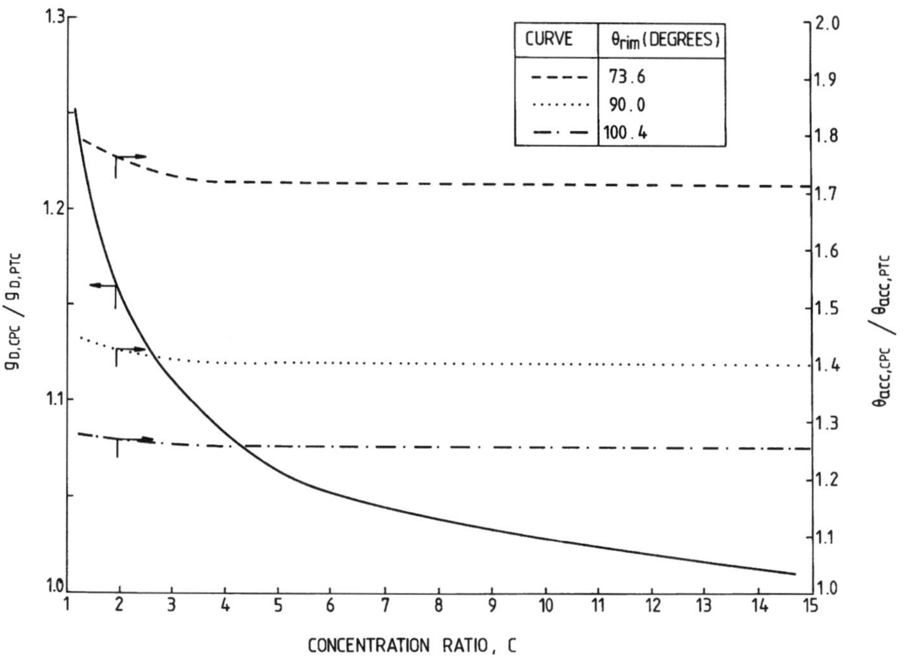

Fig. 3.18 Comparison of the exploitable fraction of incident diffuse insolation for CPC and PTC collectors

A PTC collector can be compared with a CPC collector. The CPC compared was assumed to be truncated so that its maximum concentration ratio is reduced by a factor of 0.3. Although the angular acceptance of a CPC is also affected by truncation (Carvalho et al. 1985) the design acceptance angle was not altered. It can be seen from Fig. 3.18 that a CPC collector exploits a greater part of the available diffuse insolation compared with a PTC, although this advantage diminishes as the concentration ratio increases. A CPC collector also maintains its superiority with respect to the acceptance angle.

To consider how particular skyward angular distributions of the diffuse insolation affect the performance of a PTC solar-energy collector, three distributions in the cross-section plane of the collector can be considered:

- Isotropic,
- Cosine and
- Hybrid Gaussian.

The shape of these three distributions are shown in Fig. 3.19. All distributions are normalised, i.e. the areas under the respective curves in Fig. 3.19 are the same. The hybrid Gaussian distribution combines an isotropic background with a circumsolar Gaussian part. This latter distribution is more realistic for a tracking system than use of an isotropic model that would underestimate insolation at near-zero

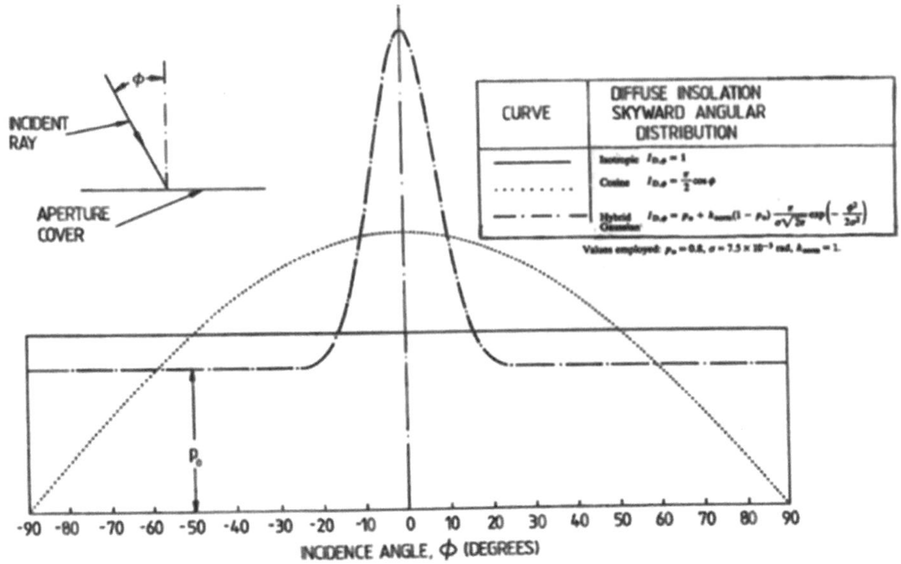

Fig. 3.19 Three angular diffuse insolation distributions

incidence angles and the cosine model that would underestimate insolation at large incidence angles. The analytical expressions for the three distributions considered are also given in Fig. 3.19.

The optical efficiency, n_{opt}, for a parabolic trough concentrator with the properties shown for the three alternative diffuse insolation distributions considered, is depicted in Fig. 3.20 for three representative values of the diffuse fraction. The hybrid Gaussian distribution yields invariably a higher optical efficiency whereas the efficiency curve given by the isotropic distribution is the lower limit. The difference in the optical efficiency for these two cases, although more pronounced for lower concentration ratio and for higher diffuse fractions is less than 2.5 % for a concentration ratio of 3 and less than 1.1 % for a concentration ratio of 10, demonstrating the weak dependence of overall PTC collector performance on the angular diffuse insolation distribution.

Real PTC collectors have an overall optical error (Rabl 1985) representing the combined individual errors as shown in Fig. 3.21 where σ represents the standard deviation for each respective error arising as indicated by each ‘subscript’. The total optical error is derived by taking into account the standard deviation of the angular intensity distribution of sun’s disc (Rabl 1985) also shown in Fig. 3.21.

The effect of the total optical error on the optical performance of a PTC collector in Fig. 3.21 has been analysed for four different total optical errors; an ideal case, a concentrator with good quality optics and two cases represent concentrators with mediocre quality optics. As can be seen from Fig. 3.21 optical efficiency is insensitive to optical errors at low concentration ratios, but as the concentration ratio increases, the optical efficiency deteriorates.

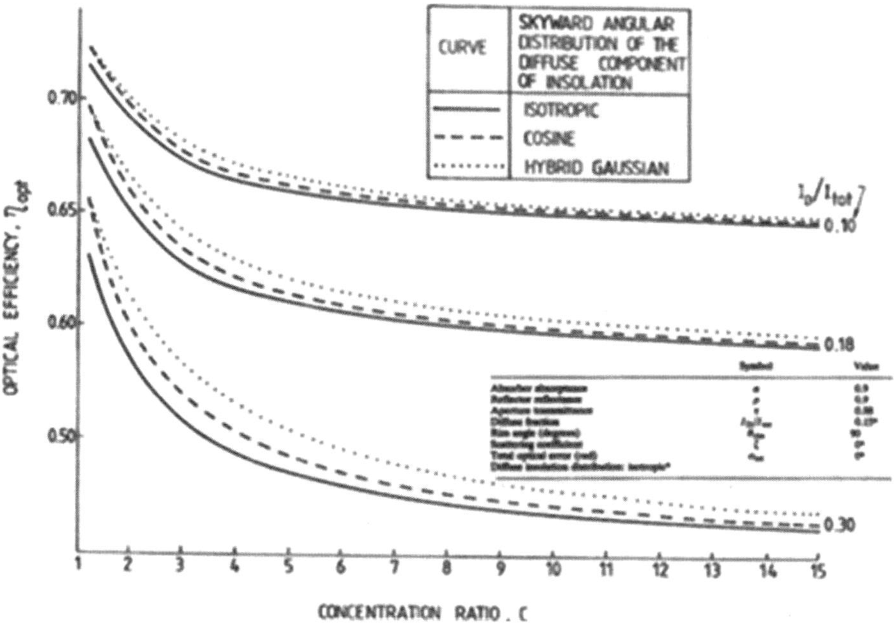

Fig. 3.20 Optical efficiency of a parabolic trough concentrator for three diffuse insolation skyward distributions for $1 < C < 15$

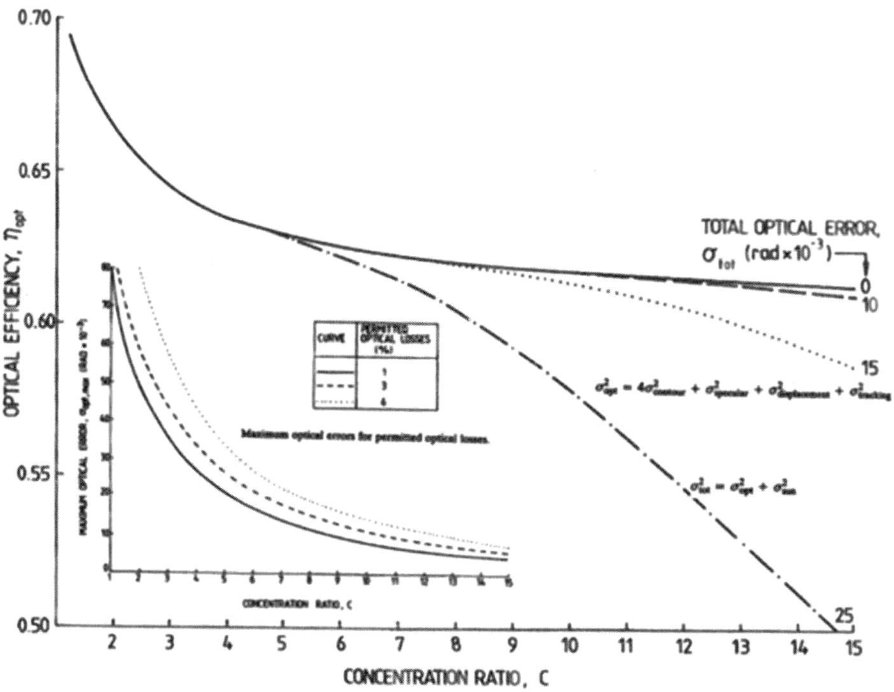

Fig. 3.21 Optical errors and optical performance for a parabolic trough collector

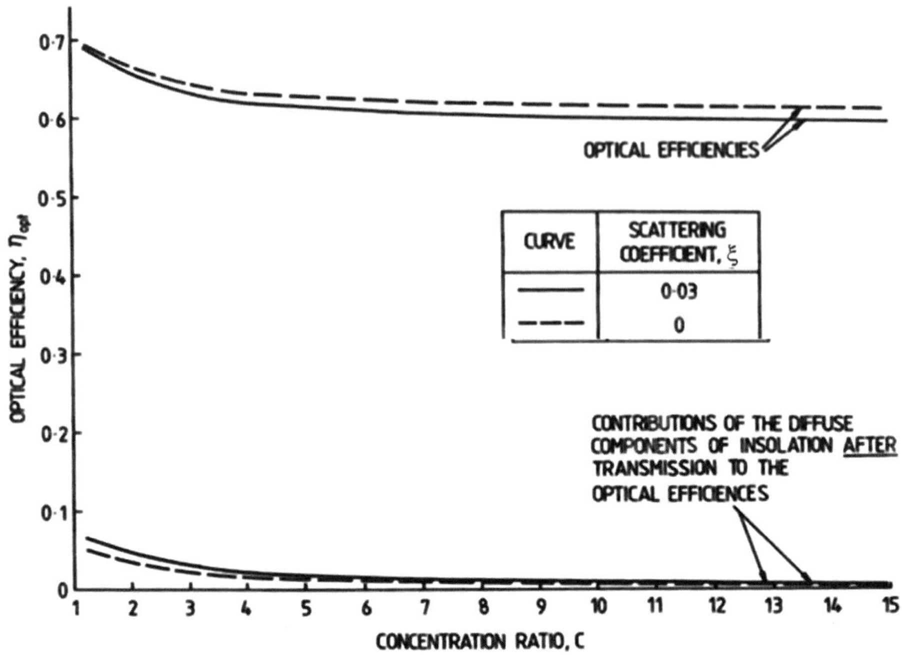

Fig. 3.22 Effect of surface scattering to dust accrual on PTC optical efficiency

For $\sigma_{sun} = 4 \times 10-3$ rad corresponding to average "clear-sky" conditions (Rabl and Bendt 1982) the maximum optical error σ_{opt} for permitted optical losses is also illustrated in Fig. 3.21, 1 %, 3 % and 6 %. As the maximum optical error is insensitive to the rim-angle of a PTC collector within the range of concentration ratio depicted, results are for a rim angle of 90°.

In passing through the transparent aperture, insolation not only suffers transmission losses but is also forward scattered. Such scattering may be increased by dust deposited on the cover. The optical performance of a concentrator is impaired by the subsequent deterioration of the collimation of the direct insolation through the concommitant forward scattering of diffuse insolation has a negligible effect on the concentrator's performance. The angular intensity of the forward scattering of direct insolation is assumed to be a cosine distribution. For an illustrative surface scattering co-efficient of $\xi = 0.03$ the effect of scattering on the optical efficiency of a PTC collector can be seen in Fig. 3.22 the optical efficiency is reduced by scattering occurs. The contribution of the diffuse component of the insolation (after transmission through the aperture cover) to the optical efficiency increases simply because the post-transmission diffuse component is augmented by forward scattered "direct" insolation.

A comparison of PTC and CPC collectors, performed on the basis of their acceptance angles and the amount of diffuse insolation exploitable shows the CPC to be the more efficient design. However, the ranges of concentration ratios found in commercially-manufactured collectors of the two types are not the same: CPC collector designers take advantage of the possibility of constructing stationary concentrators, with concentration ratios less than 2. To enable the CPC trough- axis to be orientated North–south concentration ratios as low as 1.1 +1.2 have been used. By contrast, for effective operation, PTC collectors as they track the sun are best used at higher concentration ratios. Commercially-manufactured PTC solar-energy collectors may be divided into those with

- Concentration ratio of the order of 40 that are very accurate optical devices, and
- Those with a concentration ratio in a range of 3–10, that are of moderate optical accuracy (Mills 1986).

The exploitable part of the diffuse insolation becomes negligible for high concentration ratios. When the diffuse component constitutes 10–25 % of the total hemispherical insolation then for a concentration ratio of 10 (for which the exploitable fraction of diffuse insolation is approximately 0.09) the amounts of diffuse insolation exploitable by a PTC constitutes 0.9% +2.2 % respectively of the total hemispherical insolation. A concentration ratio of 10 can be used to distinguish PTC collectors into those with

- Low concentration ratios (i.e. with $C < 10$) for which some diffuse insolation can be collected, and
- Those with high concentration ratio (i.e. with $C > 10$), for which the amount of the diffuse insolation collected is insignificant.

3.6 Heat Transfer in Line-Axis Concentrators

To simplify steady state modeling of heat transfer in concentrators the following physical assumptions are usually made:

- The absorber is a circular tube located in the line-focus of the reflector.
- The whole collector is aligned accurately with reflector surfaces are free from imperfections.
- A mean value for the reflectance of the reflector surface is used, that takes into account the multiple reflections experienced by the incident radiation, before reaching the absorber.
- The absorber may be coated with a selective material and may enclosed within a concentric transparent glass envelope as shown
- The aperture cover fitted across the top of the concentrators and glass envelope transmittance and the absorptance of all components exhibit their mean values, irrespective of the incident beam direction: This assumption, while not introducing a significant error, simplifies the expressions for the energy-exchanges between the collector components.
- Third and higher-order reflections of the incident beam are neglected, as being insignificant with respect to the overall performance.

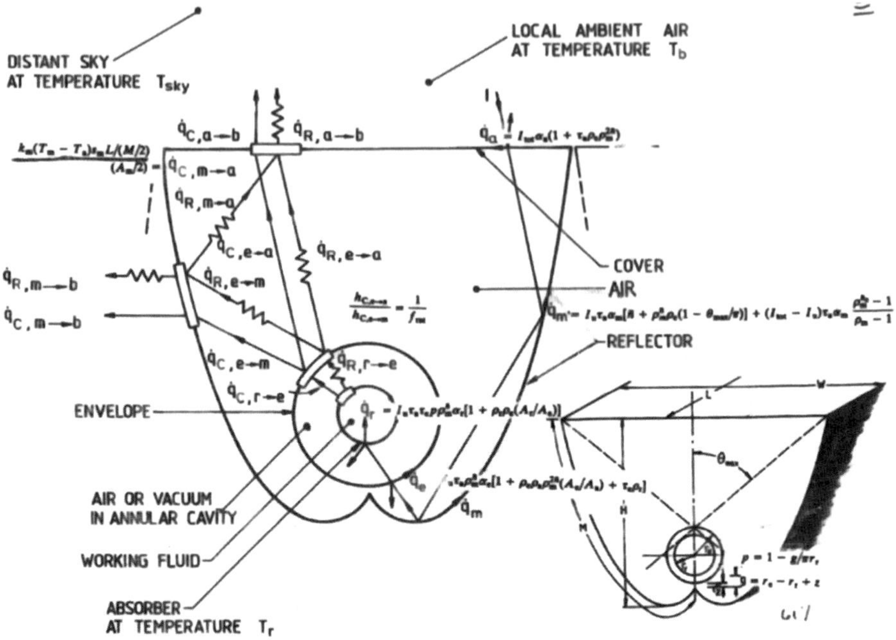

Fig. 3.23 Heat exchange processes in a line-axis compound parabolic solar energy collector

- The variation of temperature along the length of the collector is neglected. The temperature attained by a particular collector component, as a result of the heat exchanges, is considered to be the average value of the actual temperature distribution within that component.

Based on these assumptions, the various heat fluxes are shown accordingly for a CPC in Fig. 3.23 the symbols a p and T stand for the absorbances, reflectance, emittance and transmittance respectively. A and T represent area and temperature respectively, whereas the subscripts, a, e, m and r refer to the cover, envelope, reflector and absorber respectively. Part of the incident solar radiation is absorbed by the absorber envelope (Hsieh 1981), and part of incident solar radiation is absorbed by the reflector. Part of the solar radiation absorbed by the reflector would otherwise have reached the absorber. Absorbed solar radiation entering the CPC collector at an incidence angle $> \theta_{max}$ attenuated by absorption at the reflecting surfaces and the top cover, escapes out of the collector after n reflections across the reflector surfaces.

The various terms for heat exchanges between the collector components can be calculated via the expressions in Fig. 3.24. The corresponding heat transfer coefficients and thermal resistances are illustrated in Fig. 3.25. The heat-exchange rates in Fig. 3.24 (i.e. column 2 and 3) are based on the area of the component indicated first in column 1. The values of the absorbtances, reflectance and emittance for the expressions appearing in Fig. 3.24 refer to the infra-red region of the

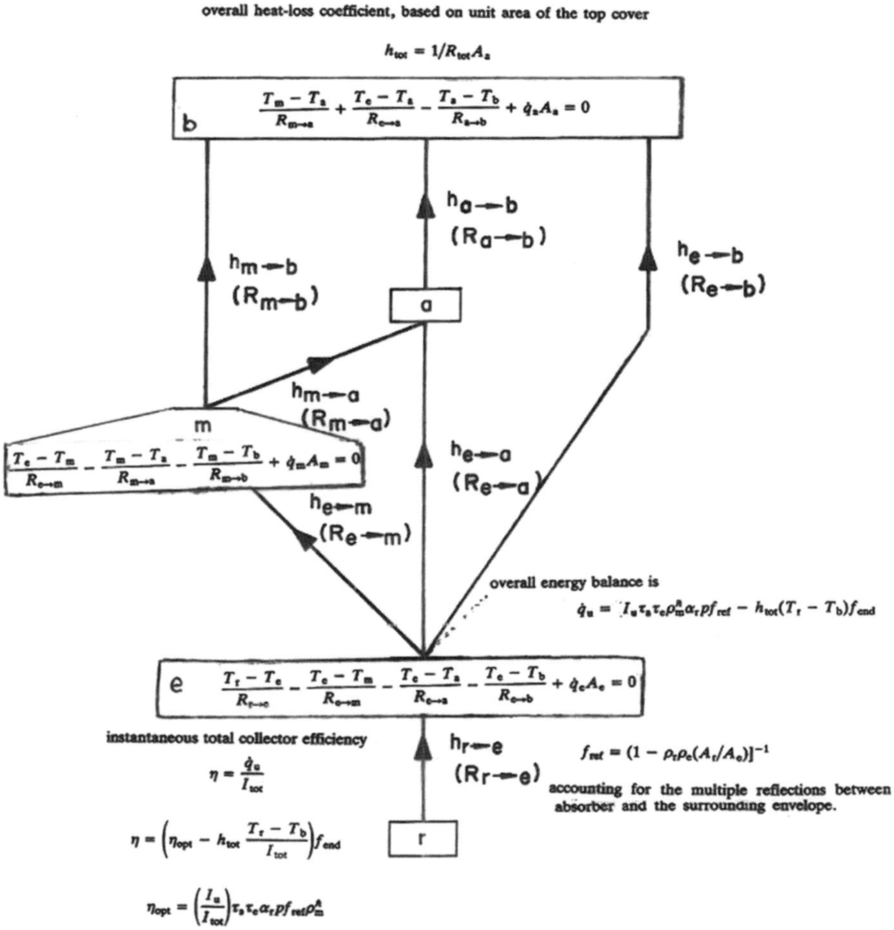

Fig. 3.24 Equation for heat exchanges in a compound parabolic concentrator

electro- magnetic spectrum, whereas in Eqs. 8.2.4, 8.2.5, 8.2.6 and 8.2.7 the values for the same parameters refer to the whole solar radiation spectrum.

The conductive heat-losses through the reflector to the top cover are weekly calculated using the reflector temperature at the middle of the reflector half-length, assuming that a linear temperature gradient exists from that point to the top cover. The expressions in Fig. 3.24 have inherent uncertainties, particularly the terms for the heat exchange between the envelope and its surroundings due to a lack of exact analytical expressions for the convective heat transfer in the particular geometry considered. The envelope is considered as an eccentrically-positioned inner cylinder in a horizontal tubular annulus, the outer cylinder being geometrically equivalent to the enclosure formed by the reflector and the top cover. The heat losses to the reflector and to the top cover are then calculated. From experimental correlations, a value of $f_{rat} = 0.55$ in Fig. 3.23 is considered appropriate (Prapas et al. 1987a).

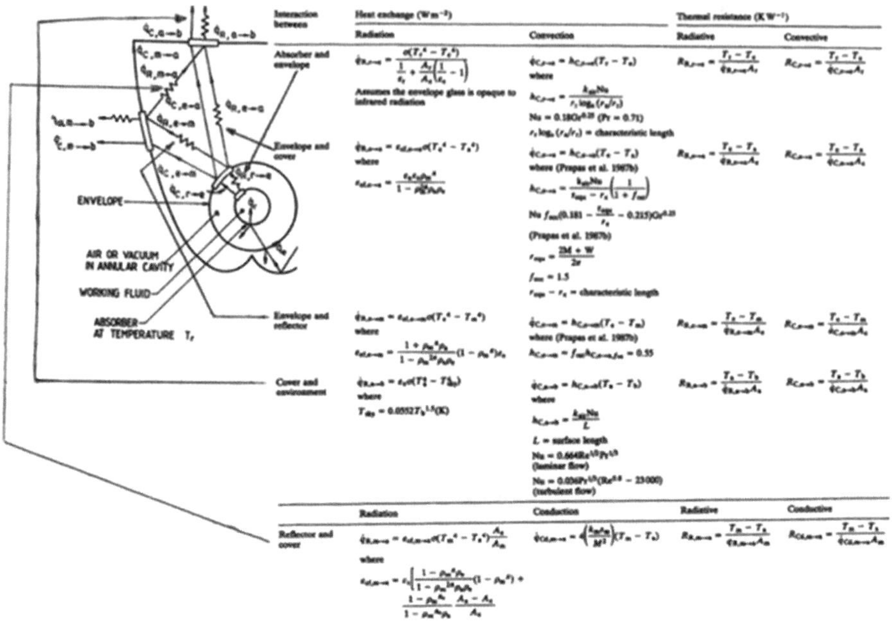

Fig. 3.25 Energy balances on overall system, the envelope, reflector and aperture cover of a concentrating collector

Heat losses from the reflector to the ambient environment are strongly dependent upon the insulation of the side-wall. These are however negligible for a properly-designed collector. The value of the heat transfer coefficient corresponding to this term in Fig. 3.25 has been derived by estimating the side-wall heat losses to be one tenth of those from the top cover of a well-insulated collector. The overall heat-transfer coefficient was found to be relatively insensitive to variations of this term; a change by a factor of four results in only an 0.5 % +14 % increase in the overall heat transfer coefficient and a consequent 0.1 % +0.5 % decrease in the overall collector efficiency.

For a concentrator without an aperture cover direct heat losses ensue from the envelope to the environment. This heat loss mechanism included in Fig. 3.25 to render the analysis applicable to open aperture concentrators.

Figure 3.25 illustrates the simultaneous radiative and convective energy balances for the envelope, the reflector and the top cover respectively. The unknown temperatures in each component of the collector are initially estimated and recalculated iteratively until the equations in Fig. 3.25 are satisfied.

Figure 3.26 illustrates Hottel-Whillier-Bliss characteristic curves for CPC solar-energy collector. The curves "cross-over" at different concentration ratio at higher temperatures with evacuation of the cavity and addition of a selective surface to the absorber. When both evacuation and a selective absorber are employed, the superior performance of the low concentration ratio collection (with C = 1.55)

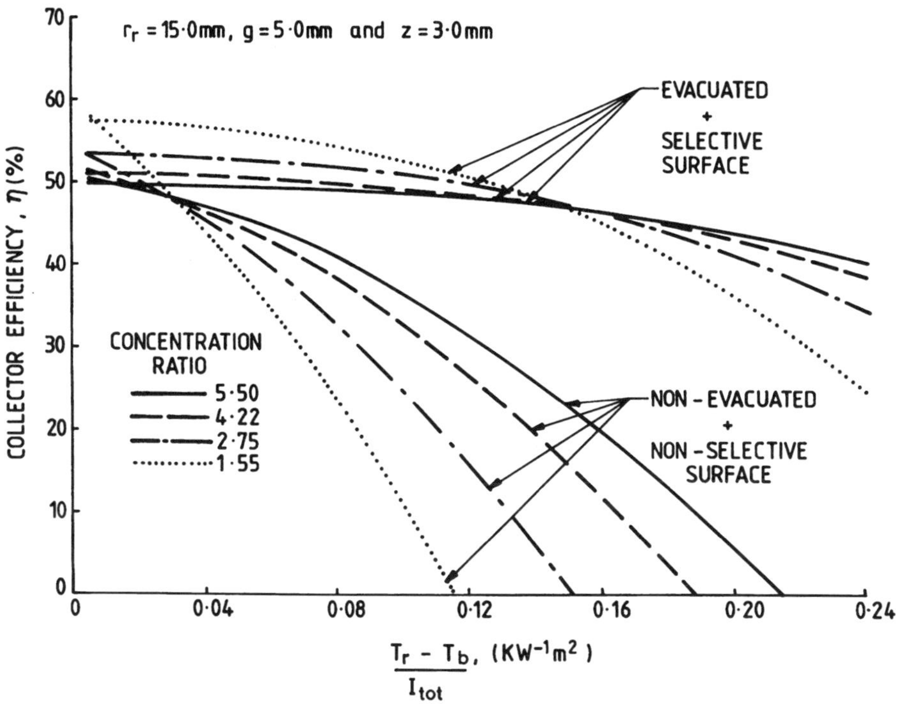

Fig. 3.26 Hottel-Whillier-Bliss collector characteristic curves for compound parabolic collectors

over those with higher concentration ratios prevails until a value of $(T_r - T_b)/I_{tot}$ = 0.15. This corresponds to a absorber temperature of 195 °C at an ambient temperature of 20 °C.

The effect of the absorption of solar radiation by various components of the collector in addition to that absorbed by the absorber on the overall performance predictions can be seen in Fig. 3.27 When this is included (i.e. the modelled system more exactly corresponds to reality), a slightly improved collector performance is predicted. This improvement compared with the situation when the absorption of the insolation by components is neglected, ranged from 0.3 % to 3 % and was more pronounced when the absorber is neither selective nor surrounded by an evacuated envelope.

As can be seen in Fig. 3.27 the evacuation of the absorber annual cavity and the application of a selective absorbing surface yielded improvements of the same order. The evacuated collector has been assumed to have no convective and conductive heat losses through the residual air. For this, the pressure in the annular space must be below 0.13 Nm−2 (Ratzel et al. 1979). The temperatures attained by various collector components versus absorber temperature are also presented in Fig. 3.27. The profound effect of the components, absorbance of the solar radiation on the attained temperature can be seen. The temperature difference between the two cases (i.e. with or without the overall absorbance included) ranged from 4 °C to 28 °C.

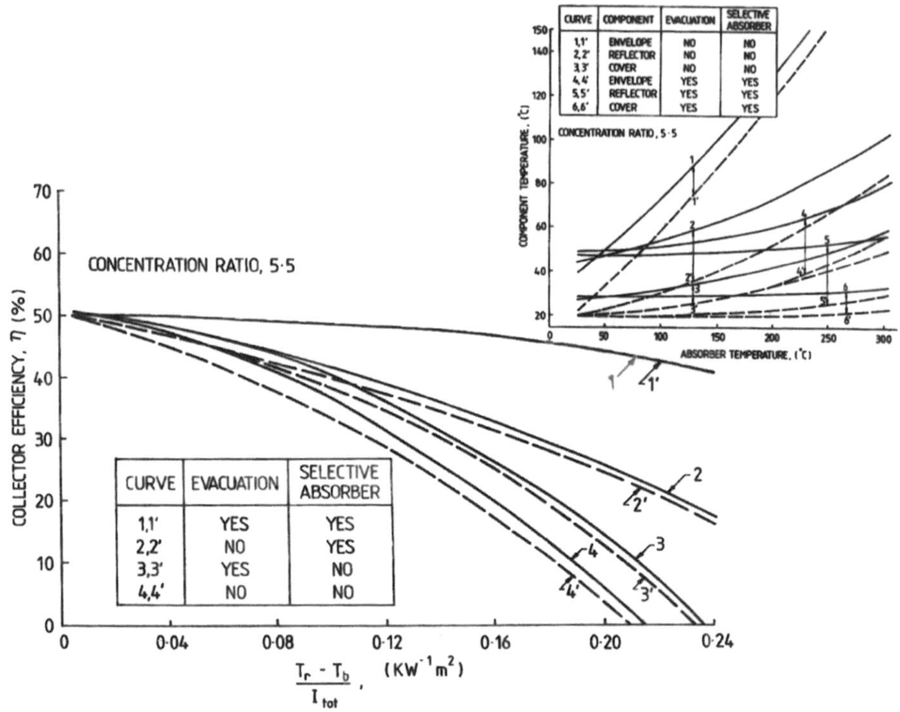

Fig. 3.27 Hottel-Whillier-Bliss characteristics and component temperatures calculated with (primed numbers) and without (unprimed numbers) the inclusion in the analysis of energy absorbed by components other than the absorber

Figures 3.28 and 3.29 show comparisons of the performances of collector configurations,

- A one-envelope absorber,
- A two- envelope absorber,
- An absorber with no envelope, and
- A single envelope absorber, with no top cover.

The superiority of the latter case with a selective surface and evacuated cavity is unfortunately hampered by operational factors, particularly the deterioration of an unprotected reflector with time and the inconvenience if surfaces have to be cleared regularly. These disadvantages have been overcome for a CPC design with an evacuated cavity, in which a single envelope serves as both the reflector and the cover (Snail et al. 1984; Duff et al. 2004). The variation of performance with inclination has also been reported (Kothdiwala et al. 1995).

The inverted absorber compound parabolic concentrator (IACPC) reflects insolation to a downward-facing absorber from which convective heat losses are thus suppressed (Mills 1978; Mills and Guitronich 1978; Kienzlen et al. 1988; Norton et al. 1991). The IACPC, in addition to use as water and air heating collectors, has applications in integrated collector-storage water heaters (Tripanagnostopoulos and

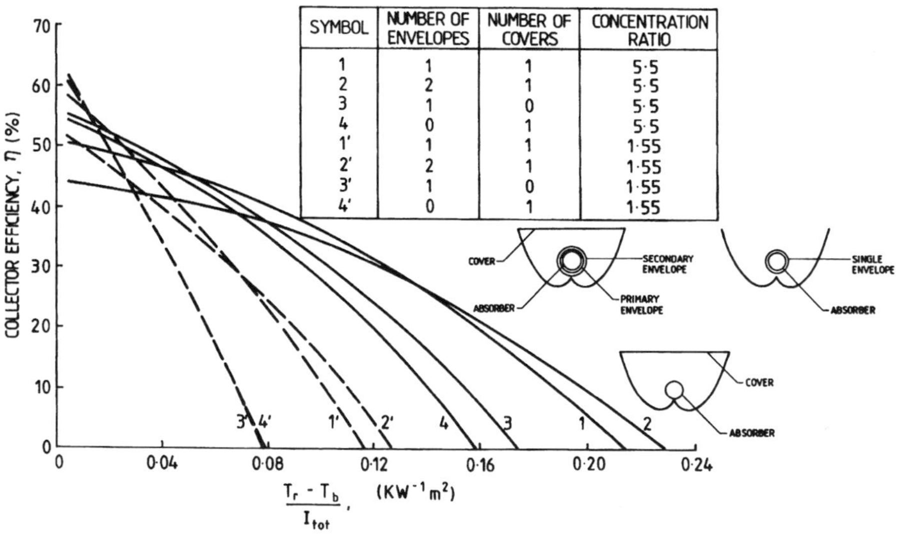

SYMBOL	NUMBER OF ENVELOPES	NUMBER OF COVERS	CONCENTRATION RATIO
1	1	1	5·5
2	2	1	5·5
3	1	0	5·5
4	0	1	5·5
1'	1	1	1·55
2'	2	1	1·55
3'	1	0	1·55
4'	0	1	1·55

Fig. 3.28 Comparison of non-evacuated compound parabolic concentrating collector configurations

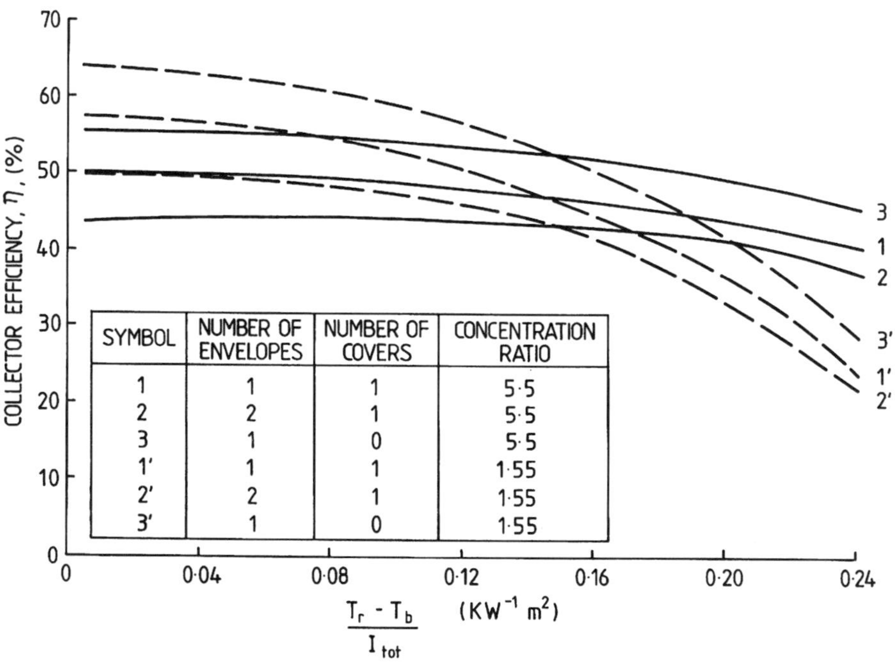

SYMBOL	NUMBER OF ENVELOPES	NUMBER OF COVERS	CONCENTRATION RATIO
1	1	1	5·5
2	2	1	5·5
3	1	0	5·5
1'	1	1	1·55
2'	2	1	1·55
3'	1	0	1·55

Fig. 3.29 Comparison of evacuated compound parabolic concentrating collector configurations

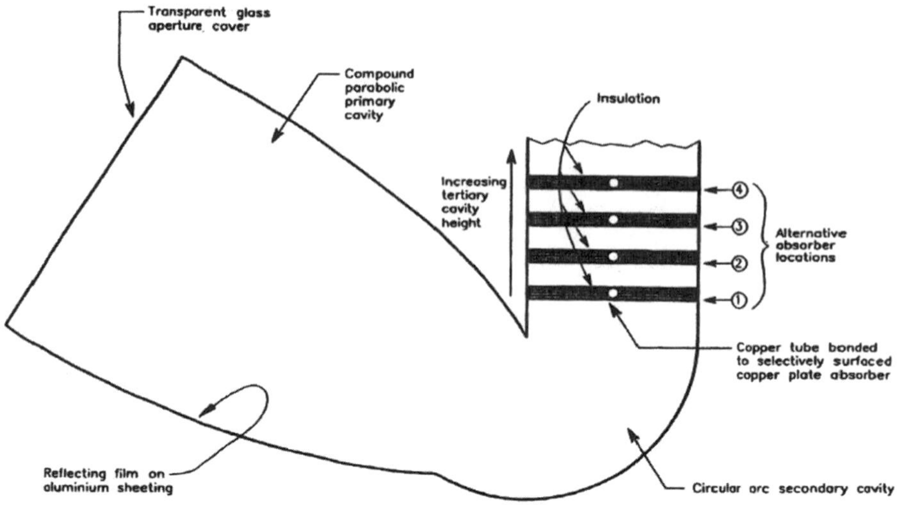

Fig. 3.30 An inverted absorber compound parabolic concentrator

Yianoulis 1992), solar distillation and solar cooling (Norton et al. 1997). A IACPC analyses is shown in Fig. 3.30 Solar radiation incident within the acceptance angle of the primary CPC reflector is concentrated onto an inverted absorber surface after reflections by secondary circular-arc and tertiary plane reflectors. Facing downwards, convection from the absorber surface is inhibited enabling medium-to-high temperatures to be achieved. The 'ideal' CPC primary reflector is often truncated (McIntire 1979) to use less materials.

A comparison (Kothdiwala et al. 1999) of a tubular absorber CPC (TACPC) and inverted absorber CPC (IACPC) is shown in Fig. 3.30. A abscissa values less than 0.04 Km^2W^{-1}, all versions of the IACPC outperform the TACPC counterpart. Above the latter value, the IACPC with absorber configuration 3 and 4 achieves superior performance over a non-evacuated IACPC absorber, configuration 3 with truncation is more efficient than a TACPC with an evacuated envelope (Fig. 3.31).

3.7 Collector Absorbers

Thermal radiation energy is emitted by matter as a result of its temperature. Thermal radiation is the energy released as a result of changes in the electron energies and in the vibrational and rotational energies of molecules and crystal structures that constitute matter. Solids, liquids and gases all both emit and absorb thermal radiation. The interaction of an oscillating electric field with the movement of the free electrons in a metallic material can be described in terms of a classical model for the behaviour of the electrons in the material (Drude 1904). By measuring the electrical properties of a material predictions can be made of the spectral optical properties. Assuming that surface scattering of electrons is negligible, spectral emittance decreases with wavelength and electrical conductivity high

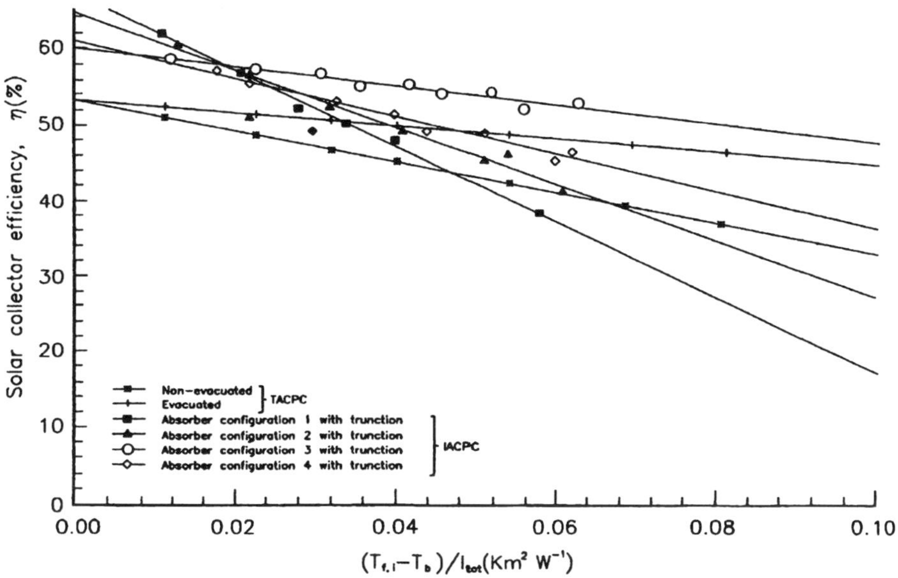

Fig. 3.31 Hottel-Whillier-Bliss comparison of inverted and tubular CPCS

efficiency thermal solar energy collection requires a large absorption of and a low emission of emitted long-wave thermal radiation. Ninty five percent of the solar radiation spectrum lies in the wavelength range of 0.32 μm; 99 % of thermal radiation at 325 K lies in the range of 3.0–30 μm. Solar selective surfaces have a high absorbance in the solar spectrum and a low emittance in the thermal spectrum. The cut-off wavelength between high and low emittance is about 2–3 μm. To reduce heat transfer between an absorber surface and a glass aperture cover in a solar collector either a low emittance absorber surface or a low emittance aperture glazing may be used. A low emittance glazing requires an infrared coating on the glass cover. Low emittance absorber surfaces are usually used however natural convection across the cavity then becomes the dominant heat losses mechanism, which introduces the necessity to consider including convection suppression devices within, or evacuate, the cavity.

Selective surfaces have a high solar absorptance and a low long-wave emittance. There are only few materials for which these optical properties are intrinsic such as silicon and germanium semi-conductors. However, these materials also have a 0.3 solar reflectance so require an anti-reflection coating. Consequently, the use of these materials as solar absorbers can be expensive. Copper, nickel and aluminium exhibit infrared reflectances $\geq$0.95, but do not have low reflectances in the solar range. Tungsten is slightly solar selective between wavelengths 0.6 and 2 μm. Selective surfaces are fabricated by placing a 0.4–1.5 μm thick layer of a material with high solar absorptance and good infra-red transmittance on metal. The solar absorptance of the resulting tandem absorber is determined by the thin-layer and the infra-red reflectance by the metal substitute. The first such tandem was black nickel or nickel coated steel (Tabor 1955). Black nickel is a nickel-zinc-sulphide complex that has an absorptance of 0.96 and an emittance typically as

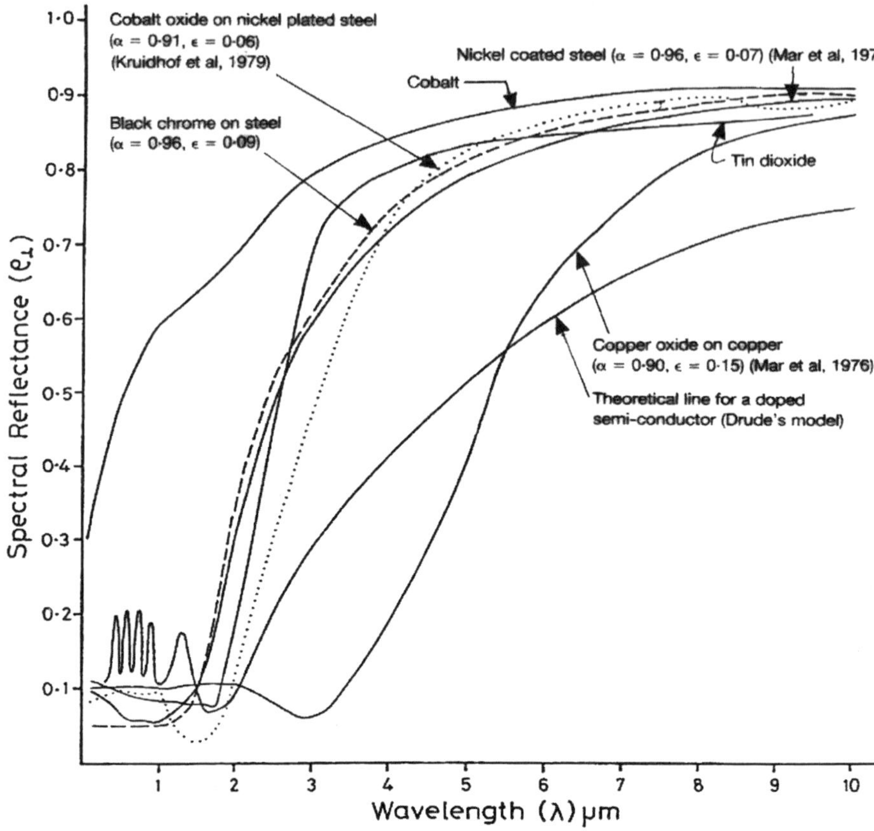

Fig. 3.32 Variation of reflectance with wavelength for a range of selective surfaces.

low as 0.08. Copper oxide on copper has absorptance and emittance of 0.9 and 0.15 respectively (Mar et al. 1976). "Black chrome" selective absorbing surfaces are common; the thin surface layer, deposited on a polished metal surface, comprises chrome particles in an amorphous chromium oxide matrix. (Lampert 1979; Lampert and Washburn 1979; Driver et al. 1975). The metal substitute is usually nickel or tin-plated steel giving solar absorptance and long-wave emittance of 0.96 and 0.1 respectively. Non-metallic materials that have a high infrared reflectance compound, to that of metallic surfaces and a low solar reflectance are termed "heat mirrors". The wavelength variations of reflectance are illustrated in Fig. 3.5 (Simonis et al. 1979) (Fig. 3.32).

Heat losses from the absorber of a collector to the surrounding environment ensue by radiation, conduction and convection. Under steady-state conditions, the interactions of these three heat-transfer modes lead to a particular temperature distribution being established, which is characteristic of the geometry and the applied temperature difference between the absorber and the ambient environment. When this pattern is known in detail, the local convective heat transfer coefficient can be estimated.

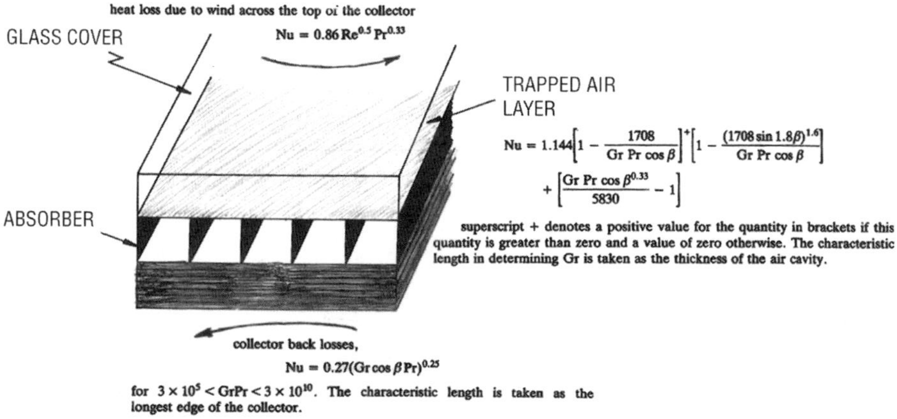

Fig. 3.33 Convective heat transfer correlations for the air-gap between absorber and aperture cover in a flat plate collector

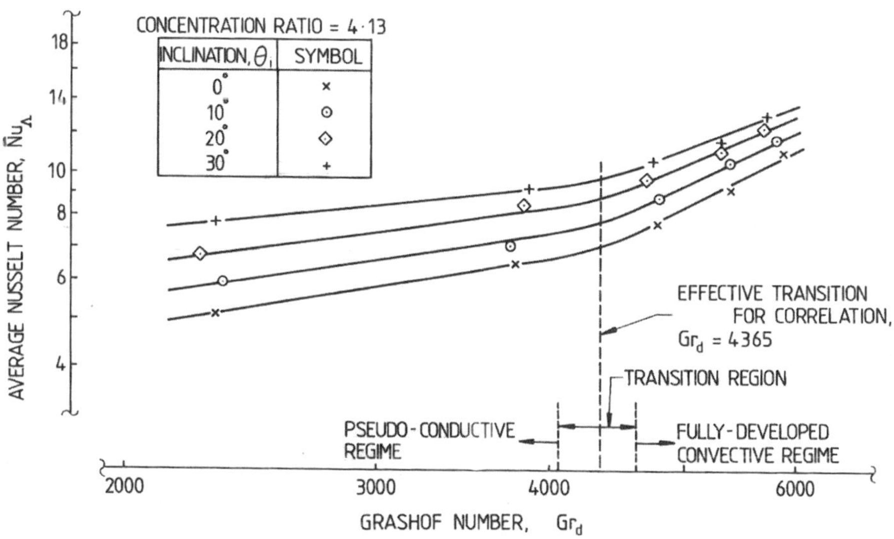

Fig. 3.34 Correlation for determining internal convective heat transfer in a CPC

To determine the heat loss by internal buoyancy driven convection between the collector plate and the glass cover in a flat plate collector, mounted at an angle between 0 and 75 from the horizontal, the correlation (Hollands et al. 1976) shown in Fig. 3.33 can be used. The characteristic length in determining Grashof number is taken as the thickness of the air cavity.

The effect of variation of the Grashof number of the rate of heat transfer within a CPC enclosure is illustrated graphically for a concentration ratio of 4.13 (Prapas et al. 1987b) in Fig. 3.34.

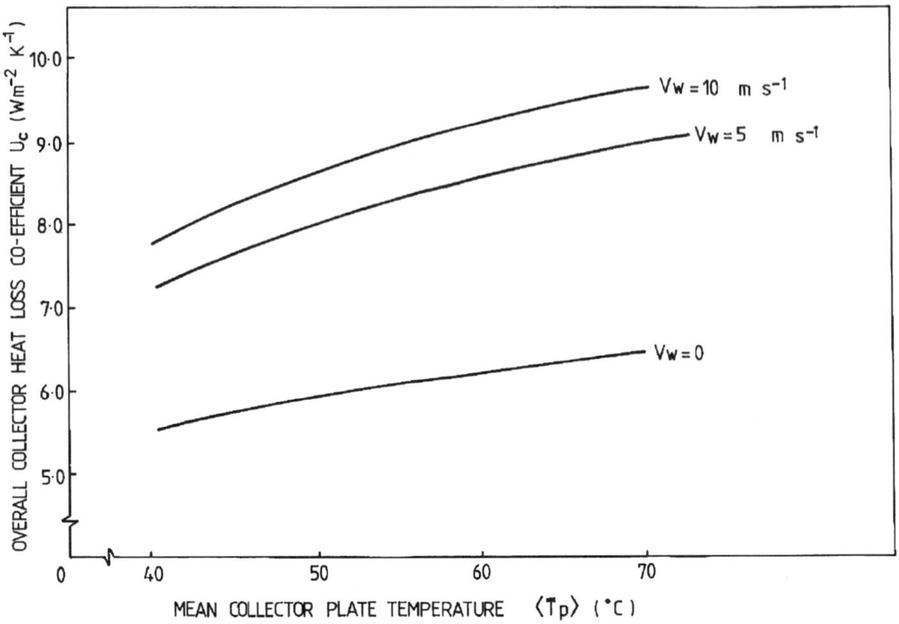

Fig. 3.35 Convective heat losses from an aperture cover due to wind

To determine the predominantly forced convective heat loss due to wind across the top of the collector (Sparrow 1979; Kothandaraman and Subramanyan 1977) Fig. 3.35 may be used. The corresponding equation is $Nu = 0.86\ Re^{0.5}Pr^{0.33}$

References

Anon (1985) ASTM E424-71. Standard test methods for solar energy transmittance and reflectance (Terrestrial) of sheet materials. American Society for Testing and Materials, Philadelphia

Anon (1977a) ASHRAE Standard 94–77. Methods of testing thermal storage devices based on thermal performance. ASHRAE, New York

Baer S (1975) "Breadbox" water heater plans. Zomeworks, Albuquerque

Baker B, McDaniels OK, Kaehn HD, Lowndes DH (1978) Time integrated calculation of the insolation collected by a reflector-collector system. Solar Energy 20:415–417

Bhaduri S, Nguyen NH (1983) Transmissivity of solar collector covers, ASME Paper 83-WA/Sol-17

Blaga A (1978) Use of plastics in solar energy applications. Solar Energy 21:331–338

Burek SAM, Norton B, Probert SD (1989) Transmission and forward scattering of insolation through transparent and semi- transparent materials. Solar Energy 42(6):457–475

Burkhard DG, Shealy DL (1975) Design of reflectors which distribute light in a special manner. Solar Energy 17:221–227

Butti K, Perlin J (1980) A golden thread. Van Nostrand Reinhold, New York

Cachorro VE, Casanova JL (1986) Optical efficiency of semistatic cylindrical-parabolic concentrator. Solar Energy 36:147–149

Carvalho MJ, Collares-Pereira M, Gordon JM, Rabl A (1985) Truncation of CPC solar collectors and its effect on energy collection. Solar Energy 35:393–399

Cheng H, Bannerot RB (1983) On the weathering of thin plastic films. ASME J Solar Energy Eng 105:149–156

Chinnery DNW (1967) Solar water heating in South Africa. SCIR Research Report, 284

Driver P, Jones EW, Riddiford CL (1975) In: Proceedings symposium on solar energy resources, ISES, Australia and New Zealand, Sect 49

Drude P (1904) Optische Eigenschaften und Electronen Theorie. An der Physik 14:936

Duff W, Winston R, O'Gallagher J, Henkel T, Berquam J (2004) Performance of the Sacremento demonstration ICPC collector and double effect chiller in 200 and 2001. Solar Energy 76:175–180

Duffie JA, Beckman WA (1974) Solar energy thermal processes. Wiley, New York

Edlin FE (1959) Plastic glazings for solar energy absorption collectors. Solar Energy 2:3–6

Evans DL (1977) On the performance of cylindrical parabolic solar collectors with flat absorbers. Solar Energy 19:379–385

Favard GJ, Nawrocki AD (1981) Preliminary optical performance study of glazing-reflector systems in breadbox water heaters. In: Proceedings of 6th national passive solar conference, Portland, pp 188–191 (September)

Fintel BW, Jakubowski GS (1985) Obtaining solar collector; cover transmissivities from a solar simulator, ASME paper 85-WA/Sol-3

Godbey LC, Bond TE, Zornig HF (1979) Transmission of solar and long-wavelength energy by materials used as covers for solar collectors and greenhouses. Trans ASAE 22:1137–1144

Grassie N (1972) Degradation. In: Jenkins AD (ed) Polymer science, vol 2. North-Holland, Amsterdam

Grassie SL, Sheridan NR (1977) The use of planar reflectors for increasing the energy yield of flat-plate collectors. Solar Energy 1(19):663–668

Grimmer DP, Zinn KG, Herr KC, Wood BE (1978) Augmented solar energy collection using various planar reflective surfaces, theoretical calculations and experimental results. Solar Energy 21:497–501

Gueymard C (1989) A simplified model for the computation of radiation transmission through a series of semi-transparent plates. Solar Energy 42:433–440

Harper CA (1975) Handbook of plastics and elastometers. McGraw-Hill, New York

Hottel HC (1976) A simple model for estimating the transmittance of direct solar radiation through clear solar atmospheres. Solar Energy 18

Hsieh CK (1981) Thermal analysis of CPC collectors. Solar Energy 27:19–29

Kaehn HD, Geyer M, Fong D, Vignola F, McDaniels DK (1978) Experimental evaluation of the reflector-collector system. In: Proceedings of the American section of the international solar energy society, Denver, vol 2(1), p 654

Kahlen S, Wallner G, Lang RW (2010a) Aging behavior and lifetime modeling of polycarbonate. Solar Energy 84:755–762

Kahlen S, Wallner G, Lang RW, Meir M, Rekstad J (2010b) Aging behavior of polymeric solar absorber materials: aging on the component level. Solar Energy 84:459–465

Kienzlen V, Gordon JM, Kreider JF (1988) The reverse flat plate collector: a stationary, non-evacuated, low-technology, medium-temperature solar collector. ASME J Solar Energy Eng 110(1):23–30

Kimball WH, Munir ZA (1978) The effect of accelerated weathering on the degradation of polymeric films. Polym Eng Sci 18:230–237

Kothandaraman CP, Subramanyan S (1977) Heat and mass transfer data book, 3rd edn. Wiley Eastern, New Delhi

Kothdiwala AF, Eames PC, Norton B, Zacharopoulos A (1999) Comparison between inverted absorber asymmetric and asymmetric tubular-absorber compound parabolic concentrations solar collectors. Renew Energy 18:277–281

Kothdiwala AF, Norton B, Eames PC (1995) The effect of variation of angle of inclination on the performance of low-concentration ratio compound parabolic concentrating solar collectors. Solar Energy 55(4):301–309
Lampert CM, Washburn J (1979) Microstructure of a black chrome solar selective absorber. Solar Energy Mater 1:82–92
Larson DC (1980) Concentration ratios for flat-plate solar collectors with adjustable mirrors. J Energy 4(4):170–175
Look DC, Sundvold PO (1983) Analysis of concentrating collectors of energy from a distant point source. Solar Energy 31:545–555
Mar HYB, Peterson RE, Zimmor PB (1976) Low cost coatings for flat plate solar collectors. Thin Solid Films 29:98–103
McDaniels OK, Lowndes DH, Mather H, Reynold J, Gray R (1975) Enhanced solar energy collection using reflector – solar thermal collector combinations. Solar Energy 17:277–283
McIntire WR (1979) Truncation of non-imaging cusp concentrators. Solar Energy 23(4): 351–355
Meinel AB, Meinel MP (1976) Applied solar energy: an introduction. Addison-Wesley, Reading
Mills DR (1978) The place of extreme asymmetrical non-focusing concentrators in solar energy utilization. Solar Energy 21(5):431–434
Mills DR (1986) Relative cost-effectiveness of periodically adjusted solar collectors using evacuated absorber tubes. Solar Energy 36:323–331
Mills DR, Guitronich JE (1978) Asymmetrical non-imaging cylindrical solar concentrators. Solar Energy 20(1):45–55
Norton B, Eames PC, Yadav YP (1991) Symmetric and asymmetric linear compound parabolic concentrating solar energy collectors. The state-of-the-art in optical and thermophysical analysis. Int J Ambient Energy 12(4): 171–190
Norton B, Eames PC, Yadav YP, Griffiths PW (1997) Solar concentrators for rural applications. Int J Ambient Energy 18(3):115–120
Oreski G, Tscharnuter O, Wallner G (2010) Determination of the solar optical properties of transparent polymer films using UV/VIS spectroscopy. Solar Energy Mater Solar Cells 94:884–891
Prapas DE, Norton B, Probert SD (1987a) Optics of parabolic trough solar-energy collectors possessing small concentration ratios. Solar Energy 39(6):541–550
Prapas DE, Norton B, Melidis PE, Probert SD (1987b) Convective heat transfers within air-spaces of compound parabolic concentrating solar-energy collectors. Appl Energy 28:123–135
Rabl A (1976) Comparison of solar concentrators. Solar Energy 18:93–111
Rabl A (1985) Active solar collectors and their applications. Oxford University Press, Oxford
Rabl A, Bendt P (1982) Effect of circumsolar radiation on performance of focusing collectors. ASME J Solar Energy Eng 104:237–250
Ranby B, Rabek JF (1975) Photodegradation, photo-oxidation and photostabilization of polymers. Wiley, New York
Ratzel A, Hickox C, Gartling D (1979) Techniques for reducing thermal conduction and natural convection heat losses in annular receiver geometries. ASME J Heat Transfer 101:108–113
Rawson H (1982) Properties and applications of glass. Elsevier, Oxford
Resch K, Wallner GM, Hausner R (2009) Phase separated thermotropic layers based on UW cured acrylic resens – effect of material formulation on overheating properties and application in a solar collector. Solar Energy 83:1689–1697
Rivero R (1958) Natural lighting – the calculation of the direct daylight factor for glazed and unglazed windows and for uniform and non-uniform skies. Building Research Station, Library Communication No. 860
Robbins FV, Spillman CK (1980) Solar energy transmission through two transparent covers. Trans ASAE 22:1224–1231
Seitel SC (1975) Collector performance enhancement with flat reflectors. Solar Energy 17:291–295

Sharafi AS, Mukminova AG (1975) Procedure for computing the reflectivity, absorptivity and transmission coefficient for radiant energy in multilayer systems with varying optical properties. Geliotekhnika 11

Shurcliff WA (1974) Transmittance and reflectance loss of multi-plate planar window of a solar-radiation collector formulas and tabulations of results for the case n = 1.50. J Solar Energy 16:149–153

Siegel R (1973) Net radiation method for transmission through partially transparent plates. J Solar Energy 15:273–276

Simonis M, v.d Leij M, Hoogendoorn CJ (1979) Physics of doped tin dioxide films for spectral selective surfaces. Solar Energy Mate 1:221–231

Snail KA, O'Gallagher JJ, Winston R (1984) A stationary evacuated collectors with integrated concentrator. Solar Energy 33:441–449

Sparrow EM, Ramsey JW, Mass EA (1979) Effect of finite width on heat transfer and fluid flow about and inclined rectangular plate. ASME J Heat Transfer 101:2

Stephenson DG (1965) Tables of solar altitude, azimuth, intensity and heat gain factor for latitudes from 43 to 55 degrees north. Solar Energy 9:81–86

Tabor H (1955) Selective radiations, wavelength discrimination. In: Transactions conference, use of solar energy, vol 2, 1A. Tuscon, pp 24–33

Touloukian YS, Dewitt DP (1972) Thermal radiative properties, nonmetallic solids, thermophysical properties of matter 8, IFI/Plenum Data Corporation

Tripanagnostopoulos Y, Yianoulis P (1992) Integrated collector-storage systems with suppressed thermal losses. Solar Energy 48(1):31–43

Wallner G, Resch K, Hausner R (2008) Property and performance requirements for thermotropic layers to prevent overheating in an all polymeric flat-plate collector. Solar Energy Mater Solar Cells 92:614–620

Weinstein A, Duncan RT Jr, Sherbin WC (1977) Lessons learned from Atlanta (Towns). Solar Energy 8:45–46

Whillier A (1953a) Solar energy collection and its utilization for house heating. ScD Thesis, MIT

Whillier A (1953b) The utilisation of solar energy in South Africa. J South African Inst Mech Eng 2:261–267

Wijeysundera NE (1978) Geometric factors for plane specular reflectors. Solar Energy 20:81–85

Williams JR, Craig JI (1976) The Shenandoah solar community sharing the Sun. In: Proceedings of the American section of the international solar energy society, vol 3. Winnipeg, pp 200–212

Winston R (1974) Principles of solar concentrators of a novel design. Solar Energy 16:89–95

Chapter 4
Storage of Solar Heat

When it was dark, you always carried the sun in your hand for me

Sean O'Casey (1928)

4.1 Overview

The key issues in the thermal storage of solar energy are;

- Storage temperature
- Duration of storage
- Magnitude of energy stored
- Charging and discharging rates
- Operation and control
- Economic optimization

In many climates heat and cooling clouds exhibit seasonal variation. The illustrative examples in Fig. 4.1 are for space and water heating loads; in this example a solar water heating system can be sized with appropriate storage to provide a hot water solar fraction all year round as water heating loads do not show usually the very strong seasonal variations associated with space heating. Space heating is not required in warm seasons and solar energy is insufficient to instantaneously satisfy space heating loads in cold seasons where at high latitudes the cold conditions are due to low insolation. Thus, as show in Fig. 2.24, without interseasonal storage, space heat loads are only satisfied in autumn and spring. Figure 4.1 illustrates that to supply an interseasonal heat storage large collector areas are obviously required; actual areas are shown in Table 8.2

Though an interseasonal heat storage system may give a larger solar function, for water heating in particular this may be only in terms of energy not exergy, that is the desired end-use temperature may not be satisfied in winter and auxiliary energy will be used.

B. Norton, *Harnessing Solar Heat*, Lecture Notes in Energy 18,
DOI 10.1007/978-94-007-7275-5_4,

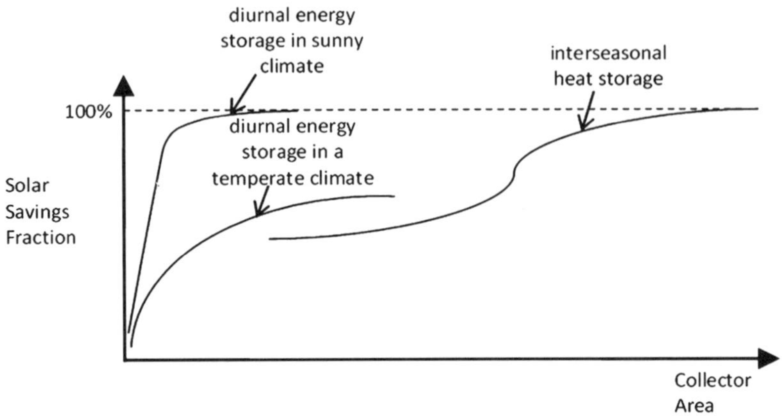

Fig. 4.1 Illustrative variation of solar savings fractions with collector area

Energy storage competes economically with the use of auxiliary heating. When a small heat store is integral to a solar water heater, the storage cost is, in reality inseparable from that of the solar water heater itself. However as a heat store becomes much larger, its cost together with its distinctive attribute of the provision of winter heat are identifiable separately from the solar collectors. In the latter case the costs of inter-seasonal storage can be compared on an economic basis with the winter fuel outlays it displaces. Long-term, and certainly, interseasonal energy storage in water requires fairly heavy insulation, for example, 0.50 m of rockwool insulation are used for the long term heat store in the Danish "zero energy house" (Korsgaard and Esbensen 1975). There are substantial advantages in using a single, large store rather than a number of small stores, as heat losses are proportional to surface area with the heat storage capacity is proportional to volume.

The choice of thermal storage media is related frequently to either the type of collector or the duration of storage. As the intermittent nature of the solar energy supply makes the provision of adequate energy storage essential for the majority of practical applications, thermal storage is needed for both low temperature and high temperature uses of solar energy. A particular storage technique has to be appropriate to both the collector outlet temperature equipment and the end use of the collected energy. To achieve high thermodynamic efficiencies for thermal power applications during periods when solar energy is not available, energy must be stored at relatively high temperatures. Storage for space heating, in contrast, requires relatively low temperatures.

Energy storage is also needed for cooling applications. For space cooling a storage temperature of about 7 °C, achieving a good coefficient of performance for refrigeration processes whilst removing energy economically from the coolth store.

Thermal energy can be stored as sensible specific heat or as latent heat of fusion. In sensible heat storage, the heat is absorbed in water, rock, brick or concrete increasing its temperature. In latent heat most of the heat transferred from the working fluid changes the enthalpy of the storage medium during a phase change process. Storage media commonly used in this type of device include inorganic salt hydrates and organic materials.

4.2 Sensible Heat Storage in Liquids

Water remains the most employed heat storage medium. Though the cost of water may be very small, a water tank, its insulation and often a support structure are required. The sensible storage of heat in water is ubiquitous because it;

- Is a commonplace material for which technical, physical, chemical and thermodynamic data is available readily with well-known viscosity, thermal conductivity and density
- Is one of the few thermal storage materials which can be used as both the collector and storage heat transfer fluid at the same time:
- Has the highest specific heat of any common generally-usable fluid.
- Is generally not a two-phase system in the medium temperature range with a vapour-liquid equilibrium temperature/pressure relationship is appropriate for non -focusing solar collectors;
- Is stable, non-toxic and non-flammable.

Its disadvantages are it:

- Is a medium for electrolytic corrosion ensuent from the use of dissimilar metals;
- Freezes and, furthermore, expands as it does so, leading potentially to damage;
- Can contain desolved oxygen, that can cause corrosion.

In sensible heat storage in liquids, it is important to minimise the mixing of hot and cold liquids to mitigate degradation of exergy. Optimal utilization of the stored heat requires maintaining a delivery temperature that satisfies or exceeds slightly the demand temperature, obviating the requirement for auxiliary heating. Though separate tanks, or movable partitions or collapsible membranes can be used for separating the hot fluid from the cold fluid, the simplest technique is the preservation of a comparatively sharp natural thermocline between hot fluid in the upper part of the tank and the cold, more dense, fluid in the lower part.

During the process of charging the heat store, the temperature to the solar collector inlet must be maintained as low as possible to retain high collection efficiency. Analytical and numerical studies have shown that thermal stratification improves the performance e.g. (Gutierrez et al. 1974; Close 1962; Sheridan et al. 1967). For a perfectly stratified storage system, the fluid at the inlet to the solar collector is always at the lowest available temperature. In comparison, a fully mixed storage system will supply fluid to the collector at progressively increasing temperatures. Vertical wall in hot water stores with copper, aluminium, steel or stainless steel liners conductance produces temperature destratisfication water storage tank depending the relative thermal conductance of the liner and the storage fluid (Miller 1977; Viskanta and Hale 1978; Sherman et al 1979; Leyers et al. 1977). Stratification of hot water stores for solar energy applications has been investigated extensively for systems where the heat input to the store occurs via the storage fluid itself, which is circulated continuously through the store, as in thermosyphon and active solar systems (Davis and Bartera 1975; Hobson and Norton 1988; Hunt et al. 1978; Koldhekar 1981; Lavan and Thompson 1977; Loehrke et al. 1978; Marshall 1981; Matsudaira and Sakakura 1972; Nakajima 1972; Phillips and Dave 1982; Sliwinski et al. 1978; Van Gallen

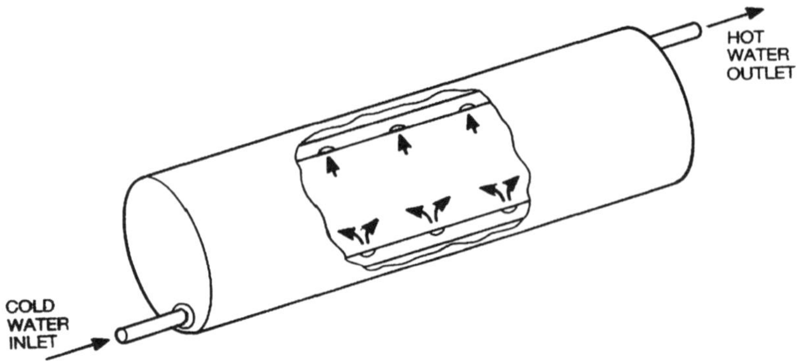

Fig. 4.2 Manifolds in a horizontal hot water store

and Van den Brink 1984; Van Koppen et al. 1978; Wood et al. 1981; Wu 1979; Jalaria and Gupta 1982). The results of these investigations cannot be applied directly to the stratification processes in an integral passive solar water heater, as the heat input comes from the insolation directly incident on the tank walls. The input and output of water from the store, and the subsequent mixing, occurs only during draw-off, and this may occur several times or only once per day depending upon the hot-water usage habits. A horizontally mounted cylindrical tank will only exhibit relatively weak thermal stratification and thus are employed usually where climatic and system sizing enable all the store to be at the demand temperature. If a horizontal store is, for most of the year, not fully heated, a lower solar fraction may be realised because of destratification. The advantage of dividing the required water volume over two or more tanks is to incorporate some inherent stratification mechanism in the design cold inlet water entering the lowest tank is prevented from mixing with the hotter water in the final input tank. Each tank is normally also fitted with inlet and outlet manifold assemblies as shown in Fig. 4.2 to further inhibit mixing in each tank. Cold inlet water enters the tank via holes the whole length of the inlet manifold, whereby the inlet velocity is so restricted to ensure that the cold water does not penetrate and mix with the warmer upper layers. Similarly, the outlet manifold collects only the hotter water from the whole length of the uppermost part of the tank, and delivers it either directly as the output or as the input to the next tank in a series-coupled system. In this way some degree of stratification is obtained in a horizontally mounted system.

Equations for fully-mixed and stratified stores operating at a time dependent temperature are shown in Fig. 4.4.

Fully-mixed hot water stores are the result of higher velocity throughputs or invariant inflowing fluid temperatures. In the absence of large mixing forces on entering, warmer fluid will seek its equilibrium in a stratum of the same density. The process of stratification is both temperature and time dependent. An exact solution for the temperature history of a stratified storage system is given by the solutions to the Fourier equation shown in Fig. 4.3 (Schmidt et al. 1960; Cabelli 1977).

A simplified approach to a stratified store (Gutierrez et al. 1974) applies an 'M' segmented tank approximation. Each segment is representative of a fully mixed sub-volume of the main tank. This is illustrated in Fig. 4.4.

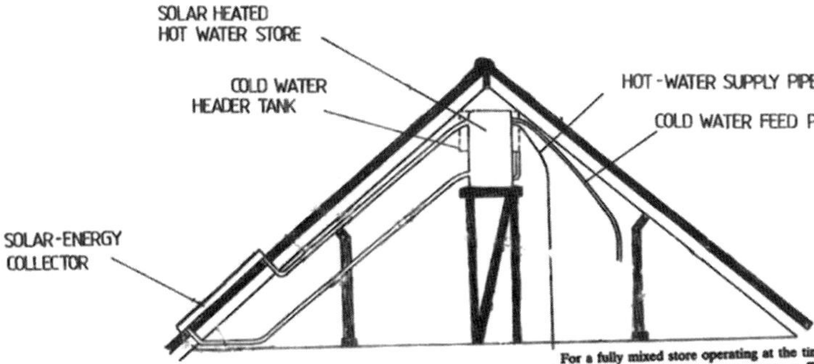

The process of stratification is both temperature and time dependent. An exact solution for the temperature history of a stratified storage system would be given by the solution to the Fourier equation,

$$(T_s - T)/T_{in} = 2/\pi^{1/2}\int_0^{2x/(\alpha T)^{1/2}} \exp(-\lambda^2)\,d\lambda$$

$$T_s(t, h) = \exp(UH/2)(\exp(-pH/2)\mathrm{erfc}(H/2\sqrt{t}) - p\sqrt{t}) + \exp(pH/2)\mathrm{erfc}(H/2\sqrt{t} + p\sqrt{t/2})$$

where U is the vertical inlet velocity to the tank and $p = U^2 + 4H$ and H is a dimensionless heat loss parameter given by

$$H = Uld^2/Ak$$

where U is the surface heat transfer coefficient, l the tank perimeter, d the tank diameter, A the base area and k the thermal conductivity.

For a fully mixed store operating at the time-dependent temperature T_s, in the presence of an ambient temperature T_a, an energy balance yields

$$M_s C_p \frac{dT_s}{dt} = \dot{Q}_u - \dot{Q}_L - (UA)_s(T_s - T_a)$$

Fig. 4.3 Analytical expressions for fully-mixed stratified hot water stores

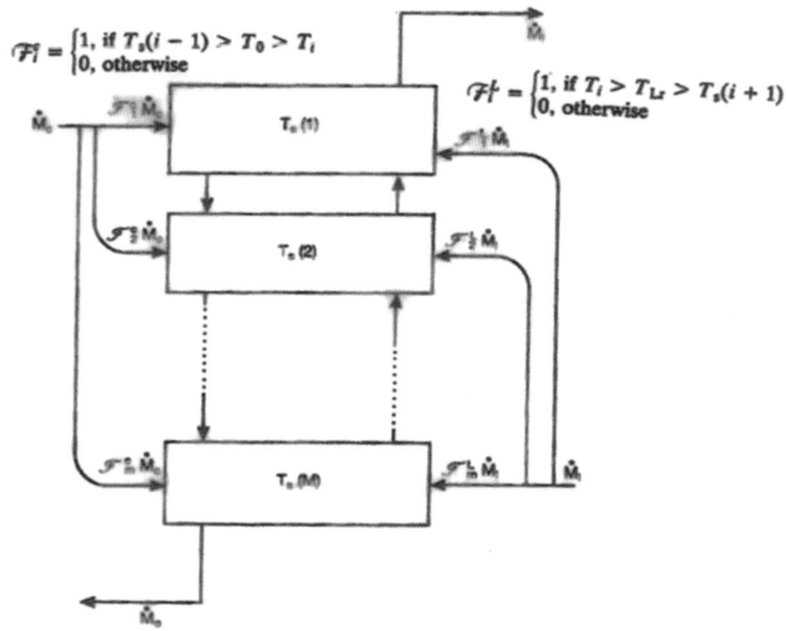

The energy balance for an "M" segmented tank

$$M_s c_p \frac{dT_s}{dt} = \dot{M}_c c_p\left[\mathcal{F}_i^c(T_0 - T_i) - (T_{i-1} - T_i)\sum_{j=1}^{i-1}\mathcal{F}_j^c\right\} + \dot{M}_l c_p\left[\mathcal{F}_i^L(T_{Lr} - T_i) + (T_{i+1} - T_i)\sum_{j=i+1}^{n}\mathcal{F}_j^L\right] + (UA)_s(T_a - T_i)$$

Fig. 4.4 Segmetial tank approach to modeling a hot water store

4.3 Sensible Heat Storage in Solids

In buildings the diurnal asynchrony between the availability of solar radiation and heating load necessitates of storing solar energy to maximise the solar contribution. Storage is provided by absorbing heat into mass of the building, the portion of the latter that participates is usually termed thermal mass. Short-term storage addresses the diurnal solar to load asynchrony, by storing solar energy collected during the middle part of the insolation period be used at night and, usually to a lesser extent, to reduce pre-heating of the building early the next day. The occupancy patterns associated with the purpose of the building – e.g., domestic, institutional, commercial – have to be considered when designing the characteristics of the storage. Thermal mass has to be coupled thermally with the passive solar energy collecting feature. The optimum size, position and coupling of thermal mass will depend upon the type of passive solar feature and its mode of use. Some passive systems have integral thermal storage, e.g. Trombe walls. In others, the provision of storage may be independent, e.g., the appropriate positioning of encapsulated phase-change materials in the walls of a direct gain room.

Three categories of thermal mass in direct gain systems can be identified:

- Primary mass – that which is insolated directly, that is falls within the diurnal motion of the sunpatch within the room. Primary mass comprises usually the internal wall and floor surfaces. For the latitudes of northern Europe, diurnal sunpatch movement constrains the location of primary mass to the floor or lower zones of the walls. Unfortunately these areas are both obstructed frequently and insulated by furnishings and carpets.
- Secondary mass – that which is irradiated by diffuse and reflected insolation and longwave thermal radiation from directly insolated surfaces. Secondary thermal mass is of more importance since (i) it can be provided by the unobstructed upper walls and ceilings (although the latter are commonly of light-weight construction) and (ii) diffuse and reflected insolation can be greater than the direct insolation.
- Tertiary mass – is that to which solar gains are transferred from the primary and secondary thermal masses via convective air movement. Tertiary thermal mass relies on a convective link, between different rooms usually otherwise it would be probably secondary thermal mass. The use of tertiary thermal mass necessitates open doorways or purpose made ducts these carry functional and economic disadvantages respectively.

The quantity of thermal mass is not related solely to the total amount of actual mass, but to the association between mass and surface area (Mazria 1980). For example, a dense concrete wall does not increase significantly effective thermal mass after the wall thickness is increased beyond about 100 mm. Thus low mass elements such as double-sided plaster-board partitions, and furnishings make a significant contribution in spite of their low actual mass, due to their large surface area to mass ratio. It should be noted that high levels of thermal mass in intermittently heated buildings are likely to be a disadvantage.

Indirect passive solar features present the option of providing storage either within the feature itself, or within the heated part of the building. Storage within the building is essentially tertiary storage linked to the collecting element by convection, and thus must be well distributed with maximum surface area. Generally conventional building materials together with the building contents will be sufficient for this. To provide primary thermal storage in conservatories is relatively easy since typically a conservatory is furnished sparsely and hard, heavyweight finishes are acceptable, and indeed to be encouraged in order to emphasise the periodically-habitable "transition from indoors to outdoors" nature of such features. Otherwise the user may occupy the conservatory in winter providing auxiliary heating to the space. The effectiveness of storage located in conservatories is much reduced due to the high conductance to ambient.

The effect of the storage will be dependent upon the predominant mode of operation. For a conservatory relying on either the natural or forced circulation of air between the conservatory and the heated building only rarely will useable solar energy be carried over to the evening, due to the rapid cooling of the thermal mass. Indeed thermal mass is likely to reduce the solar performance in these modes since it will reduce the total time that the conservatory temperature is above the threshold required for a net heat gain to ensue. The main advantage of thermal mass in the conservatory, is to elevate minimum temperatures. This is more significant than energy saving, since the conservatory may contain plants which have to be frost-protected, or may form a glazed street. A temperature above ambient of 3 °C is sufficient. Low minimum temperatures might tempt occupants to partially heat such areas. Another advantage is that due to the ease by which large areas of primary mass can be provided, significant reductions in peak temperature can be attained, although this should never be considered to remove the need for adequate shading and ventilation. Solar ventilation pre-heating performance is relatively independent of the level of thermal mass in the conservatory.

4.4 Latent Heat Storage

A material undergoing a phase change absorbs or releases latent heat at a relatively constant temperature as shown in Fig. 4.5.

The absorption of large amounts of energy at a constant temperature is one of the main attractions of phase change materials (PCM) (Fath 1998). Phase transformation of materials have been employed in a variety of temperature stabilization applications (Ismail and Goncalves 1999; Abhat 1983) including widely in this temperature control of electronics (Wirtz et al. 1999).

PCM used for solar energy storage applications have been investigated widely both experimentally and analytically (Baetens et al. 2010; Banaszek et al. 1999; Kenisarin et al. 2007). In solar energy heating systems a PCM is placed in a container, through which is changed by a either water or air as a heat transport fluid passes transferring heat to the PCM which melts. To discharge when the stored

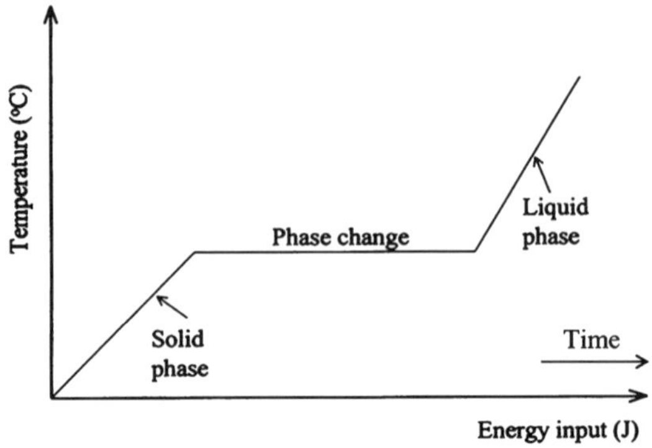

Fig. 4.5 Schematic process of PCM heating

heat is required the process is reversed. The thermal performance of latent heat storage systems during the charging and discharging process have been studied theoretically and experimentally (Esen 2000; Fath 1998; Comakli et al. 1993; Kaygusuz 1995). A PCM should be selected firstly on the basis of its melting temperature, rather than its latent heat capacity as the melting temperature has the most significant effect on system performance (Jurinak and Abdel-Khalik 1978, 1979). PCM stores have been included in the complex dynamic systems such as solar-assisted heat pumps and storage tanks used for drying grain (Kaygusuz et al. 1991).

PCM storage elements can be installed within floor, wall and roof constructions, in plaster board and concrete, to increase the thermal mass of a building. Absorbed solar energy is stored in the walls during high daytime temperatures and subsequently released to the room in the evening. As a result, the more stable room temperature will be more comfortable with energy consumption for both air conditioning and heating decreased (Salyer and Sircar 1997). For double glazing filled with different thickness of PCM (Ismail and Henriquez 1997) transmittance and reflectance tests indicate large reductions in the infrared and ultraviolet radiation while maintaining good visibility.

Thermal energy storage in the form of latent heat involves phase transformation which is associated with the absorption and storage of heat (Jotshi et al. 1992). In order to undertake a long-term performance analysis of a specified solar heating system, an adequate model of the heat storage unity is required (Klein et al. 1976). The design of the solar thermal energy storage unit should take into account both structural and thermal system requirements. Structurally, the storage unit must be leak-proof, able to withstand long exposure to temperature cycles, endure pressure from both static and dynamic loads, and be corrosion resistant. Thermally, it must be optimised both in size and shape in order to accommodate the required storage capacity, the change in volume and to minimise heat losses. The most-analysed latent heat energy storage unit is the shell-and-tube unit with the PCM filling the shell with a heat transfer fluid flowing through the tubes (Lacroix 1993; Bansal and Buddhi 1992; Esen et al. 1998; Ismail and Alves 1986) or, vice versa, with PCM filling the tube and the heat transfer fluid flowing parallel to it (Esen et al. 1998; Ghoneim 1989). The shape of a PCM energy store seeks generally to minimize both heat losses and construction materials cost. Being cheap to fabricate, vertical circular cylinders are most common, although such tanks require more insulation, especially at the top and base. Spherical units offer minimal heat loss and construction material requirement but are difficult to fabricate, require special supports and consume more volume in a building. Rectangular containers require nearly half the melt time compared to cylindrical containers of the same volume and heat transfer area (Zivkovic and Fujii 2001). Packing the PCM in small mass-produced rectangular containers enables modular construction of PCM thermal energy storage units. Complete melting of large PCM masses is difficult in a shell-and-tube heat exchanger unit (Zivkovic and Fujii 2001). Encapsulation of phase change materials must accommodate the expansion and contraction that accompanies melting and solidification. The provision of voids for expansion leads to a larger reduction in heat transfer surface area. Leakage and possible corrosion over repeated heating and cooling cycles, can be avoided by proper selection of container construction materials. Galvanised steel, stainless steel and aluminium are used as containers for PCM storage systems as they all have high strength-to-weight ratios, are corrosion resistant, and have relatively good thermal conductivities. Although stainless steel is heavier, it is strong enough to accommodate large volume changes (Lee et al. 1979).

The PCM melting point should be higher than the average ambient environmental temperature for solar thermal storage system so that under most circumstances no melting occurs due to the ambient thermal loads. In addition to the phase transition temperature, a high latent heat per unit mass is required. A phase change process should be congruent, so that the molten PCM and the solid PCM are identical in composition. Supercooling during freezing, should be avoided or minimized (Abhat 1981). Supercooling is defined as the difference between the melting point and the temperature at which solidification initiates. To catalyse the process of solidification, solidification promoting agents can be added. For many PCMs usually used in conjunction with metallic fillers, the fillers themselves promote the initiation of solidification. A PCM should be chemically stable, should not undergo any chemical reaction with the container material and should have a flash point considerably higher than the operating maximum temperature of the system and it should be non-flammable and non-explosive. Requirements are summarized in Table 4.1.

Differential thermal analysis and computer-driven differential scanning calorimetry are the two main techniques for the measurement of PCMs thermo-physical properties. The difference in the heat flow to or from a sample and to or from a reference is monitored as a function of temperature or time, while the sample is subject to a controlled temperature (Abhat and Malatidis 1981).

Currently available solid–liquid PCMs for thermal energy storage applications are either; inorganic compounds, organic compounds, and eutectics of inorganic and/or organic compounds. Inorganic compounds include salt-hydrates, salts, metals, and alloys; whereas organic compounds are comprised of paraffins, non-paraffins, and polyalcohols. Their ranges of use are summarised in Fig. 4.6

Salt hydrate PCMs have been studied extensively for latent heat storage in space heating and cooling applications (Yanadori and Masuda 1986; Kimura and Kai

Table 4.1 Selection criteria for a phase change material (Pal and Joshi 1999)

Properties	Requirements
Thermal	High latent heat of fusion
	High specific heat capacity
	Good thermal conductivity
	Reversible phase change transition on repeated cycling
Physical	Total congruent melting
	Low volume change
	High density
Kinetic	No supercooling
	Good crystallization rate
Chemical	Long-term chemical stability
	Non-corrosive
	Non-flammable
	Non-explosive
	Non-toxic
Economic	Available, low initial cost and cost effective in use

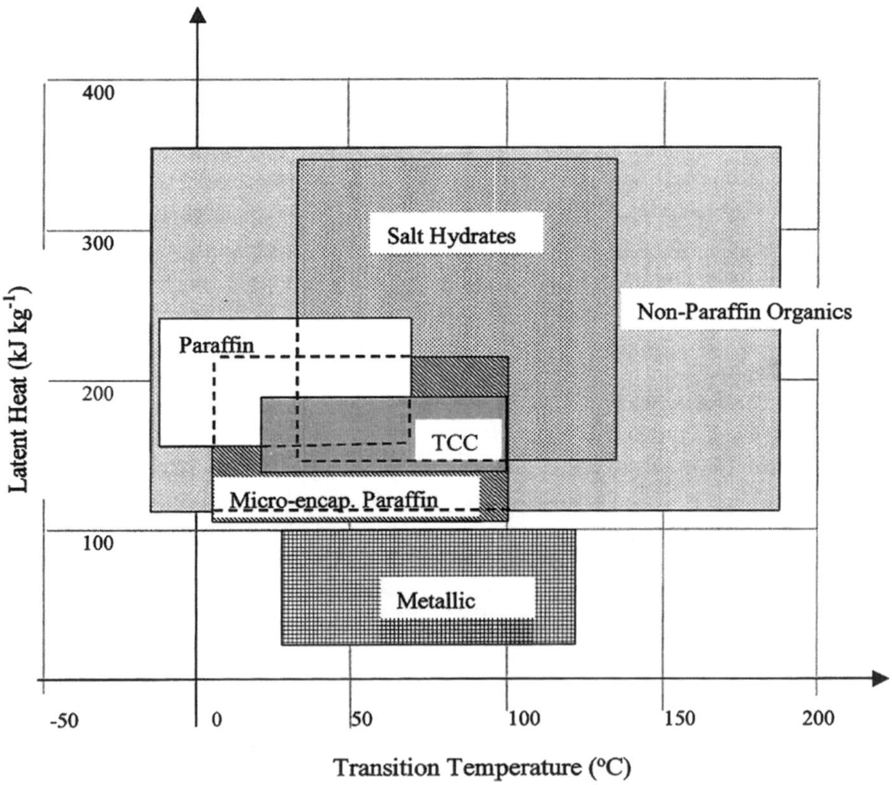

Fig. 4.6 Ranges of latent heat capacities and transition temperatures for generic type of phase change energy storage materials

1988; Salyer and Sircar 1997; Fath 1991). Salt hydrates have a high heat of fusion per unit volume, high thermal conductivity (0.4–0.6 $Wm^{-1}K^{-1}$ for $MgSO_47H_2O$, almost double that of organic paraffins) and a small volume change (<1 %) on melting. These materials are compatible with plastics, but are usually toxic. Salt hydrates are generally corrosive, absorb and loose water during phase transition, and with a complex melting behavior (Jotshi et al. 1992) tend to form partially hydrated crystals, that effects performance adversely (Antohe et al. 1996). These disadvantages mean salt hydrates are in general thus unsuitable for building applications.

For an effective and uniform absorption of heat from the heat source to the PCM, high thermal conductivity is a desirable property. Metallic alloys usually have high thermal conductivity (4–18 $Wm^{-1}K^{-1}$) and on a unit volume basis, their latent heats are larger than hydrocarbons (Pal and Joshi 1997). Metallic alloys have been employed in some high-performance systems like electronic chip cooling system (Antohe et al. 1996; Pal and Joshi 1997; Ishizuka and Fukuoka 1992). Metallic PCMs are denser than organic PCMs resulting in relatively heavy systems. For example, the density of the alloy Bi/Pb/Sn/In is about 12 times greater than n-eicosane (Pal and Joshi 1997).

Waxes and paraffins are organic materials obtained from refining crude oil or produced synthetically (Mortimer 1986) and are inexpensive and widely available (Leoni and Amon 1997). Paraffins exhibit many desirable characteristics suitable for use in temperature control and thermal storage applications (Salyer and Sircar 1997) though, if not encapsulated fully, can be a fire hazard. Low melting temperature paraffins with phase change in the near ambient temperature range about 25 °C, are used for space and greenhouse heating, (Benard et al. 1978). For use at temperatures above 0 °C and below 100 °C, paraffins have good overall cost performance, and to be more versatile than other PCMs (Salyer and Sircar 1990). Although paraffin wax has many desirable properties suitable for temperature control and thermal storage applications, low thermal conductivity (about one-half that of salt hydrates), large volume change during melting and freezing (approximately 10 % by volume expansion or contraction) leading to leakage in the liquid phase, high wetting ability and flammability are significant disadvantages. To improve the low thermal conductivity of paraffin-based PCMs, metallic fillers, metal matrix structures (honeycomb or thin strips), finned tube and aluminium shavings can be used. To overcome the volume change on melting and freezing, elastic containers and different container geometries may be used. Pure paraffin waxes are very expensive, hence only technical grade paraffins are used for latent heat storage. Technical grade paraffins are usually mixtures of many hydrocarbons and therefore do not have sharp melting points. Paraffins are stable chemicals but should be protected from air while hot to preclude slow oxidation (Jotshi et al. 1991).

The paraffin PCM storage container must be designed to avoid leakage. Chemical compatibility and corrosion determine life cycle, thus compatibility of the PCM and the containment material is essential for a reliable system. PCMs are compatible with the metals aluminium and copper (Abhat 1981, 1983).

Non-paraffin organics include a wide variety of organic materials such as fatty acids, esters, alcohols, and glycols. About 70 non-paraffin organics that have melting points in the range of 7–187 °C (Hale et al. 1971). Some of the features of non-paraffin organics are: high heat of fusion, low thermal conductivity, low flash point, inflammability, varying levels of toxicity and instability at high temperatures (Mortimer 1986). Their melting points of 31–38 °are suitable for heating applications and the heat of their fusion values is comparable to those of paraffins and salt hydrates. They exhibit excellent melting/freezing characteristics without any supercooling (Jotshi et al. 1991). Their major drawback is cost which is about 2–2.5 times greater than paraffins. A eutectic mixture of 49.3 % $MgCl_2.6H_20$ and 50.7 wt% $Mg(NO_3)_2.6H_2O$ containing additives to initiate crystallization and prevent supercooling has been marketed commercially (Lane 1983). This eutectic is stable over 1,000 heating and cooling cycles. It is compatible with plastics or mild steel (preferably coated), but not stainless steel, since it is attached by $MgCl_2.6H_2O$. Aluminium is corroded severely by both components.

Binary mixtures of fatty acids such as lauric-capric, lauric-palmitic, lauric-stearic and palmitic-stearic systems possess good chemical stability (Feldman et al. 1989). However their long term stability and that of their eutectic mixtures has not been verified by extensive thermal life cycle testing.

PCMs, in general have low thermal conductivities in the range (0.33–0.18 $Wm^{-1}K^{-1}$) so materials are introduced to improve heat transfer during the melting and solidification, charging and discharging, processes.

For organic and inorganic PCMs, a filler material can enhance heat transfer. This may be achieved by using aluminium in the form of powder, foam, wool or a honeycomb (Pal and Joshi 1999). The use of a filler material increases the weight of the heat sink, but reduces the latent heat storage volume. An optimum filler volume to provide the best thermal performance depends on the application.

The common types of phase change material used for energy storage are from solid to liquid. The liquid phase presents fluid containment problems, due to both the presence of the liquid and volume expansion during transition to the liquid phase. "Dry" material systems to avoid these problems include micro-encapsulated solid–liquid phase change composites and solid-solid organic phase change compounds. Micro-encapsulated phase change composites in powder form, consist of tiny beads that contain the phase change material inside thin polymer shells (Fossett et al. 1998). Individual capsules range from 10 μ to 1 mm in diameter with impermeable, semi-rigid shell walls of typically less than 1 μ thickness. The core phase change material, which is usually a blend of paraffin compounds, comprises 80–85 % of the composite mass. Core PCM's are blended to establish the phase transition temperature. Because the core material is a blend of materials, transition actually occurs over a relatively wide temperature interval, typically about 10 °C (Wirtz et al. 1999).

Materials that undergo reversible solid-state crystal structure transitions at temperatures ranging from ambient up to about 100 °C have significant heat of transition (Wirtz et al. 1999). Transition temperatures can be selected by forming solid solutions from different compounds. Transitions can occur in an interval of less than 5 °C, but due to the PCM mass being in granular form, its effective thermal conductivity is less than the thermal conductivity of the base material. The reduction in thermal conductivity depends on the packing arrangement of the particles, particle shape, and particle contact areas. Typically, overall thermal conductivity is 10–25 % of the particle thermal conductivity (Kaviany 1995).

In a phase change process, complex solid and liquid phase regions exist simultaneously. Most work on PCM heat storage has been experimental or has used simplified theories to underpin the mathematical modeling approach adopted (Costa et al. 1998). Rigorous detailed theories have not been developed due to the non-linear nature of phase change at the moving liquid–solid interface and the different thermophysical properties of the two phases. A numerical model for a phase change energy storage system coupled to air and water-based solar heating systems should include the effects of the solar savings fraction (the fraction of the load met by solar energy) and the required storage capacities(Ghoneim 1989). In a parametric study to determine the effects of various thermal and geometric parameters of a storage tank and PCMs on the quantity of solar thermal energy that could be stored (Esen and Ayhan 1996) found smaller cylinder radii (<0.024 m) should be used with fins, lamellae or matrices structured into PCM to increase thermal conductivity. Robin et al. (1995) presented a one-dimensional transient heat conduction simulation of a salt-hydrate PCM suitable for an integrated solar collector storage system in which convective heat transfer in the liquid phase was neglected in these models.

Accurate determination of the temperature distribution in a PCM is essential to understand the thermal performance of a PCM system. The temperature profile in ice was found to be linear in the early stages of the experiment during which no convection occurred (Tien and Yen 1966). However as the thickness of the molten layer increased, the temperature in the liquid phase became uniform due to convective mixing. To fully understand the role of convection in such processes it is necessary to undertake detailed temperature measurements in both phases (Hale and Viskanta 1980). In a latent heat thermal energy storage system heat transfer can be enhanced using an externally finned tube with the heat transfer fluid contained inside a cooling tube (Lacroix 1993; Zhang and Faghri 1996; Sasaguchi et al. 1998). A packed bed of encapsulated phase change material will not have energy densities as high as a fully solid phase change material, however, it has the advantage of increased heat transfer effectiveness as particle size becomes small added to the sensible heat content of the water which usually surrounds the phase change capsules.

References

Abhat A (1981) Low temperature latent heat thermal storage. In: Beghi G (ed) Ispra courses on energy systems and technology. D. Reidel, Holland, pp 33–91

Abhat A (1983) Low temperature latent heat thermal energy storage: heat storage materials. Solar Energy 30:313–332

Abhat A, Malatidis M (1981) Determination of properties of heat of fusion storage materials for low temperature applications. In: Millhone JP, Willis EH (eds) New conservation technologies and their commerciasation, vol 1. Springer, Berlin, pp 847–856

Antohe BV, Lage JL, Price DC, Weber JL (1996) Thermal management of high frequency electronic systems with mechanically compressed microporous cold plates. Thermal management of commercial and military electronics. In: Proceedings of the ASME national heat transfer conference, Houston, pp. 179–186

Baetens R, Jelle BP, Gustavsen A (2010) Properties, requirements and possibilities of smart windows for dynamic daylight and solar control in buildings: a state-of-the-art review. Solar Energy Mater Solar Cells 94:87–105

Banaszek J, Domanski R, Rebow M, El-Sagier F (1999) Experimental study of solid-liquid change in a spiral thermal energy storage unit. Appl Thermal Eng 19:1253–1277

Bansal NK, Buddhi D (1992) An analytical study of a latent heat storage system in a cylinder. Energy Converse Manage 33:235–242

Benard DG, Levesque D, Wirgin A (1978) Theoretical and experimental analysis of latent heat storage system. In: Proceedings of the ISES Congress vol 1. New Delhi, pp 524–530

Cabelli A (1977) Storage tanks – a numerical experiment. Solar Energy 19:45–54

Close DJ (1962) The performance of solar water heaters with natural circulation. Solar Energy 6:33–40

Comakli O, Kaygusuz K, Ayhan T (1993) Solar-assisted heat pump and energy storage for residential heating. Solar Energy 51:357

Costa M, Buddhi D, Olivia A (1998) Numerical simulation of a latent heat thermal energy storage system with enhanced heat conduction. Energ Convers Manage 39:319–330

Davis ES, Bartera R (1975) Stratification in solar water heater storage tanks. In: Proceedings of the workshop on solar energy storage subsystems for the heating and cooling of buildings, Charlottesville

Esen M (2000) Thermal performance of a solar-aided latent heat store used for space heating by heat pump. Solar Energy 69:15–25

Esen M, Ayhan T (1996) Development of a model compatible with solar assisted cylindrical energy storage tank and variation of stored energy with time for different phase change materials. Energy Convers Manage 37:1775–1785

Esen M, Aydin D, Ayla D (1998) Geometric design of solar-aided latent heat store depending on various parameters and phase change materials. Solar Energy 62:19–28

Fath HES (1991) Heat exchanger performance for latent heat thermal energy storage system. Energy Conversion Manag 31:149–155

Fath HES (1998) Technical assessment of solar thermal energy storage technologies. Renew Energy 14(1–4):35–40

Feldman D, Shapiro DB, Fuks CJ (1989) Fatty acids and their mixtures as phase change materials for thermal energy storage. Solar Energy Mater 18:210–217

Fossett AJ, Maguire MT, Kudirka AA, Mills FE, Brown DA (1998) Avionics passive cooling with microencapsulated phase change materials. Transactions of the ASME. J Electron Packaging, 120, pp 238–242

Ghoneim AA (1989) Comparison of theoretical models of phase-change and sensible heat storage for air and water-based solar heating systems. Solar Energy 42:209–220

Gutierrez G, Hincaple F, Duffie JA, Beckman WA (1974) Simulation of forced circulation water heaters; effects of auxiliary energy supply, load type and storage capacity. Solar Energy 15:287–298

Hale DV, Hoover MJ, O'Neill MJ (1971) Phase change materials handbook. Marshall Space Flight Center, Alabama

Hale NW Jr, Viskanta R (1980) Solid–liquid phase change heat transfer and interface motion in materials cooled or heated from above or below. Int J Heat Mass Transfer 23:283–289

Hobson PA, Norton B (1988) Verified accurate performance simulation model of direct thermosyphon solar energy water heaters. ASME J Solar Energy Eng 11:282–292

Hunt BJ, Richtmyer TE, Hill JE, Franklin AE (1978) Testing of water tanks for thermal storage according to ASHRAE Standard 94–77. NBS report no. NBSIR-78-1548, Washington DC

Ishizuka M, Fukuoka Y (1992) Development of a new high density package cooling technique using low melting point alloys. In: Proceedings, ASME/JSME thermal engineering joint conference, vol 2, pp 375–380

Ismail KAR, Alves CLF (1986) Analysis of shell-tube PCM storage systems. In: Proceeding of the eight international heat transfer conference, San Francisco, pp 1781–1786

Ismail K, Goncalves M (1999) Thermal performance of a PCM storage unit. Energy Convers Manage 40:115–138

Ismail K, Henriquez JR (1997) PCM glazing systems. Int J Energy Res 21:1241–1255

Jalaria Y, Gupta SK (1982) Decay of thermal stratification in a water body for solar energy storage. Solar Energy 28(2):137–143

Jotshi CK, Goswami DY, Tomlinson JJ (1991) Thermal energy storage in phase change materials for temperature application zero to 200C. Report no SEEC 91–1. Solar energy and energy conversion lab, University of Florida, Gainesville

Jotshi CK, Tomlinson JJ, Tomlinson JJ (1992) Solar thermal energy storage in phase change materials. In: Proceedings of the 1992 annual conference AMER SOLAR Energy Society

Jurinak JJ, Abdel-Khalik SI (1978) Properties optimization for phase change energy storage in aire-based solar heating systems. Solar Energy 21:377

Jurinak JJ, Abdel-Khalik SI (1979) Sizing phase change energy storage units for air-based solar heating systems. Solar Energy 2:355

Kaviany M (1995) Principles of heat transfer in porous media, 2nd edn. Verlag, New York

Kaygusuz K (1995) Experimental and theoretical investigation of latent heat storage for water based solar heating systems. Energy Convers Manage 36:315

Kaygusuz K, Comakli O, Ayhan T (1991) Solar-assisted heat pump systems and energy storage. Solar Energy 47:383

Kenisarin M, Mahkamov K (2007) Solar energy storage using phase change materials. Renew Sustain Energy Rev 11:1913–1965

Kimura H, Kai J (1988) Mixtures of calcium chloride hexahydrate with some salt hydrates or anhydrous salts as latent heat storage materials. Energy Convers Manage 28:197–200

Klein SA, Beckman WA, Duffie JA (1976) A design procedure for solar heating systems. Solar Energy 18:113–127

Koldhekar SM (1981) Temperature stratification in hot water solar thermal tanks. Report no AIAA-81-0368. Proceedings of the 19th AIAA Aerospace sciences meeting, St. Louis, pp 12–15

Korsgaard V, Esbensen TV (1975) Nul-energihusprojeckted ved DTH. DTH Lyngby, Denmark

Lacroix M (1993) Numerical simulation of a shell-and-tube latent heat thermal energy storage unit. Solar Energy 50:357–367

Lane GA (1983) Solar heat storage and latent heat of materials, vol 1. CRC Press, Boca Raton

Lavan Z, Thompson J (1977) Experimental study of thermally stratified hot water storage tanks. Solar Energy 19:519–524

Lee C, Taylor DJ, Heibein S (1979) Solar energy storage options. Workshop on thermal energy storage for solar heating and cooling, San Antonio

Leoni N, Amon CH (1997) Transient thermal design of wearable computers with embedded electronics using phase change materials, HTD – 343. National heat transfer conference, 5, ASME 1997, pp 49–56

Leyers HJ, Scholez F, Tholen A (1977) Analytical and experimental determination of the temperature distribution in stratfield hot water stores. International seminar on heat transfer in buildings, Dubrovnik, pp 683–693

Loehrke RI, Gari HN, Sharp MK, Haberstcoh RD (1978) A passive technique for enhancing thermal stratification in liquid storage tanks, ASME paper no 78-HT-50. In: Proceedings of the 2nd AIAA/ASME thermophysics and heat transfer conference, Palo Alto, 24–26 May 1978, pp 2–8

Marshall RH (1981) Experimental experience with ASHRAE/NBS procedures for testing a phase change thermal storage device. In: Proceedings of the 1st BHRA international conference on energy storage, Brighton

Matsudaira H, Sakakura Y (1972) Analysis on the thermal capacity of heat storage water tanks, part I. J Soc Heat Air-Cond Sanit Eng Jpn 46(4):11–17

Mazria E (1980) Solar home book. Rodale Press, Emmaus

Miller CW (1977) The effect of a conducting wall on a stratified fluid in a cylinder. In: 12th AIAA thermophysics conference, Albuquerque

Mortimer CE (1986) Chemistry, 6th edn. Wadsworth, Belmont

Nakajima Y (1972) Studies on the weighting functions of thermal storage water tanks. Part I Trans Jpn Archit Soc 199:37–47

Pal D, Joshi YK (1997) Application of phase change materials for thermal control of plastic quad flat packages (PQFP): a computational study. Numer Heat Transfer Part A 30:19–34

Pal D, Joshi YK (1999) Thermal control of horizontally mounted heat sources using phase change materials. EEP – 26–2, Advances in electronic packing-1999, 2, ASME 1999, pp 1625–1630

Phillips WF, Dave RN (1982) Effects of stratification on the performance of liquid-based solar heating systems. Solar Energy 29(2):111–120

Robin Y, Bar-Niv I, Korin E, Mikic B (1995) Integrated solar collector storage system based on a slat-hydrate phase change material. Solar Energy 55:435–444

Salyer IO, Sircar AK (1990) Phase change materials for heating and cooling of residential buildings and other applications. In: Proceedings of the 25th intersociety energy conversion engineering conference, vol 1. Reno, 12–17 Aug 1990, pp 236–241

Salyer IO, Sircar AK (1997) A review of phase change materials research for thermal energy storage in heating and cooling applications at the University of Dayton from 1982–1996. Int J Global Energy Issues 9

Sasaguchi K, Yoshida M, Nakashima S (1998) Heat transfer characteristics of a latent heat thermal energy storage unit with a finned tube (effects on fin configuration). Trans JSME 55:475–482

Schmidt AF, Bircell JR, Wilson WA, Smith V (1960) An experimental study concerning the pressurization and stratification of liquid hydrogen. Adv Cryogenic Eng 5

Sheridan NR, Bullock KJ, Duffie JA (1967) Study of solar processes by analog computer. Solar Energy 11:69

Sherman C, Wood BD, Mason J (1979) Effect of vertical wall conductance on temperature relaxation in thermally stratified liquid thermal storage tanks. In: Proceedings of the ISES conference, Atlanta, pp 591–595

Sliwinski BJ, Mech AR, Shih TS (1978) Stratification in thermal storage during charging. In: Proceedings of the 6th international heat transfer conference, vol 4. Toronto, pp 149–154

Tien C, Yen YC (1966) Approximate solution of a melting problem with natural convection. AIChE Chem Eng Prog Symp Ser 62:166–172

Van Gallen E, Van den Brink GJ (1984) Recommendations for European solar storage test methods. Final report no 303, 212, TNO, Delft

Van Koppen CWJ, Fischer LS, Dijkmans A (1978) Stratification effects in the short and long term storage of solar heat. In: Proceedings of the ISES conference, New Delhi, pp 554–558

Viskanta R, Hale NW (1978) Free convection circulation in a thermal energy storage system driven by heat loss from the wall. In: Proceedings of the 14th southeastern seminar on thermal sciences, North Carolina State University

Wirtz RA, Zheng N, Chandra D (1999) Thermal management using "dry" phase change materials. In: Proceedings of the fifteenth IEEE semiconductor thermal measurement and management symposium, San Diego, IEEE#99CH36306, 9–11 Mar 1999, pp 74–82

Wood RJ, Al-Muslahi SM, O'Callaghan PW, Probert SD (1981) Thermally stratified hot water storage systems. Appl Energy 9:231–242

Yanadori M, Masuda T (1986) Heat transferrential study on a heat storage container with a phase change material. Solar Energy 36:169–177

Zhang Y, Faghri A (1996) Heat transfer enhancement in latent heat thermal energy storage system by using an external radial finned tube. J Enhance Heat Transfer 3:119–127

Zivkovic B, Fujii I (2001) An analysis of Isothermal phase change of phase change material within rectangular and cylindrical containers. Solar Energy 70:51–61

Chapter 5
Flat-Plate and Evacuated Tube Collectors

The sun has got his hat on
Hip hip hip hooray!
The sun has got his hat on,
and he's coming out to play

Gay and Butler (1932)

5.1 Overview

Solar-energy collectors are devices employed to gain useful heat energy from the incident solar radiation. They can be of the concentrating or the flat-plate type. A simple flat-plate solar energy collector consists basically of an absorbing surface which absorbs the insolation and transmit it in form of heat to a working fluid commonly air or water. A general steady-state analysis for flat-plate collectors is shown in Fig. 5.1.

Instantaneous efficiencies are plotted against $(T_{fii}-T_a)/I$ and the intercept and the slope determined (7.1.5). The long-term performance of many solar heating collectors can be characterized by a thus determined intercept and slope as shown by the illustrative examples for different collector types given in Fig. 5.2. These parameters are not constant but depend on temperature and wind speed. Heat removed is also a weak function of overall heat loss. To account for these effects non-linear equations have been developed and together with an angular dependent "incident angle modifier" are used in various versions as the basis for rating tests. These are shown in Fig. 5.3.

The absorber material in a flat-plate collector, in addition to having a high absorbance of the incident radiation should also have a low emissivity, good thermal conductivity, and be stable thermally under temperatures encountered during operation and stagnation. It should also be durable, have low weight per unit area and, most importantly, be cheap. Black coated metal sheets (in air heater) and fins in water heaters are used frequently as absorbers due to their effectiveness, ease of use, availability and relatively low cost. Wisely or not, a multitude of

B. Norton, *Harnessing Solar Heat*, Lecture Notes in Energy 18,
DOI 10.1007/978-94-007-7275-5_5,

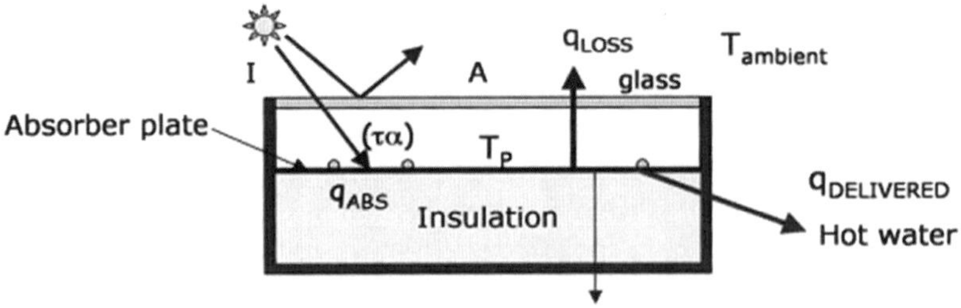

$$q_{DEL} = q_{ABS} - q_{LOSS}$$

$$q_{DEL} = IA(\tau\alpha) - U_L A\,(T_P - T_{AMB})$$

$$\eta = \frac{q_{DEL}}{q_{INCIDENT}} = \frac{IA(\tau\alpha) - U_L A(T_P - T_{AMB})}{IA} = (\tau\alpha) - U_L\left(\frac{T_P - T_{AMB}}{I}\right)$$

Fig. 5.1 Heat balance of a flat plate solar collector

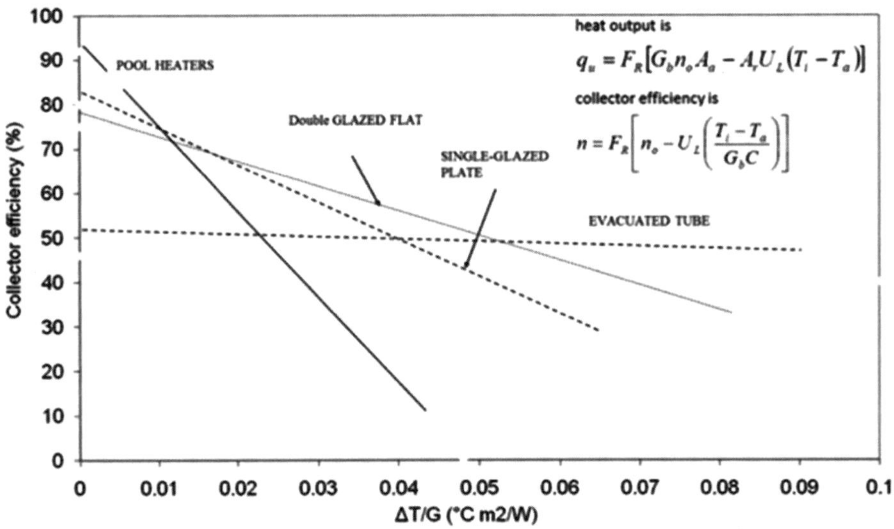

Fig. 5.2 Hottel-Whillier-Bliss characteristics of non-concentrating collector types

$$n = F_R(\tau\alpha)_n k_{\alpha\tau} - c_1\frac{(T_i - T_a)}{G_t} - c_2\frac{(T_i - T_a)^2}{G_t}$$

For single glass cover where

$$k_{\alpha\tau} = 1 - b_o\left(\frac{1}{\cos(\theta)} - 1\right) - b_1\left(\frac{1}{\cos(\theta)} - 1\right)^2$$

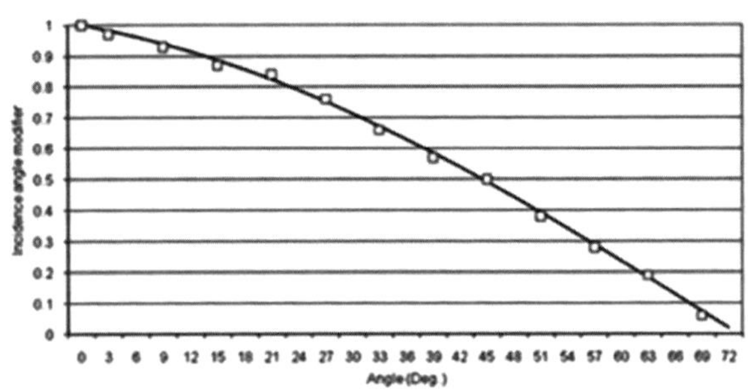

Fig. 5.3 Non-linear collector efficiency equation including incidence angle modifier

different absorber materials have been tried for both air and water heating solar collectors, absorber materials used have included black plastic sheets, black-painted rocks and charcoal.

5.2 Air Heating Flat-Plate Collectors

Solar air heating is employed both in buildings and a wide variety of drying and other process heat applications (Hastings and Morck 2000) and many collector types are available commercially (Hastings et al. 1998). A bare plate solar air heater is the simplest type of flat-plate collector. It consists simply of an air duct, the uppermost surface of which acts as the solar heat absorber plate with the rear surface insulated is shown in Fig. 5.4. Bare-plate collectors are widely used in crop drying operations (both for natural and forced convection systems).

Corrugated sheet roofs of buildings have been adapted as bare plate collectors for the heating of the air space within the building as in some large solar storage barns (Bailey and Williamson 1965). Generally, optical losses due to transmission reduction of the incoming solar radiation by transparent covers in solar-energy collectors are roughly 10 % (Sodha et al. 1987). Thus the efficiency attainable at very low collector temperature rises (<10 °C above ambient) by covered plate solar-energy collectors is thus 10 % less than that of bare-plate collectors. However, though they can experience minimal optical losses, bare-plate solar energy collectors have large thermal losses through the exposed surface. Consequently they have low thermal efficiencies (Ferguson and Bailey 1979; Kreider and Kreith 1977) at moderate elevated temperatures, and thus are only suited for low temperature drying applications. The poor performance of bare-plate collectors is however compensated for by their simple low cost construction.

Upward heat losses from most solar air heaters are minimised by the use of one or more transparent cover materials above and usually parallel to the absorber plate. Common cover materials used are glass, twin-walled polycarbonate sheet and clear plastic. The cover material prevents convective heat losses from the absorbing plate and reduces long-wave radiative heat loss. Appropriate cover materials must thus be reasonably opaque to long-wave radiation. Covered solar air heaters operate at higher efficiencies than bare-plate solar air heaters at moderate temperature elevations, but fabrication cost is increased and since the cover materials are potentially vulnerable to breakage, running costs are also increased marginally. Covered plate solar air heaters

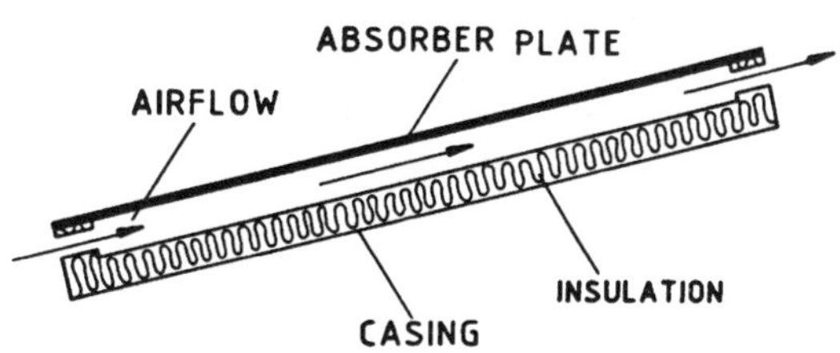

Fig. 5.4 A bare-plate air heater

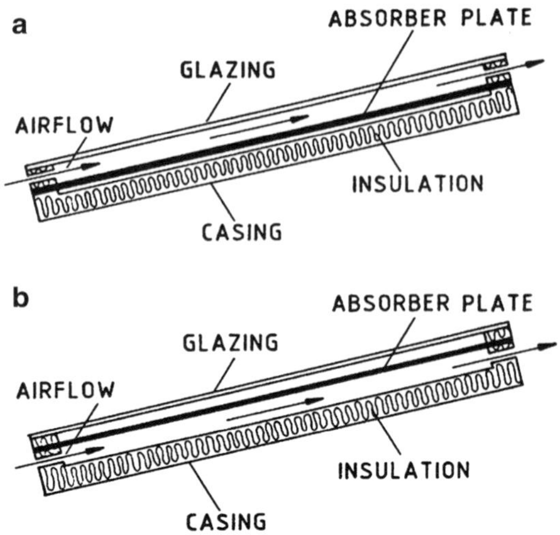

Fig. 5.5 A back-pass and front-pass solar air heaters

are generally recommended for temperature elevations of 10–35 °C above ambient. Covered plate solar air heater types include:-

- Front Pass Covered Solar Air Heaters: Here the air to be heated passes through the duct between the cover material and the absorber plate (the back side of which is insulated) – shown in Fig. 5.5. Heat transfer to the air stream is thus from the front side of the absorber plate.
- Back Pass Covered Solar Air Heaters: The absorber plate here is placed directly behind the transparent cover with a contained layer separating it from the cover. The air to be heated flows between the inner surface of the absorber plate and the layer of insulation with heat transfer via the rear side of the absorber (see Fig. 5.5b). Back pass solar air heaters have generally been have generally been found to be more efficient than the front pass types (Close and Yusoff 1978). To ensure an even flow distribution, an arrangement of parallel ducts as shown in Fig. 5.6 may be employed.
- Suspended Plate Covered Solar Air Heaters: The absorber plate here, is fixed between the cover material and the backing layer of insulation. The air to be heated thus flows on either side of the absorber plate thus increasing the heat transfer area. The absorber plate is thus at a lower temperature and consequently re-radiates less heat. The two different configurations of the suspended plate air heater are the parallel pass solar air heaters (Fig. 5.6) and the double pass solar air heaters shown in (Fig. 5.5). The suspended plate solar air heaters thus operate at higher efficiencies than the bare plate, front or back pass solar air heaters (Ferguson and Bailey 1979). A cross-section of another variant of the collector type is shown in Fig. 5.7.
- Matrix Solar Air Heaters: This is a modified form of the suspended plate collector, with a porous high surface area absorber between the air and the absorber, (Macedo and Altemani 1978) as shown in Fig. 5.8 to increase the heat transfer surface.

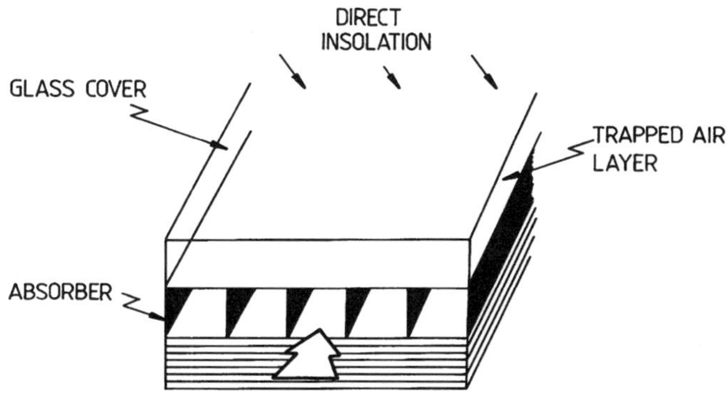

Fig. 5.6 A back-pass solar air heater with multi parallel ducts

Fig. 5.7 A parallel-pass suspended-plate solar air heaters

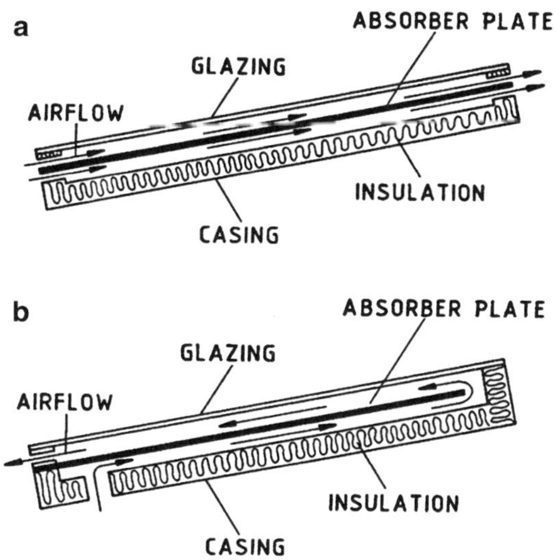

Fig. 5.8 A perinatal matrix air solar air heater

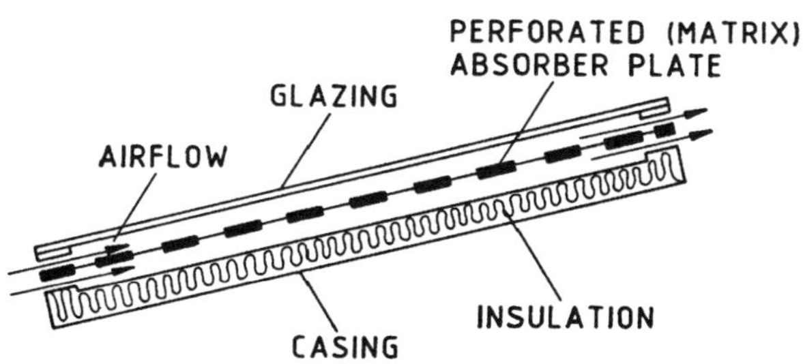

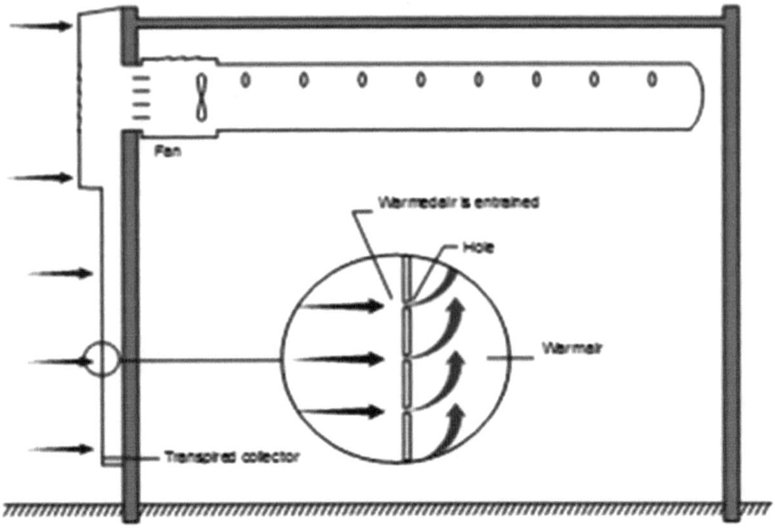

Fig. 5.9 Transpired solar air heater

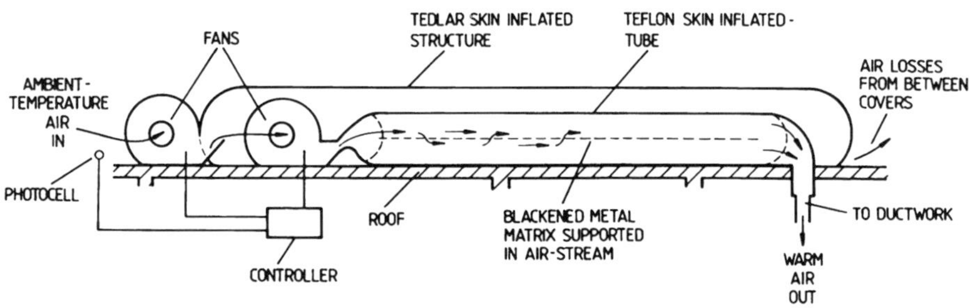

Fig. 5.10 Roof mounted inflated air-heating solar energy collector

- Transpired air heating solar collectors entrain convective heat transfer via perforations in the base absorber plate (Hollick 1990) as show in Fig. 5.9. The performance of transpired air-heating solar collectors is dependent on the;
  - Absorber material
  - Perforation size and spacing (Tsilingiris 2000) (Leon and Kumar 2007; Motahar and Alemrajabi 2010)
  - Air flow rate with efficiency improving at higher flow rates (Khatlab 2000) (Kutscher et al. 1993; Liao et al. 2007; Metwally et al. 1997).
  - Air channel depth (Njomo and Daguenet 2006)
  - Prevailing wind speed (Kutscher et al. 1993)
  - Insolation (Yildiz et al. 2002)
- Inflatable solar air heaters as shown in Fig. 5.10

- For applications which require temperature elevations of <40 °C above ambient, single glazed solar air heaters are adequate. For higher temperature rises, the reduction of heat losses from the absorber plate becomes necessary by the use of additional glazing.

5.3 Water Heating Flat-Plate Collectors

The water heating flat-plate collector shown in Fig. 5.11 is the most commonly found in thermosyphon solar water heaters. In order to predict the fluid outlet temperature a collector heat removal factor is employed. This factor gives the overall heat transfer characteristics, usually of a fin and tube arrangement, under steady state conditions and is the ratio of the actual amount of heat transferred to the collector fluid, to the heat which would be transferred if the entire collector plate was at the fluid inlet temperature as also shown in Fig. 5.11.

The heat removal factor represents the actual amount of that heat absorbed by the fluid at a particular point along the collector, to the amount of heat that would be absorbed were the whole fin at the local fluid temperature. The collector tubes can therefore be considered to be divided up into a series of lengths in each of which heat removal is assumed to be constant for each of these lengths. A transient

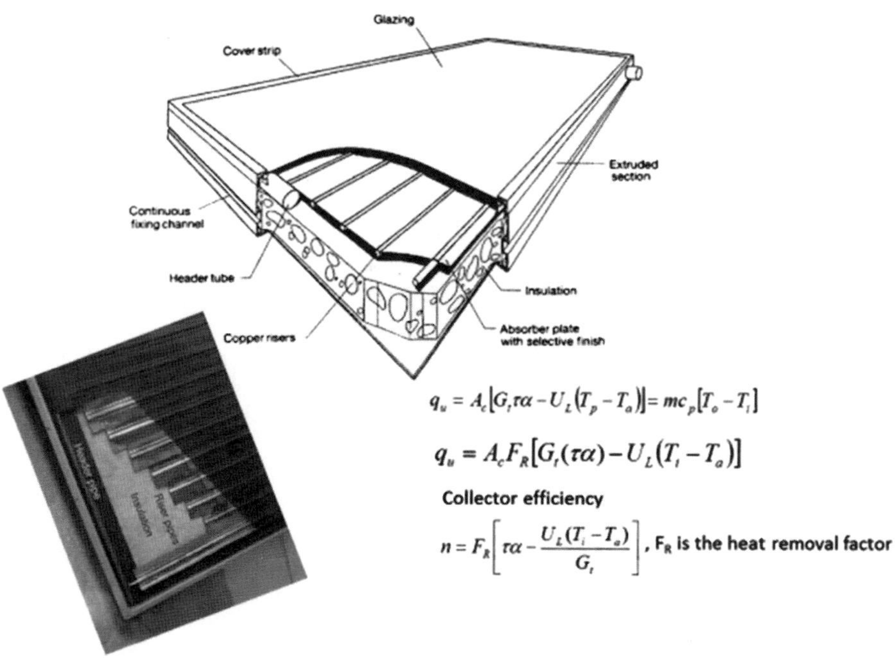

Fig. 5.11 A water heating flat plate collector

thermal analysis may then be carried out on the fluid within each length or node, the heat transfer from the absorber plate to the fluid being calculated using the local heat removal factor. This provides the additional advantage that since the fluid temperature of each node is known, the contribution due to buoyancy forces within the collector can be calculated accurately when determining the mass flow rate in a thermosyphon system, without having to assume a linear temperature profile between the inlet and outlet. The time lag between maximum plate and fluid temperatures attributed in part to the transient behaviour of the collector plate can have a significant effect on the collector outlet temperature (Shitzer et al. 1979).

When comparing the Hottel-Whillier-Bliss model with a steady-state finite difference model of a flat plate collector good agreement between the respective predicted water temperature distributions is found at high mass flow rates (Kirchhoff and Billups 1976). However, the Hottel -Whillier-Bliss model predicts too high efficiencies at low thermosyphonic mass flow rates. This discrepancy was attributed to heat flux in the collector plate, parallel to the risers becoming significant at low flow rates. This is not accounted for in the Hottel-Whillier-Bliss analysis.

Thus for accurate performance prediction of many water-heating flat-plate solar collectors in thermosyphon systems a two dimensional transient finite difference model is required. The main advantages of this approach over the use of Hottel-Whillier-Bliss relationship are that (i) the fluid, collector plate and glass cover assume individual thermal masses and temperature profiles, and (ii) fewer assumptions are necessary concerning heat transfer processes within the collector plate and between the collector plate and fluid. However, even in a detailed collector thermal simulation model, assumptions remain to avoid predicting quantities that neither undergo a spatial nor a temporal variation. For a collector plate of a fin and parallel tube configuration, such assumptions usually include

- For the purposes of radiative heat exchange the glass cover and collector plate are treated as two large, parallel, grey bodies.
- The glass cover is represented by a single node and is therefore considered to be at a single uniform temperature at each moment in time.
- The glass cover is opaque to long-wave radiation.
- Because the fin material is thin, the temperature gradient through the fin is assumed constant; i.e. two dimensional heat flow only.
- Conduction within the collector fluid in the direction of flow is negligible.

The energy balance for the two-dimensional plate temperature distribution is shown in Fig. 5.12.

The boundary condition connecting the fluid temperatures to the plate temperatures is obtained from a heat balance on an incremental volume of fluid within the riser pipes. The pipe wall is assumed to take on the same local temperature as that of the fin. The rate of increase in internal energy plus the heat convected into the incremental fluid volume from the "down stream" direction equals the convective heat transfer between the fin and fluid. To avoid the use of complex shape factors, the glass cover is assumed usually to be at a single uniform temperature at any instant in time is used and the cover temperature is obtained in the

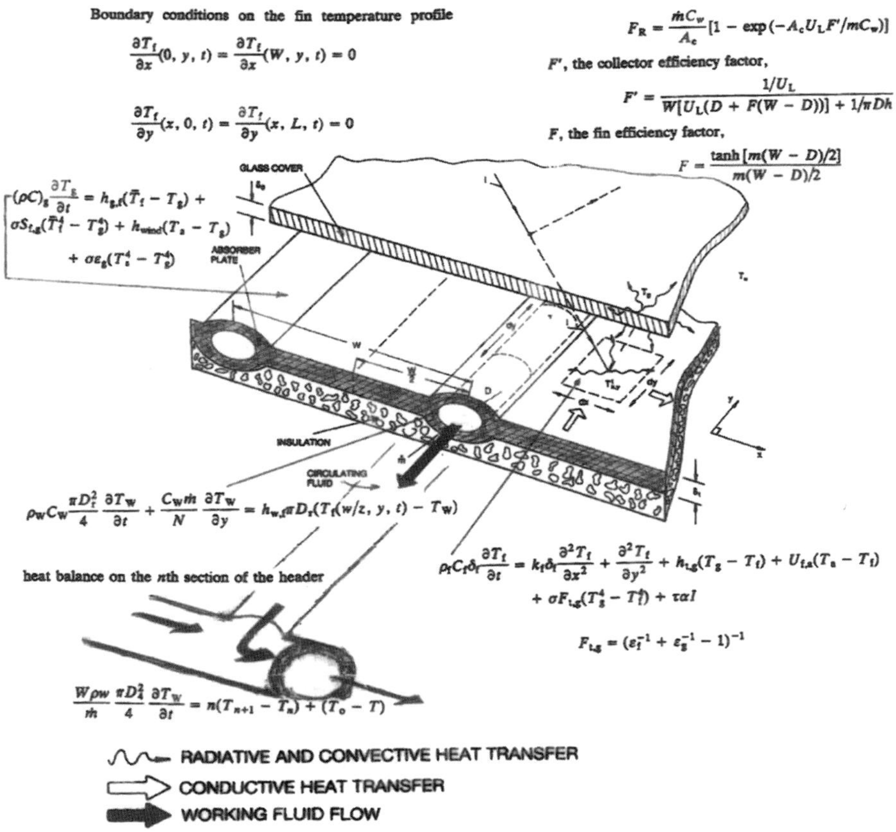

Fig. 5.12 Energy balance of solar collector absorber

following energy balance. The mean temperature of the collector plate is employed to determine the heat transfer between plate and cover. That is the rate of increase in internal energy of the glass cover equals the sum of (i) the heat transferred by convection between the absorber plate and cover, (ii) radiative heat transfer between plate and cover, (iii) convective heat transfer between the outer surface of the glass cover and surroundings (due predominantly to wind effects) and (iv) radiative heat transfer between glass cover and sky).

In order to be able to simulate accurately the transient response time of the collector, it is necessary to take into account the thermal mass of water in the collectors header pipes. The transient response of fluid in this component has been shown to account for the experimentally observed retardation time, that is the time lag between a change in the temperature of the fluid leaving the end of the riser pipes and the time when this change appears at the outlet of the header pipe. The mass flow rates in the sections of the header pipes associated with each of the riser pipes will each have different values. It is therefore necessary to consider each rise tube and fin individually as shown in Fig. 5.13. Non-uniformities can

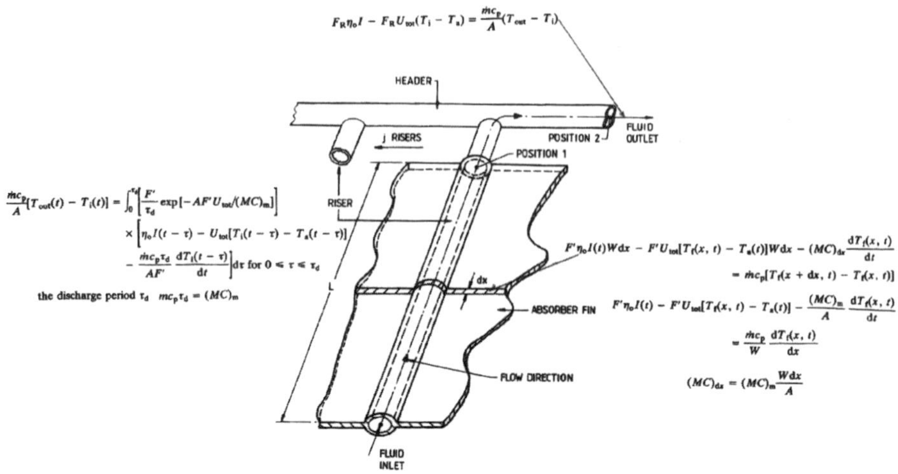

Fig. 5.13 Energy balance on an absorber tube and fin section

arise in the flow distribution through the risers of header-and-riser type collectors (Jones and Lior 1978). Each collector riser then provides different individual contributions to the resulting outlet temperature.

The difference between the rates of the heat *absorbed* by the collector and *lost* to the environment is that removed by the working fluid. (An additional term, accounting for the heat stored (or released) by the collector's components, would appear in the corresponding equation for transient conditions). For a discrete element of a collector in a one-dimensional model a single value for the effective thermal capacitance can be used, and the rate of conduction of heat along the fluid flow path can be ignored. Assuming homogeneity of the collector in the direction of the fluid flow, the effective thermal capacitance of the considered element $(MC)_{dx}$ can be expressed as shown in Fig. 5.13.

The discharge period (Rogers 1980) represents the time interval within which the volume of the working fluid removed from the collector presents a thermal capacitance equivalent to the effective thermal capacitance of the collector system. The rate of heat output from the collector system can be expressed as a convolution integral (Kuo 1966), which accounts for the delays caused in the outlet temperature by the effective thermal capacitance of the collector. This is shown on the left hand side of Fig. 5.13. The first part in brackets on the right-hand side is then the response function of the collector system, and the second part in brackets represents the rate of energy input to the working fluid under the considered generalised conditions. Thus the response function of the collector may be expressed as shown in Fig. 5.14.

The response period represents the elapsed time for the response function of the collector to achieve 0.368 of its initial value. The response period corresponds to the standard definition of the response period for a first-order system that is that period within which the transient phase would be completed if the rate of variation of the associated variable was held constant at its initial value (Atkinson 1980). The response function of a collector under steady-state conditions degenerates to a single value, namely the heat-removal factor.

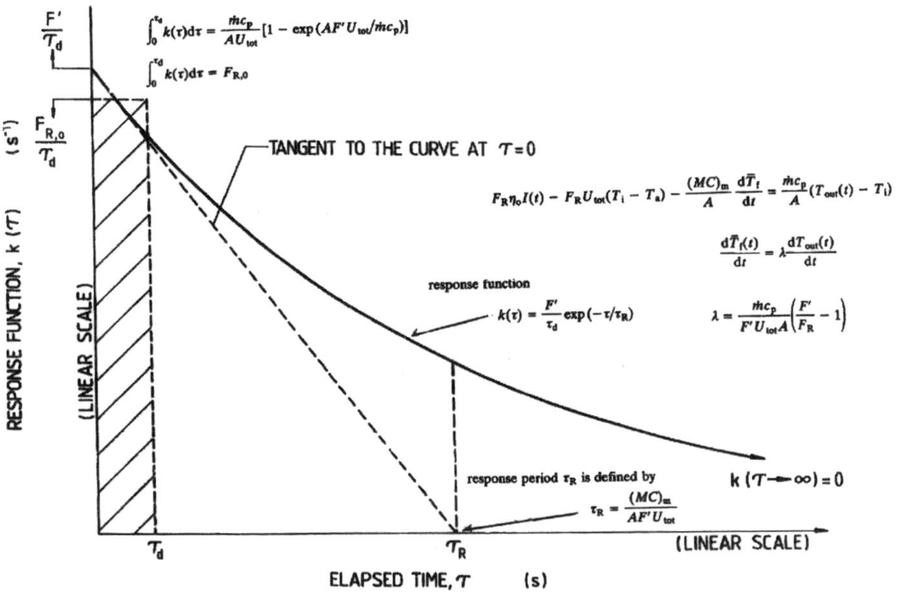

Fig. 5.14 Response function for solar collectors

The dynamic effects under transient operation of a collector have been also analysed theoretically (De Ron 1980; Kamminga 1985; Chakraverty et al. 1987; Wang et al. 1987). Simplified models such as the equation shown on the top right-hand side of Fig. 5.14 (Hill et al. 1977; Simon 1975; Gillet and Moon 1985) permit a satisfactory long-term prediction of collector's performance under transient conditions to be obtained. They have used extensively to produce expressions for various characteristics of a solar-energy collector, such as the response period (Hill et al. 1977) and the effective thermal capacitance (Simon 1975). However, because of the simplifying assumptions they incorporate, they cannot adequately predict short-term behavior, e.g. the minute-by-minute thermal behavior during intermittently-sunny conditions.

Assuming a single value for the effective thermal capacitance and that the rate of conduction of heat along the fluid flow path can be neglected, then an exact analytical solution of can be obtained for a generalised temporal pattern of the insolation intensity if the insolation variation with time is specified. For example, consider a step-change in the insolation intensity from an initial value of zero. The ambient and inlet temperatures are assumed to be equal and to remain invariant during the course of the transient phase. Then, for this particular case, the solution shown in Fig. 5.15 can be obtained. Figure 5.15 also shows perspective view of the evolution of the temperature profiles during the transient phase.

Figures 5.15 and 5.16, show the elapsed time for the transient phase to be completed with various segments along the fluid-flow path attaining steady-state temperature earlier, depending on their location. The position nearest to the inlet of the collector is the first to attain a steady-state as the steady-state "front" travels

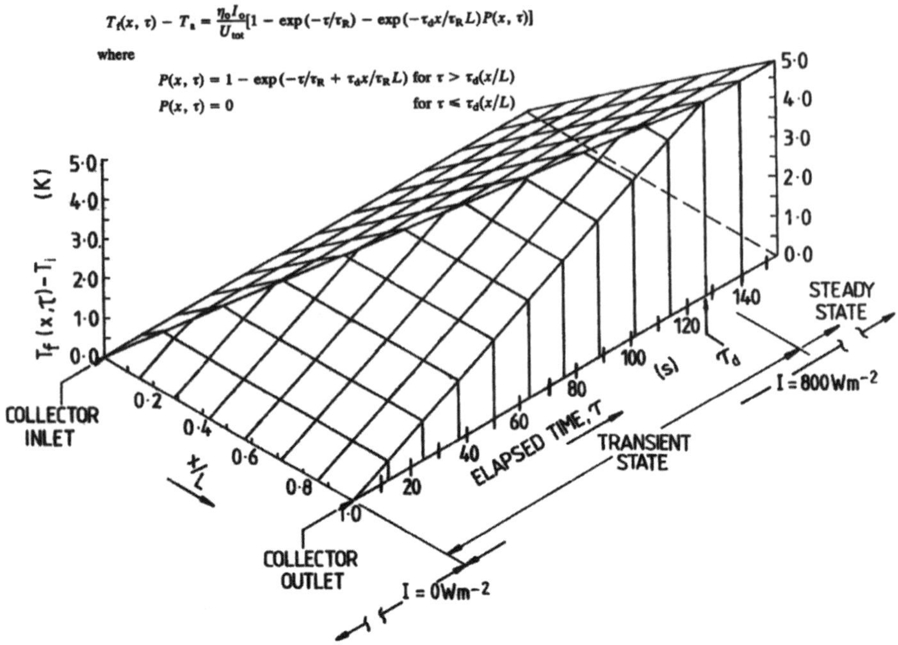

Fig. 5.15 Response of collector absorber temperature

towards the outlet of the collector. This front reaches the outlet after an elapsed period equal to the discharge time that would be derived from a steady-state analysis (Klein et al. 1974).

During transient conditions the local thermal behavior in a solar-energy collector is not uniform. Different locations along the fluid-flow path as shown in Fig. 5.16 respond in a different manner, to the changes in their temperature encountered. A response function would be required that accounts for both the time and position dependences of a collector outlet temperature. However, such a response function is, neither analytically nor experimentally, derived readily. If it is assumed that a sequence of transient operating conditions can be approximated by a series of discrete step-changes, each one of a duration of not less than the discharge time of the collector enables the dependence of the outlet temperature on the distributed nature of a collector system to be taken as an average effect over each step change. Then the thermal behaviour of the collector can be described by a one-variable response function that can be derived for the outlet temperature values shown in Fig. 5.16 (solid line). As a result of taking into account the overall effect of the distributed nature of the collector this response function as depicted in Fig. 5.17 can be seen to present an almost square shape (compared to the curve in Fig. 5.14) with the exponential nature of the curve being very gradual.

The behaviour of the effective thermal capacitance cannot be described by a single value under transient conditions (Prapas et al. 1988a, b). It can be expressed approximately as shown on the top left hand side of Fig. 5.17.

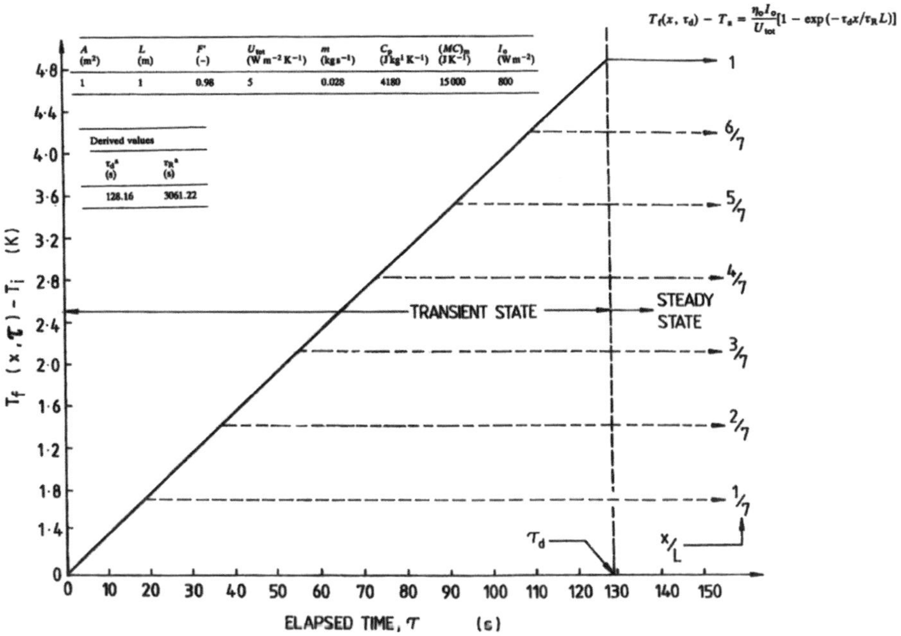

Fig. 5.16 Evolution of the transient thermal state in a collector absorber

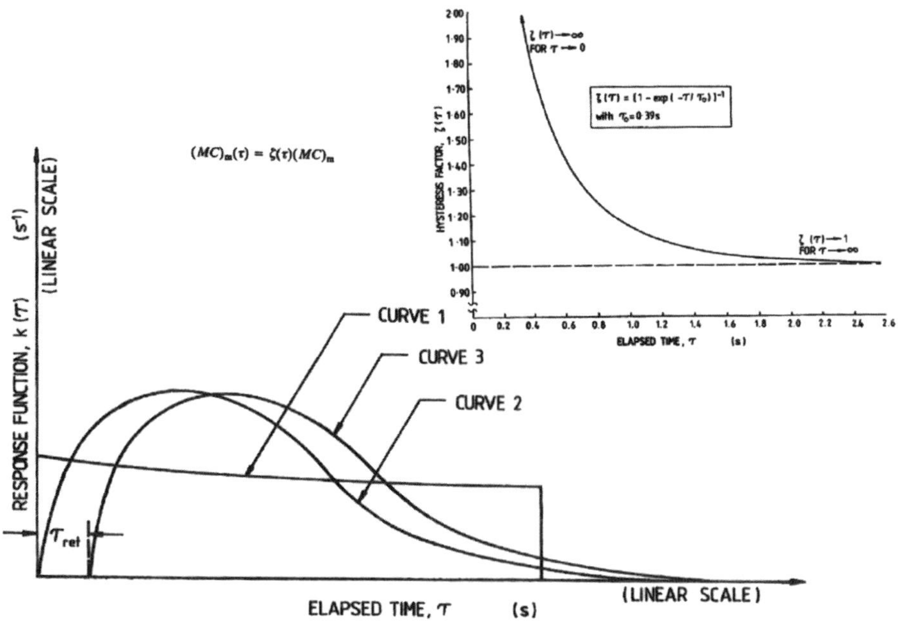

Fig. 5.17 Response functions and hysteresis factors for a solar collector

The hysteresis factor in Fig. 5.17 corresponds to an elapsed time from the onset of the transient phase. The effective thermal capacitance during a transient phase is greater than the steady-state. During transient conditions, heat is stored in the collector absorber in advance of delivery to the working fluid. For steady-state thermal capacitance values close to the asymptote are attained by most metal absorbers within 10 s.

If the effective transient thermal capacitance is employed in a collector then the heat balance equation becomes non-linear. The response function can then only be readily obtained approximately by linearization of that equation. However, the response period of a absorber-working fluid sub-system of about 0.40 s is much smaller normally than the discharge times of a collector with a typical metal absorber of 130 s. Assuming that step-wise transient changes are of duration of longer than the discharge time enables the thermal capacity effects to be averaged out over the transient interval. So, in spite of the mathematical prohibitions which inhibit obtaining a general analytical solution, simplifications can always be used to obtain a "response function" experimentally for a collector system. By incorporating the average effects of both the distributed nature of the collector and the effective thermal capacitance, this latter function can describe adequately the transient performance of the collector system as long as the operating conditions change step-wise with a time interval between two successive changes longer than the discharge time. This experimentally-obtained response function is different in form from that derived with the assumption of constant effective thermal capacitance (see curve 1 in Fig. 5.17). The shape of this function is shown qualitatively by curve 2 in Fig. 5.17.

If zero thermal capacity is assumed for the pipework between positions 1 and 2 in Fig. 5.13, then the response function associated with the outlet temperature measured at the second position is identical in form to the curve 3 in Fig. 5.17 corresponding to the response function (i.e., curve 2) associated with the position 1 but for a retarded by a time that can be predicted from the dimensions of the pipework and the fluid flow-rate. In a typical header-and-riser type flat-plate collector absorber riser-and-fin assembly, when the flow distribution through the collector risers is uniform, the overall response function of the collector system can then be derived from a linear superposition of the individual response functions of each of the collector risers. The resultant overall response function is illustrated qualitatively in Fig. 5.18. The overall response can be seen to be dependent on the pipe-work geometry within the collector and the choice of the position at which the outlet temperature is to be measured.

5.4 Evacuated-Tube Flat-Plate Collectors

Vacuum tube (also referred to as evacuated tube) solar collectors are comprised of an evacuated tubular annulus surrounding and absorber. The absorber can be;

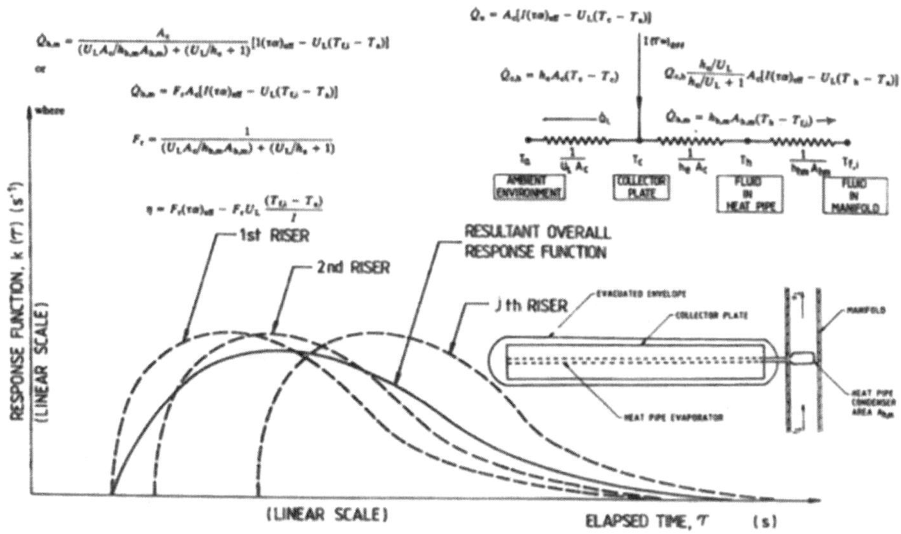

Fig. 5.18 Overall response function of a collector absorber

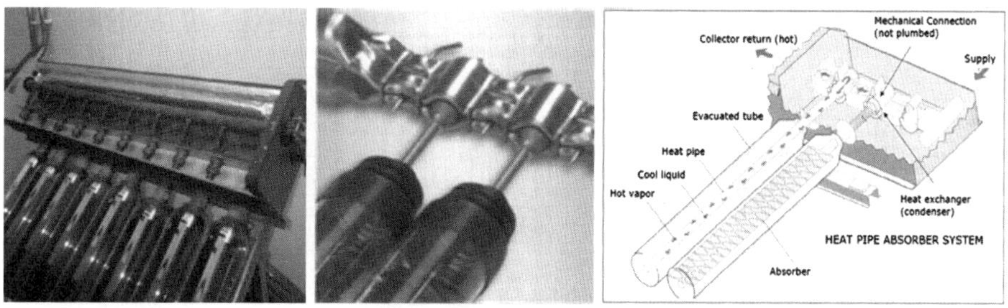

Fig. 5.19 Manifold arrangements for heat-pipe evacuated tube solar collectors

- In an integral all-glass collector, the solar-selective coated inner surface of a continguous inner glass tube within which the heat transfer fluid flow (Wang et al. 1987, 1989; Budihardjo et al. 2007; Li et al. 2010)
- In a fluid-in-metal collector, a metal (usually copper) inner tube containing the heat transfer fluid (Window and Harding 1983; Window 1983; Zhigiang and Harding 1984, 1985; Harding and Zhiqiang 1985a, b)
- A closed heat pipe (Tabassaun et al. 1988). Several manifold arrangements for heat pipes are shown in Fig. 5.9

All-glass vacuum tube collectors have a higher heat extraction efficiency than fluid-in-metal units (Wang et al. 1987, 1989). However the aspect ratio of the inner tube should not be so large that no flow is entrained at the lowest end of the collector (Morrison et al. 2004, 2005; Budihardjo et al. 2007; Shah and Furbo 2007).

In a dry type evacuated-tube collector, shown in Fig. 5.19, each absorber fin is enclosed in a separate cylindrical glass envelope. The latter generally has only one

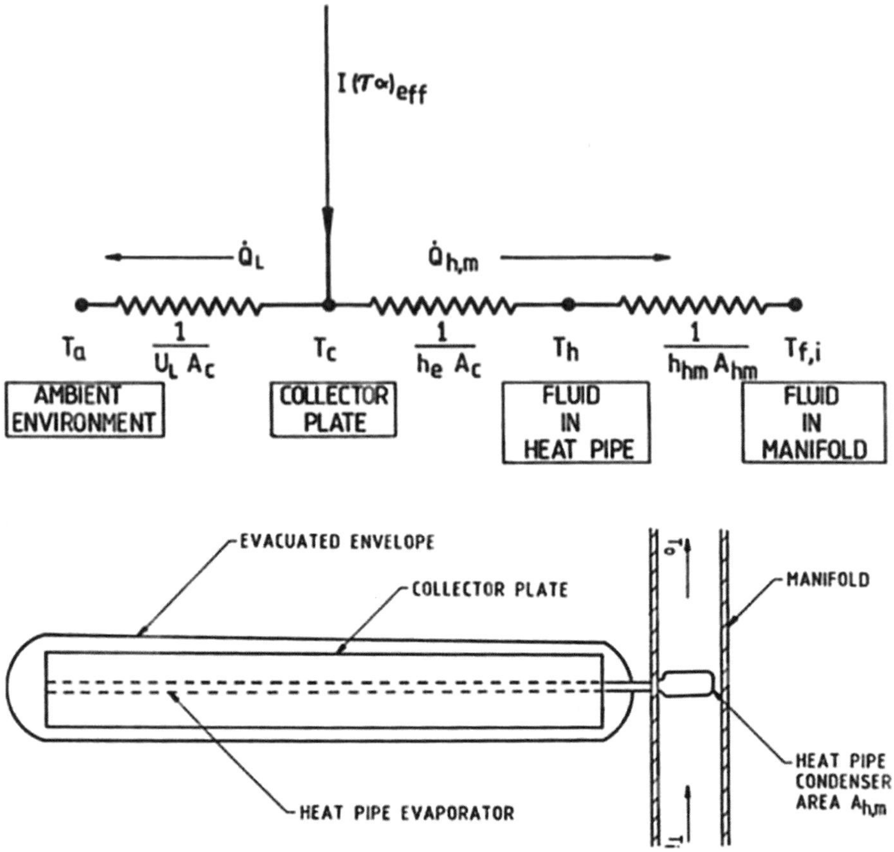

Fig. 5.20 Heat transfer in a heat-pipe evacuated tube collector

protrusion of pipework, against which the glass is sealed, through which heat is removed. Evacuation of the envelope prevents convective heat loss from the absorber plate to the glass. The transfer of heat from the collector is usually accomplished by employing a heat-pipe absorber.

Inside the condenser and the evaporator, the heat transfer phenomena are especially complicated. They system considered here consists of a single fin-and-tube heat-pipe absorber within an evacuated glass sheath. The fin-plate usually has a selective surface and the condenser of the heat-pipe transfers heat to the fluid via a manifold. A simple mathematical model has been developed to study the effects on performance of variations of heat-transfer areas and coefficients of the condenser and the evaporator.

Figure 5.20 shows a simplified resistance network for a heat-pipe solar-energy collector. The rate of radiative heat loss is low (because of the use of selective surfaces in such collectors) compared with the rate of convective heat loss. The steady-state rates of useful heat removal by the solar-energy collector plate

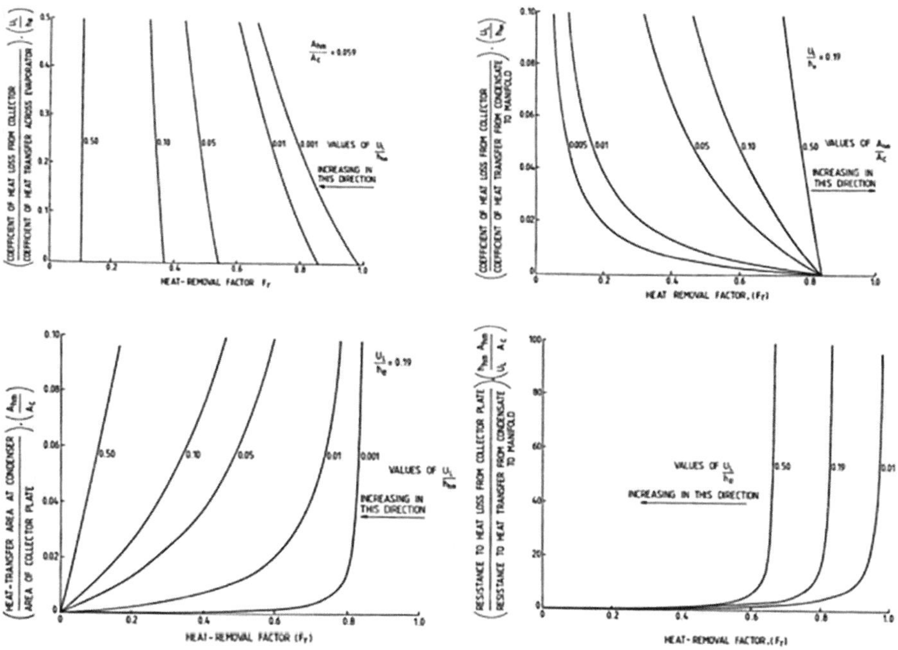

Fig. 5.21 Dependency of heat pipe absorber performance on effective condensation in the manifold

for ratios of critical parameters are shown in Fig. 5.21. The collector heat-loss coefficient can be regarded as constant over the working range. Thus the heat removal factor has a great influence on the shape of the efficiency curve.

A heat-pipe absorbers heat removal factor is dependent on the condenser transferring heat to the fluid in the manifold at a rate that enables sufficient fluid to return to the absorber to effect efficient evaporative heat removal (Kamminga 1986) as shown in Fig. 5.20.

In a "wet" type of evacuated tube, water circulates in a thermosyphon comprised of the evacuated tube and the store. Figure 5.22 shows a cross-section through a wet-type evacuated tube collector. Flow-through "wet" evacuated tube collectors avoid the heat exchanger efficiency losses associated with closed "dry" collectors as they are inserted directly into sockets in the store as show in Fig. 5.23.

There are estimated in 2010 to be over 15 million thermosyphon installations in China alone with the wet type of evacuated tube. Though various brands have been exported from China to other regions, internationally such systems have not yet displaced close-coupled thermosyphon units comprising separate flow and return pipes to a flat-plate collector.

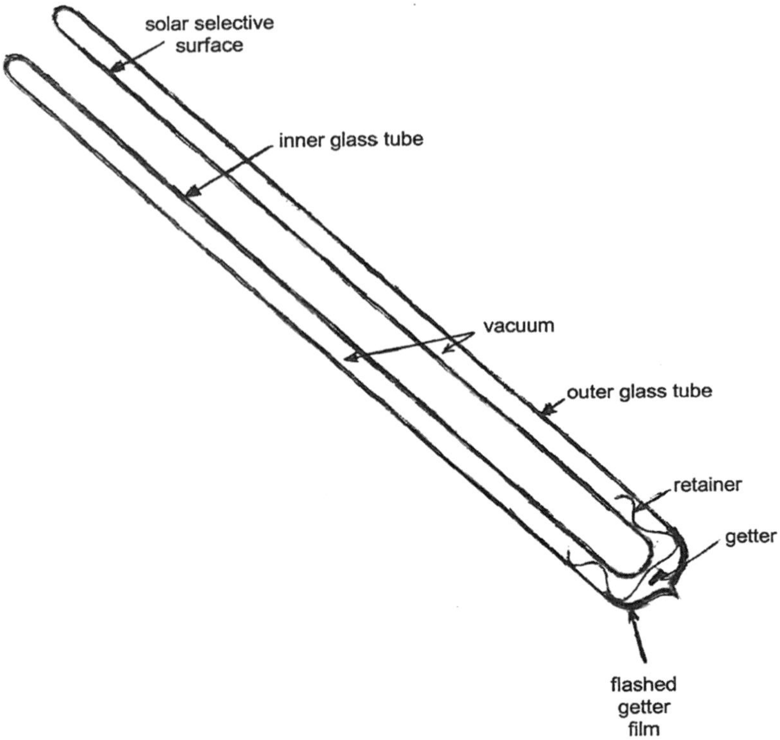

Fig. 5.22 An all-glass evacuated tube collector

5.5 Rating Tests for Solar Collectors

Since the early work (e.g., Whillier and Richards 1961; Anon 1977a, b), much research has been completed in order to develop and refine methods for determining the performances of solar-energy collectors. Testing procedures are available for both liquid-heating and air-heating collectors, which may be of either flat-plate or concentrating types. Transient test methods are available (Anon 1986), that employ experimental measurements are acquired during actual insolation conditions. In the analysis of the behaviour of a collector under either steady-state or transient conditions, the Hottel-Whillier-Bliss characteristics of the collector namely the optical efficiency and for the overall heat-loss coefficient) are obtained directly as lumped values. Transient tests can be used for both the "heating" and "cooling" behaviours of a collector can be used to determine its response function (Prapas et al. 1989). Collectors are also rated for their durability. Limited work has been undertaken to relate the outputs of such test to each of the solar collector failure sources shown in Fig. 5.24.

Fig. 5.23 Socket-fitting of an evacuated tube collector directly into the hot water store

5.6 Architectural Integration of Solar Collectors

Little attempt is made or is seemingly necessary to integrate domestic solar collectors even when they are included in a new construction. Wall-mounted collectors have on occasion been installed as façade features as shown in Fig. 5.25. For façade-mounted collectors, the reflection from the collector of light in the visible range of the spectrum gives a collector with a coloured rather than black appearance. If the visible spectral range selected is a sufficient narrow wavelength band then collector absorptance may remain acceptable (Munari-Probst and Roeker 2010; Mertin et al. 2011).

Larger-scale solar energy collector installations mounted on a flat roof are often conceded when ground by a judiciously designed parapet. For example in Fig. 5.26

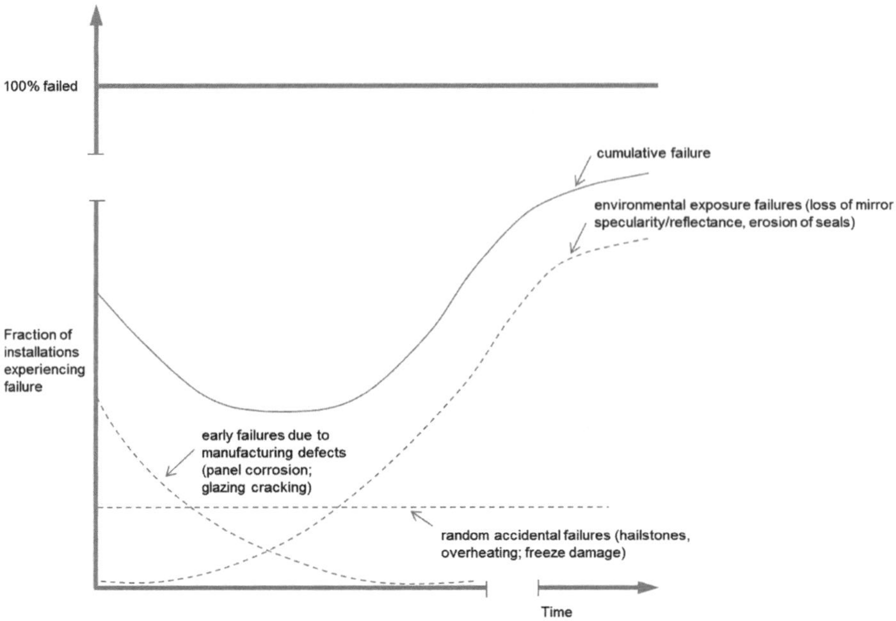

Fig. 5.24 Illustrative prevalence of failure modes over time for solar thermal collectors

Fig. 5.25 Horizontally mounted collectors on a façade in Malmo, Sweden

Fig. 5.26 Roof-mounted collectors shown on the *left* hidden from ground level view as shown on the *right* by a parapet in Jiuquan, China

which shows two photographs of a hotel, in Jiuquan, China, the evacuated tube collector away that provides service hot water is not visible when viewed from ground level.

References

Anon (1986) Methods of test for the thermal performance of solar collectors, British Standard 6757. British Standards Institution, London

Anon (1977a) ASHRAE Standard 94-77. Methods of testing thermal storage devices based on thermal performance. ASHRAE, New York

Anon (1977b) Methods of testing to determine the thermal performances of solar collectors, AS RAE Standard 93-77. American Society of Heating, Refrigeration and Air-conditioning Engineers, New York

Atkinson P (1980) Feedback control theory for engineers. Heinemann Educational Books, London

Bailey PH, Williamson WF (1965) Some experiments on drying grain by solar radiation. J Agr Eng Res 10:191–196

Budihardjo I, Morrison GL, Behnia M (2007) Natural circulation flow through water-in-glass evacuated tube solar collectors. Solar Energy 81:1460–1470

Chakraverty S, Bansal NK, Garg HP (1987) Transient analysis of a CPC collector with time dependent input. Solar Energy 38:179–185

Close DJ, Yusoff MB (1978) The effect of air leaks on solar air collector performance. Solar Energy 20:459–465

De Ron AJ (1980) Dynamic modelling and verification of a flat-plate solar collector. Solar Energy 24:117–128

Ferguson WE, Bailey PH (1979) Solar air heater for near ambient crop drying; description test methods and 1977 results. The British Society for Research in Agricultural Engineering, Scottish Institute of Agricultural Engineering departmental note, SIN/275

Gillett WB, Moon JE (1985) Solar collectors- test methods and design guidelines. D. Reidel, Boston, pp 123–130

Harding GL, Zhiqiang Y (1985a) Thermosiphon circulation in solar water heaters in corporating evacuated tubular collectors and a novel water-in-glass manifold. Solar Energy 34:13–18
Harding GL, Zhiqiang Y (1985b) Heat extraction efficiency of a concentric glass tubular evaluated collector. Solar Energy 35:71–79
Hastings SR (ed) (1998) Solar air systems, built examples. James and James, London
Hastings SR, Morck O (2000) Solar air systems: a design handbook. Earthscan, London
Hill JE, Jenkins JP (1977) The application of ASHRAE 93–77 standard to concentrating collectors. In: Proceedings of the US energy research and development administration conference, concentrating solar collectors, Atlanta, pp 1–8
Hollick PR (1990) Method and apparatus for reheating ventilation air for a building. US patent 4, 899, 728
Jones GF, Lior N (1978) Isothermal flow distributions in solar collectors and collector manifolds. In: Proceedings of the annual meeting of the American Section of ISES, Denver, pp 362–367
Kamminga W (1985) The approximate temperatures within a flat- plate solar collector under transient conditions. Int J Heat Mass Transfer 28:433–440
Kamminga W (1986) The testing of an evacuated tubular collector with a heat pipe using the Fourier frequency domain. Int J Heat Mass Transfer 29:83–90
Khatlab HM (2000) Evolution of perforated plate solar air heater. Int J Solar Energy 21:46–62
Kirchhoff RH, Billups M (1976) A two dimensional heat transfer model of a flat plate collector. ASME paper 76-WA/Sol-2
Klein SA, Duffie JA, Beckman WA (1974) Transient considerations of flat-plate solar collectors. ASME J Eng Power 96A:109–113
Kreider JF, Kreith G (1977) Solar heating and cooling, engineering practical design and economics. McGraw-Hill, New York
Kuo FF (1966) Network analysis and synthesis, chapters 1 and 7. Wiley, New York
Kutscher CF, Christiansen CB, Barker GM (1993) Unglazed transpired solar collectors; heat loss theory. ASME J Solar Energy Eng 115:12–188
Leon MA, Kumar SC (2007) Mathematical enabelling and thermal performance analysis of unplugged transpired air collectors. Solar Energy 81:62–75
Li Z, Chen C, Luo H, Zhang Y, Xue Y (2010) All-glass vacuum tube collector heat transfer model used in forced-circulation solar water heating system. Solar Energy 84:1413–1421
Liao AK, Athienitis L, Candonedo L, Park K-W, Poissant Y, Collins M (2007) Numerical and experimental study of heat transfer in a BIPV-thermal system. ASME J Solar Energy Eng 129:423–430
Macedo I, Altemani CAC (1978) Experimental evaluation of natural convection solar air heaters. Solar Energy 20:367–369
Mertin S, Le Caër V, Joly M, Scartezzini J-L, Schueler A (2011) Coloured coatings for glazing of active solar thermal facades by reactive magnetron sputtering. In: Proceedings of the CISBAT, Lausanne
Metwally AM, Abou-Ziyan HZ, El-Leathy AM (1997) Performance of advanced corregated duct solar air collector compared to five conventional designs. Renew Energy 10:519–537
Morrison GL, Budihardjo I, Behnia M (2004) Water-in-glass evacuated tube solar water heater. Solar Energy 76:135–140
Mortimer CE (1986) Chemistry, 6th edn. Wadsworth, Belmont
Motahar S, Alemrajabi AA (2010) An analysis of unglazed transpiral solar collectors based on exergetic performance criteria. Int J Thermodynamics 13:153–160
Munari-Probst MC, Roeker C (2010) Architectural integration of solar thermal systems. Detail 1, 10, pp. 42–45
Njomo D, Daguenet M (2006) Sensitivity analysis of thermal performances of flat plate solar air heaters. Heat Mass Transfer 42:1065–1081
Prapas DE, Norton B, Milonidis E, Probert SD (1988a) Response function for solar-energy collectors. Solar Energy 40:371–383

Prapas DE, Norton B, Probert SD (1988b) Variations of the effective thermal capacitance in a solar-energy collector. Int J Ambient Energy 8:59–72

Prapas DE, Norton B, Probert SD (1989) Transient performance testing of a parabolic trough concentrating solar energy collect. Proc Inst Mech Eng 203:21–29

Rogers BA (1980) Transient testing of collectors. In: Proceedings UK_ISES conferences, solar energy codes of practice and test procedures, London, pp 45–55

Shah J, Furbo S (2007) Theoretical flow investigations of an all glass evacuated tubular collector. Solar Energy 81:822–828

Shitzer A, Kalmanoviz D, Zvirin Y, Grossman G (1979) Experiments with a flat-plate solar water heating system in thermosyphonic flow. Solar Energy 22:27–36

Simon FF (1975) Flat-plate solar–collector performance evaluation with a solar simulator as a basis for collector selection and performance prediction. NASA Report No TMX-71793, also presented at the ISES meeting, Los Angelas, 29 July–3 Aug 1975

Sodha MS, Bansal NK, Kumar K, Bansal PK, Malik MAS (1987) Solar crop drying. CRC Press, Florida

Tabassaun SA, Norton B, Probert JD (1988) Heat removed from a solar energy collector with a heat pip absorber. Solar Wind Technol 5:141–145

Tsilingiris PT (2000) Heat transfer analysis of low thermal conductivity solar energy absorbers. Appl Therm Eng 20:1297–1314

Wang LT, Weide L, Zhai Z (1987) Experimental study on the flow pattern inside the horizontally disposed all glass evacuated solar collector with visual technique. Acta Energiae Solaris Sinica 8:177–184

Wang, LT, Weide, L, Zhai, Z (1989) Experimental study on the natural convective heat transfer in horizontally disposed all glass evacuated solar collector tube. Acta Energiae Solaris Sinica 10:176–185

Weinberger H (1964) The physics of the solar pond. Solar Energy 8:45–56

Whillier A, Richards S (1961) A standard test for solar- energy water heaters. In: Proceedings of the U.N. conference on new sources of energy, paper S/97, Rome

Window B (1983) Heat extraction from single ended glass absorber tubes. Solar Energy 31:159–166

Window B, Harding GL (1983) Buoyancy effects and the manifolding of single ended absorber tubes. Solar Energy 31:53–157

Yildiz C, Togrul IT, Sassilmaz C, Pehliean D (2002) Thermal efficiency of an air solar collector with extended absorption surface and increased convection. Int Commun Heat Mars Transferm 29:831–840

Zhiqiang Y, Harding GL (1984) Optical properties of dc reactively sputtered thin films. Thin Solid Films 120:81–108

Zriken Z, Bilgen E (1987) Theoretical study of Trombe-Michel wall solar collector system. Solar Energy 39:409–419

Chapter 6
Use of Heat From, and Thermal Management of, Photovoltaics

The Sun is shining, but the ice is slippery

Gibson (1931)

6.1 Overview

A photovoltaic-thermal (PV/T) is a photovoltaic panel from which heat is removed.
The three attributes of PV/T collectors are:

- PV efficiency can be improved by heat removal as cell operating temperature is lower for mono and polycrystalline silicon PV-cells, whose efficiency decreases with increasing temperature by approximately 0.5 %°C.
- Solar heat and electricity are both provided (Van Helden et al. 2004) PV modules absorb 80 % of the incident solar radiation but convert only often 15–20 % of this to electrical energy, the remainder being dissipated as thermal energy.
- The, usually, roof area dedicated available for solar energy devices can be employed more optimally by combining both systems into a single unit.

In flat plate PV/T collectors air or water are used for heat removal. Heat removal from concentrated photovoltaic systems has been accomplished using fluids such as air and water (Royne et al. 2003; Saki et al. 1996; Du et al. 2012; Teo et al. 2011) or by heat pipes (Anderson et al. 2008). PV/T collectors and systems for the thermal management of photovoltaics can be distinguished by the latter seeking solely to maximise photovoltaic output. A PV thermal management system thus collects solar heat but dissipates it to an ambient heat sinks. If the heat is used from a PV thermal management system this is likely to be asynchronous with insolation (Fig. 6.1).

B. Norton, *Harnessing Solar Heat*, Lecture Notes in Energy 18,
DOI 10.1007/978-94-007-7275-5_6, © Springer Science+Business Media Dordrecht 2014

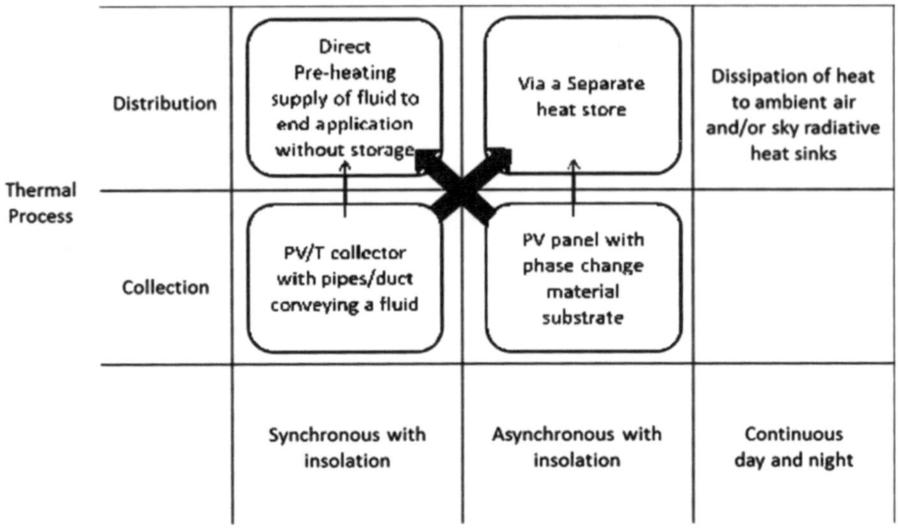

Fig. 6.1 Distinction between PV/T and PV/PCM solar collectors and associated storage systems

6.2 PV/T Water Heating Collectors

Photovoltaic thermal water heating (Ji et al. 2006) collectors continue to be examined for use in both pumped and in thermosyphon- solar water heaters (Chow et al. 2008; Soulites et al. 2011).

A 2 m^2 domestic solar water heater to generate a useful amount of electricity when integrated with a series of PV cells (Garg and Agarwal 1995; Garg et al. 1991). In a water heating PV/T collector the PV module replaces the collector absorber plate (Cox and Raghuraman 1985). A theoretical analysis of a PVT solar collector can take the form of a modified version of the Hottel-Whillier Bliss equation (Florschuetz 1979). The costs of heat and electrical energy displaced and of the PV/T collector itself affect the economic viability of PVT particularly for water heating (Andrews 1981). Algorithms have been developed to calculate both the thermal and electrical efficiency of a PVT system (Bergene and Lovvik 1995). They suggested that mush systems might be useful as pre-heaters for domestic hot water. For PV/T water heating collectors using polycrystalline-Silicon and amorphous-Silicon PV cells Tripanagnostopoulos et al. (2002) found that the cooling provided in improved PV cells efficiency by approximately 10 %. Water provided better cooling than air, with air heating collectors, glazing the collectors improved thermal performance but reduced the electrical efficiency.

A system a thermosyphon can be used to circulate the cooling water in a PV/T, He et al. (2006) found that such system showed combined efficiency of 50 %, with the thermal efficiency contributing approximately 40 %. Though less hot water was produced than a conventional thermosyphon solar water heater, due to the electricity purchased the overall energy production was greater. Due to a larger temperature

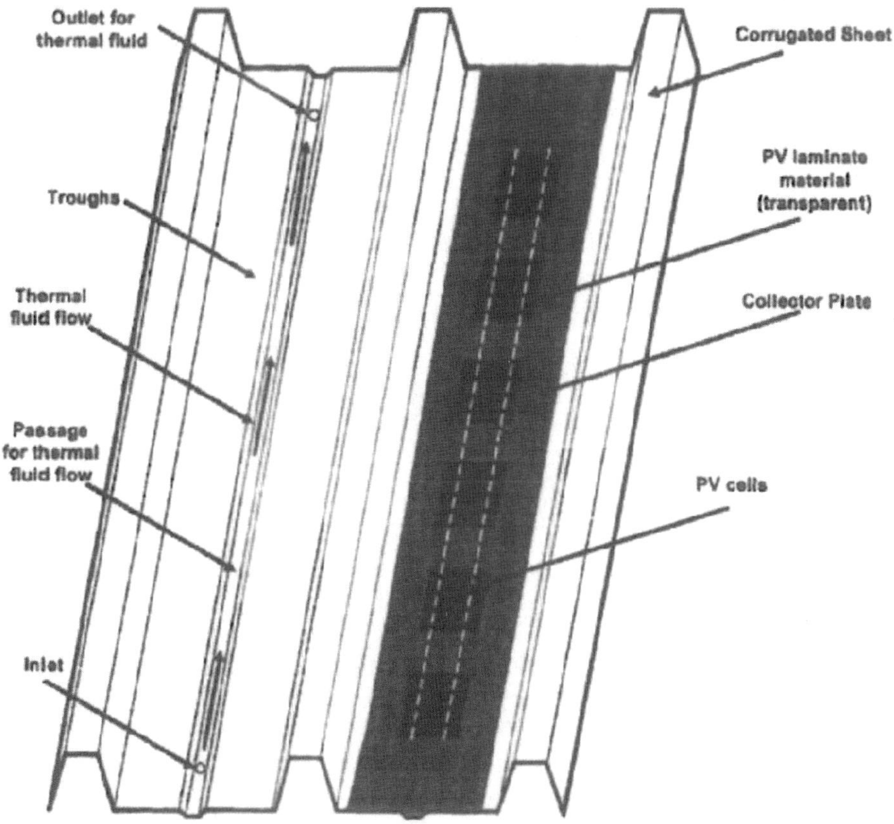

Fig. 6.2 A PV/T roofing system (Anderson and Duke 2007)

gradient between the water entering the thermosyphon tubes and the PV cells in the lower portion of the collector, locating the PV cells in the lower part of the collector improves electrical and thermal efficiency (Chow et al. 2006). Ji et al. (2006) and Chow et al. (2007) both examined a PVT system for integration as building walls in Hong Kong and Anderson and Duke (2007) have proposed a novel collector that integrates photovoltaic cells with aluminium or coated steel sheet metal roofing, as shown in Fig. 6.2.

PVT is economically viable in industrial applications in a Mediterranean environment (Kalogirou and Tripanagnostopoulos 2007). PVT systems based on amorphous silicon technology, although having lower electrical efficiencies, would have shorter payback times.

PVT system using concentration by linear Fresnel reflectors with a concentration ratio of 11 gave a maximum thermal efficiency of approximately 60 % with no electrical load (Rosell et al. 2005). There have been several proposals a CPC reflector designs with a PVT module mounted at its focus (Brogren et al. 2001). One design with concentration ratio of 37 had a maximum reported combined efficiency of 69 %

(Coventry 2005). Though the system had a low thermal efficiency, heat losses where also low due to its smaller heated area. Imperfections in the concentrator shape resulted in non-uniform illumination affecting adversely the electrical performance. A PVT system using a small-scale parabolic dish concentrator provides sufficiently high temperatures to drive an absorption cooling system (Kribus et al. 2006).

Garg and Adhikari (1999) demonstrated the use of several truncated CPCs with a concentration ratio of 3, in a single PVT module for air heating. A similar system (Othman et al. 2005) utilised a double pass with a rear finned surface to improve heat transfer on the rear face of the PV module.

6.3 PV/T Air Heating Collectors

In an air heating PVT collector (Bhargava et al. 1991) air circulates in a plenum behind a PV module (Brinkworth et al. 1997). Heat transferred from the panel heats the air for air heating PV/T collectors with a glass aperature cover mounted above the PV module to form a second air gap. The best compromise between electrical and thermal performance is provided when air passed both between the back surface of the module and the insulation layer and between the top surface of the module and the glass cover (Hegazy 2000). For an unglazed PVT air collector with a maximum thermal efficiency of 38 %. Adding glazing or a static reflector (with concentration ratio of approximately 1.3) increased this efficiency to approximately 60 % (Tripanagnostopoulos et al. 2002). By adding both glazing and the reflector thermal efficiency increased to 75 % however, increased optical losses resulted in decreased electrical efficiency. The addition of fins to the rear of the PV module improved the electrical and thermal efficiency of PVT systems as did the inclusion of a thin metal sheet in the air passage behind the PV module (Tonui and Tripanagnotopoulos 2007a, b).

6.4 Thermal Management of Photovoltaics

Elevated operating temperatures reduce the solar to electrical conversion efficiency of silicon photovoltaic devices (Weakliem and Redfield 1979). Active heat dissipation using air or water cooling introduces costs associated with both pumping and increased system maintenance although these costs can be offset at least in part if the generated thermal energy can be utilised in the building. The most common passive approach to heat dissipation involves a duct arranged behind the PV panel or its mounting system allowing natural convection and wind induced air-flow across the back of the PV panel. This is inherently limited due to the uncertainty of wind direction and pressure.

At an insolation of 750 Wm^{-2} and ambient temperature of 23 °C, the PV cell temperature can rise to over 45 °C (Huang et al. 2006). When the average ambient temperature is over 23 °C with incident solar radiation above 400 Wm^{-2}, conventional air-cooling arrangements are insufficient to maintain good PV performance. Concentrating PV systems have additional cooling requirements beyond these limits.

Phase change materials (PCM) can absorb/discharge a large amount of energy over a limited temperature range during phase change. PCMs are of interest for use in applications as diverse as thermal energy storage and the thermal management of systems, as well as for active and passive cooling of electronic devices. The capacity of a PCM for energy storage and temperature control depends on its properties, heat transfer methods and system configuration. The properties of various PCMs, methods of heat transfer enhancement, design configurations of heat storage facilities as part of solar passive and active space heating systems, thermal energy storage systems and building applications have been the subject of recent reviews (Kenisarin and Mahkamov 2007; Agyenim et al. 2010). Paraffin and paraffin wax are both non-toxic and inert with most materials and do not pose a danger to either health or the environment. The more important technical barriers when using paraffin and paraffin wax for energy storage and temperature regulation are their low thermal conductivities (~0.2 $Wm^{-1}K^{-1}$) and the presence of impurities that cause heterogeneous nucleation over a temperature range. The low thermal conductivity frequently makes an anticipated level of thermal storage untenable within an acceptable time period.

An investigation of a PV/PCM system with a selection of metal fin configuration in the PCM to limit passively the temperature rise of BIPV has been conducted. The system is compare with conventional air-cooling arrangements both experimentally and using numerical simulation (Huang et al. 2004, 2006). It has been found that using a PCM "RT27" with metal fins can significantly moderate the temperature rise of the PV in a PV/PCM system. In order to optimise the system, a detailed study to provide insight into the effect of fins on natural convection inside the system is required. Venkatesana et al. (2005) simulated the paraffin crystal structure formed when was deposits on an inner pipe wall. It was found that the crystallisation of paraffin is determined by both the shear rates and cooling rates. With higher wax contents, the introduction of oscillatory motion not only promotes was deposition, but also accelerates the crystal growth to achieve full was deposition (Lukman et al. 2008). The effect of convection and crystalline segregation on the heat transfer efficiency within a PV/PCM system for PV temperature regulation has been evaluated experimentally and the thermal performance of bulk PCM with crystallisation segregation for different internal fin arrangements is also presented (Huang et al. 2011). The selection of the thermal management approach employed depends on costs and the prevailing insolation as illustrated in Fig. 6.3.

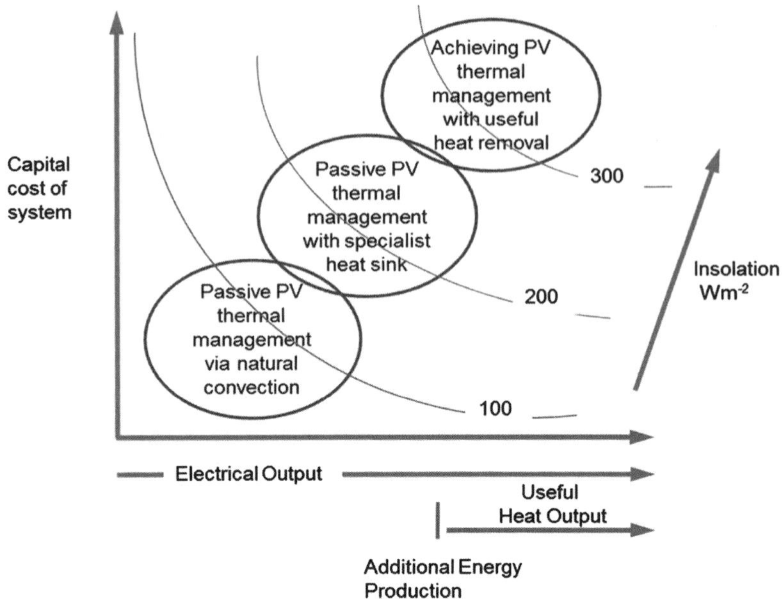

Fig. 6.3 Indicative criteria for use of different techniques for the thermal management of photovoltaics

References

Agyenim F, Hewett N, Eames P, Smyth M (2010) A review of materials, heat transfer and phase change problem formulation for intent heat thermal energy storage systems (LHTESS). Renew Sustain Energy Rev 14:615–628

Anderson TN, Duke M (2007) Analysis of a photovoltaic/thermal solar collector for building integration. In: Proceedings of the SB07, New Zealand sustainable building conference, Auckland

Anderson WG, Dussinger PM, Saraff DB (2008) Heat pipe cooling of concentrating photovoltaic cells. In: Proceedings of the 53rd IEEE photovoltaic specialists conference, pp. 1–6

Andrews JW (1981) Evaluation of flat-plate photovoltaic/thermal hybrid systems for solar energy utilization, U.S. Department of Energy, Report no BNL-51435

Bergene T, Lovvik OM (1995) Model calculations on a flat-plate solar heat collector with integrated solar cells. Solar Energy 55:453–462

Bhargava AK, Garg HP, Agarwal RK (1991) Study of hybrid solar system-solar air heater combined with solar cell. Energ Convers Manage 31(5):471–479

Brinkworth BJ, Cross BM, Marshall RH, Hongxing Y (1997) Thermal regulation of photovoltaic cladding. Solar Energy 61:169–178

Brogren M, Nostell P, Karlsson B (2001) Optical efficiency of a PV-thermal hybrid CPC module for high latitudes. Solar Energy 69:173–185

Chow TT, He W, Ji J (2006) Hybrid photovoltaic-thermosyphon water heating system for residential applications. Solar Energy 80:298–306

Chow TT, He W, Ji J (2007) An experimental study of façade-integrated photovoltaic/ water-heating system. Appl Therm Eng 27:37–45

Chow T-T, He W, Ji J (2008) Hybrid photovoltaic-thermosyphon water heating system for residential application. Solar Energy 80:298–306

Coventry JS (2005) Performance of a concentrating photovoltaic/thermal solar collector. Solar Energy 78:211–222

Cox CH, Raghuraman P (1985) Design consideration for flat-plate photovoltaic thermal collectors. J Solar Energy 35:227–241

Du B, Hu E, Kolle M (2012) Performance analysis of water cooled convertrated photovoltaic (CCPV) system. Renew Sustain Energy Rev 16:6732–6736

Florschuetz LW (1979) Extensive of the Hottel-Whillier model to the analysis of combined photovoltaic/thermal flat plat collectors. Solar Energy 22:361–366

Garg HP, Adhikari RS (1999) Performance analysis of a hybrid photovoltaic/thermal (PV/T) collector and integrated CPC troughs. Int J Energy Res 23:1295–1304

Garg HP, Agarwal RK (1995) Some aspects of a PV/T collector/forced circulation flat plate solar water heater with solar cells. Energy Convers Manage 36:87–99

Garg HP, Agarwal PK, Bhargava AK (1991) The effect of plane booster reflectors on the performance of a solar air heater with solar cells suitable for a solar dryer. Energy Convers Manage 32:543–554

He W, Chow TT, Lu J, Pei G, Chan L (2006) Hybrid photovoltaic and thermal solar-collector designed for natural circulation of water. Applied Energy 83:199–210

Hegazy AA (2000) Comparative study of the performances of four photovoltaic/thermal solar air collectors. Energy Convers Manage 41:861–881

Huang MJ, Eames PC, Norton B (2004) Thermal regulation of building-integrated photovoltaics using phase change materials. Int J Heat Mars Transfer 47:2715–2733

Huang MJ, Eames PC, Norton B (2006) Experimental performance of phase change materials for limiting temperature rise in building integrated photovoltaics. Solar Energy 80:1121–1130

Huang MJ, Eames PC, Norton B, Hewitt NJ (2011) National convection in an internally formed phase change material heat sink for the thermal management of photovoltaics. Solar Energy Mater Solar Cells 95:1398–1603

Ji J, Hun J, Chow T-T, Yi H, Lu J, He W, Sun W (2006) Effect of fluid flow and packing factor on energy performance of wall-mounted hybrid photovoltaic/water heating collector system. Energy Build 38:1380–1387

Kalogirou SA, Tripanagnostopoulos Y (2007) Industrial application of PV/T solar energy systems. Appl Therm Eng 1259–1270

Kenisarin M, Mahkamov K (2007) Solar energy storage using phase change materials. Renew Sustain Energy Rev 11:1913–1965

Kribus A, Kaftori D, Mittelman G, Hirshfeld A, Flitsanov Y, Dayan A (2006) A miniature concentrating photovoltaic and thermal system. Energ Convers Manage 47:3582–3590

Lukman I, Westncott RE, Xiongwei N (2008) On the effect of wax content on paraffin wax deposition in a batch oscillating baffled tube apparatus. Chem Eng J 137:205–213

Othman MYH, Yatin B, Sopian K, Bakar MNA (2005) Performance analysis of a double-pass photovoltaic/thermal (PV/T) solar collector with CPC and fins. Renew Energy 30:2005–2017

Rosell JI, Vallvendu X, Lechon MA, Ibanez M (2005) Design and simulation of a low concentrating photovoltaic/thermal system. Energy Convers Manage 46:3034–3046

Royne A, Dey CJ, Mills DR (2003) Cooling of photovoltaic cells under concentrated; illumination; a critical review. Solar Energy Mater Solar Cells 86:451–481

Saki G, Arboiro JC, Lugwe A (1996) The EUCLIDES prototype; an efficient parabolic trough for PV concentration. In: Proceedings of the 25th IEEEE photovoltaics specialists conference, pp 1207–1210

Soulites M, Quinlan P, Tripanagnostopoulos Y, Smyth M, Zacharopoulos A, Ramirez U, Yianoulis P (2011) Heat retaining integrated collector storage solar water heater with asymmetric CPC reflector. Solar Energy 85:2474–2487

Teo HG, Lee PS, Hawlader MNA (2011) An active cooling system for photovoltaic module. Appl Energy 90:309–315

Tripanagnostopoulos Y, Nousia T, Souliotis M, Yianoulis P (2002) Hybrid photovoltaic/thermal solar systems. Solar Energy 72:217–234

Van Helden WGJ, van Zolingen RJC, Zondag HA (2004) PV thermal systems: PV panels supplying renewable electricity and heat, progress in photovoltaics. Res Appl 12:415–426

Venkatesana R, Magarajanb NR, Pasoa K, Yic YB, Sastrye AM, Foglera HS (2005) The strength of paraffin gels formed under static and flow conditions. Chem Eng Sci 60:3587–3598

Weakliem HA, Redfield D (1979) Temperature dependence of the optical properties of silicon. J Appl Phys 50:1491–1493

Chapter 7
Solar Thermal Power Generation and Industrial Process Heat

In time, manufacturing will to a great extent follow the sun

C.G. Abbot (1928)

7.1 Overview

High temperature heat production requires a concentrator, an absorber and a heat transfer fluid, with the addition of an engine if electricity is to be generated. Reflector materials need to retain high specular reflectance under often harsh conditions. Specialised high-temperature selective multiple cermet layer coatings prepared by physical vapour deposition are necessary for operating temperatures above 400 °C (Kennedy and Price 2006). Selective coatings for medium-temperature applications such as metal-dielectric cermet composites of metal particles in a ceramic matrix are not stable at high operating temperatures.

Energy from concentrating insolation when converted to a high temperature fluid can either drive an engine for electricity generation or be used directly for industrial or thermochemical processes. For both applications, heat storage may be both feasible and economically viable. There are a range of combinations shown in Fig. 7.1.

Solar heat can also be converted to electricity using Seebeck thermoelectric effect devices operating under high solar energy concentration. Solar thermoelectric electricity generation has been achieved with an efficiency of 4.6 % for an air-mass 1.5 solar spectrum at 1,000 Wm^{-2} (Kraemer et al. 2011).

Solar thermal power generation technology generally refers to a power generation system that involves collecting solar radiation through concentrated collectors to an absorber surface which will heat a carrier fluid to a high temperature. Solar tracking mode are shown in Fig. 7.2, heliostat mirrors in central receiver systems and parabolic dish systems fully track the Sun. Parabolic trough and linear Fresnel lens systems generally employ E-W polar tracking mode.

B. Norton, *Harnessing Solar Heat*, Lecture Notes in Energy 18,

DOI 10.1007/978-94-007-7275-5_7,

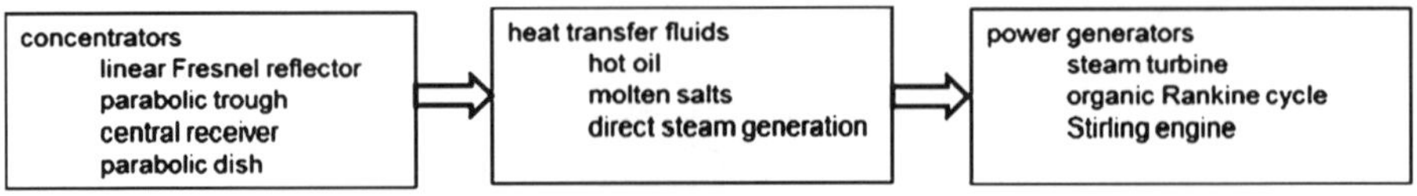

Fig. 7.1 Generic component options in solar thermal power generation systems

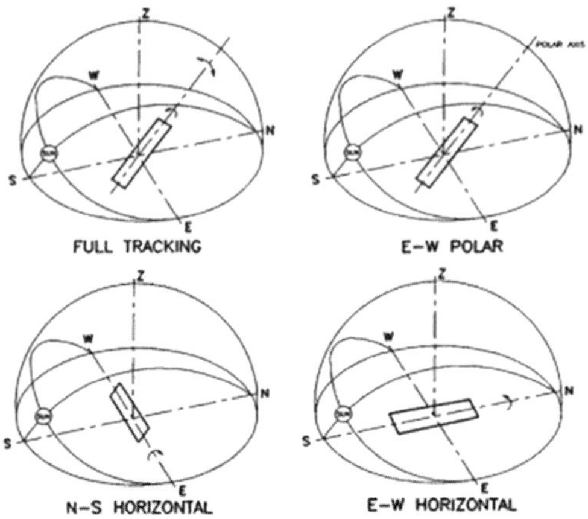

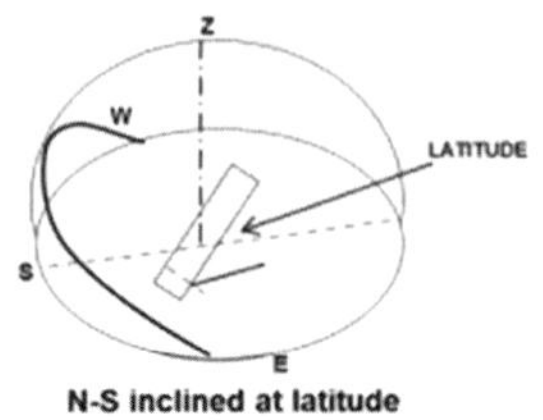

		Direct component insolation incident on aperture relative to that on a fully tracking aperture		
		Equinoxes	Summer Solstice	Winter Solstice
Tracking mode	Full tracking	1.00	1.00	1.00
	E-W Polar	1.00	0.92	0.92
	N-S Horizontal	0.74	0.74	0.86
	E-W Horizontal	0.89	0.98	0.61
	N-S Inclined at latitude	0.87	0.68	0.70

Fig. 7.2 Solar tracking modes

Thermochemical hydrogen production process is based on the use of concentrated solar radiation as the energy source of high-temperature process heat for driving an endothermic chemical transformation (Seinfeld 2005; Rodat et al. 2009). Large-scale concentration of solar energy is mainly based on three optical configurations using parabolic reflectors, namely: trough, tower, and dish systems. At higher concentration ratios lower heat losses from smaller area absorbers ensue giving higher temperatures at the receiver.

Through a piping and boiling system the hot fluid will be able to generate steam to power a turbine. Using concrete to store sensible heat for parabolic trough power plants with synthetic oil as the heat transfer fluid has been investigated as has a graphite-based, high-temperature sensible heat storage system. Solar thermochemical processes convert solar energy into chemical energy, with the absorbed, concentrated, solar radiation driving an endothermic chemical reaction. Concentrated solar radiation is used as the energy source for high temperature process heat to drive chemical reactions for the production of storable fuels. Solar heat can also be used to drive a reversible reaction in a solar chemical reactor; after transporting the latter to where the energy is to be used, an exothermic reverse reaction yields the stored energy.

Using concrete to store sensible heat for parabolic trough power plants with synthetic oil as the heat transfer fluid has been investigated as has a graphite-based, high-temperature sensible heat storage system. Solar thermochemical processes convert solar energy into chemical energy, with the absorbed, concentrated, solar radiation driving an endothermic chemical reaction. Concentrated solar radiation is used as the energy source for high temperature process heat to drive chemical reactions for the production of storable fuels. Solar heat can also be used to drive a reversible reaction in a solar chemical reactor; after transporting the latter to where the energy is to be used, an exothermic reverse reaction yields the stored energy.

The first patent for a solar powered engine was granted to Mouchout (1869, 1879) for glass-enclosed iron sphere through which concentrated insolation was transmitted from a two-axis sun tracking mirror to heat water to operate a steam engine. Adams (1878) developed an improved mirror array. From at least 1868 until 1889, Ericsson (1868, 1884; Church 1907) developed a solar-powered steam engine that was the first to include a parabolic trough collector. Tellier (1889) designed of the first solar engine powered by a flat plate collector using pressurised Ammonia working fluid to drive a water pump. In marked contrast, a 10 m diameter conical dish reflector comprised 1,788 individual mirrors concentrated solar energy onto an absorber to generate steam that was transferred to an engine that drive an irrigation water pump (Eneas 1901); the structure supporting the large weight of the reflector proved unable to cope successfully with strong winds and hailstorms. Shuman and Boys (1917) used sun-tracking line-axis parabolic troughs to focus insolation on a tube surrounded by a glass envelope to drive a steam engine (Shuman 1911). In 1912 it was used to power an irrigation plant near Cairo, Egypt. In a parabolic trough solar power plant use a large field of parabolic trough track the sun during the day to concentrate insolation onto a receiver tube located at the focus of the parabolic mirrors. A heat transfer fluid passes through the receiver and is heated to temperatures required to generate steam that in most systems drives a conventional Rankine cycle steam power plant.

7.2 Parabolic Trough Systems

7.2.1 *Overview*

Parabolic trough reflectors concentrate sunlight onto receiver tubes in which a transfer fluid is circulated as shown in Fig. 7.3. Heated to approximately 400 °C, this fluid, usually a specialist oil, is then pumped through heat exchangers to produce superheated steam. The steam is converted to electrical energy in steam turbine generator, the latter can be integrated into a combined steam and gas turbine cycle.

Fixed-Mirror Tracking absorber systems employ segmented or parabolic reflector as shown in Fig. 7.4. Concentration ratios of up to 50 can be obtained with an average optical efficiency of about 60 % (Pujol et al. 2011).

Fig. 7.3 Parabolic trough concentrators

Fig. 7.4 Tracking-absorber fixed-reflector system (Pujol et al. 2011)

7.3 Fresnel Mirror System

A linear concentrating Fresnel solar concentrator is an planar array of linear mirror strips that reflect sunlight onto a stationary thermal receiver as shown in Fig. 7.5. The operation temperature range is 100–400 °C.

In a linear Fresnel reflector, a single-axis tracking of an array of mirror strips focuses insolation onto a linear receiver. Linear Fresnel reflector systems have relatively low initial cost as inexpensive planar mirrors are driven by a simple tracking system and the area required is used efficiently by the mirror strips for one absorber being interspaced with those for the adjacent absorber. The fixed absorber tube obviates the need for flexible high pressure joints or thermal expansion bellows. As the reflector strips are ground-mounted, wind loads on the reflector strips are low; the reflector width per absorber tube can therefore be three times that of a parabolic trough. When used for direct steam generation, no heat exchanger is

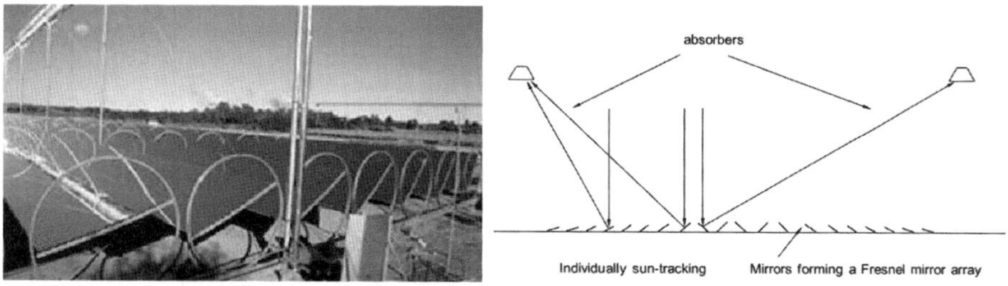

Fig. 7.5 A compact linear Fresnel reflector system

required. A disadvantage of linear Fresnel systems is a reduction in the efficiency compared to parabolic troughs, which has to be compensated for by a lower investment cost in the solar field. These cost reductions can come from economies of scale and design optimisation of the collector and there are also potential savings offered by lower operation and maintenance costs.

7.4 Heliostat Field Central Receiver Systems

Solar power tower systems consist of a field of heliostat sun-tracking mirrors that reflect direct insolation to a receiver located at the top of a tall tower as shown in Fig. 7.6. A molten salt heat-transfer fluid, heated in the receiver is used to generate the steam required for a steam turbine-generator to produce electricity. Both parabolic trough and solar power tower systems have inherent economy of scale as they are coupled with turbine generators whose minimum economic capacity is usually over 50 MW.

Each heliostat consists of a tracking unit with drive motors, controls as well as the reflector. Being dual-axis tracking, heliostats are adjusted constantly. A circular or semi-circular array as shown in Fig. 7.7 of large individually-tracking heliostat mirrors concentrate sunlight onto the central receiver.

Water, steam, molten salts, liquid sodium and air are employed as intermediate heat transfer fluids. With concentration ratios exceeding 500, power towers deliver solar heat at temperatures of over 500 °C and over 1,000 °C for steam cycles and gas turbines respectively. Modern central receiver technology was demonstrated during the 1980s with the "Solar 1" and "Solar 2" facilities in California. "PS-10", an 11 MW central receiver system near Seville, Spain, commenced operation in 2007. The heliostat field of "Solar Tres", a 17 MW steam-generating central receiver system is three times the size of Solar Two comprising 2,493 96 m^2 glass-metal heliostats with 16 h of heat storage provided by a 600 MWh 6,250 tonne molten nitrate salt thermal energy store.

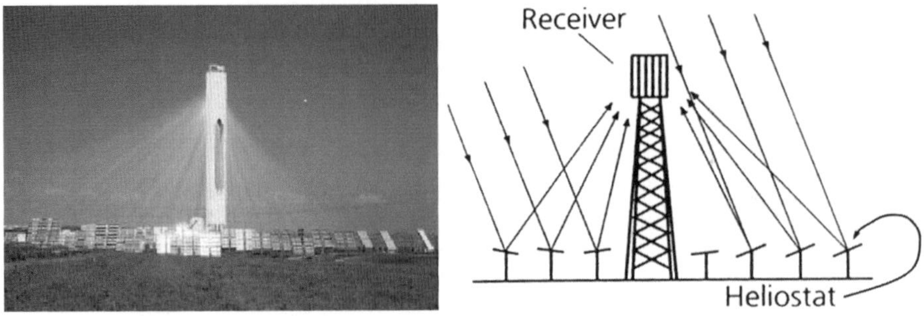

Fig. 7.6 11MW PS10 heliostat and central receiver system near Seville, Spain

Fig. 7.7 Large scale heliostat arrays

7.5 Parabolic Dish Systems

Parabolic dishes concentrate insolation onto an absorber as shown in Fig. 7.8. The latter can be a Stirling engine where air is heated and transferred from the hot cylinder to the cold cylinder of the engine; the working fluid movement is used to generate electricity.

As parabolic dishes are always aligned to the solar position, they operate at their highest optical efficiency. This optical efficiency is considerably higher than that of trough, linear Fresnel receiver or central receivers where cosine losses ensue.

Solar Parabolic dishes are usually composed of mirror facets that focus solar energy onto a receiver as shown in Fig. 7.9. A heated working fluid such as hydrogen drives a turbine or, more usually, a Stirling engine.

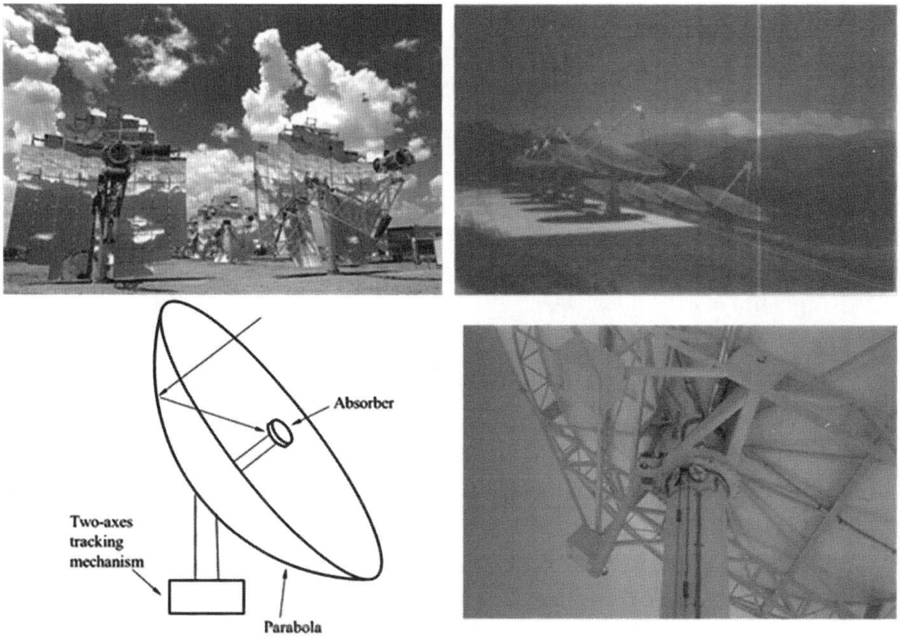

Fig. 7.8 Parabolic dish reflectors, schematic diagram and photographs

Fig. 7.9 Parabolic dish reflectors fabricated from mirror segments

7.6 Characteristics of Concentrating Solar Thermal Electricity Generating Systems

Solar thermal power generation can be operated:

- To generate solely when direct insolation is sufficient
- As a "solar-assist" to provide steam to an existing generating plant as it integrate well with conventional thermodynamic cycles and power generation equipment

Table 7.1 Thermal photovoltaic and thermoelectric conversion of solar heat to electricity

	Parabolic trough	Central receiver	Parabolic dish
Typical power (MW)	30–320	10–200	5–25
Operating temperature (°C)	390	560	750
Annual capacity factor (%)	23–50	20–77	25
Peak efficiency (%)	20	23	29
Net annual efficiency (%)	11–16	7–20	12–25

- As dispatchable power when integrated with thermal storage and/or gas co-firing giving good matching between insolation and where electrical demand is driven by summer air conditioning.

Economic viability is likely to continue to improve with;

- Increasing plant sizes to reduce operating costs through economies of scale as both collectors and systems are fabricated using established manufacturing process with mostly readily available materials
- Development of economically viable thermal storage
- Siting solar thermal electricity generation near gas sources or pipelines to utilise gas as a co-firing fuel (Table 7.1).

Heat engines are employed in all current practical systems for the conversion of solar energy to electricity. Two other approaches are

- Thermophotovoltaics; solar photons heat an intermediate solar selecture absorber that re-emits photons whose energy matches the band-gap of a thermophotovoltaic (TPV) cell. TPV cells have a theoretical maximum efficiency of 85.4 %, similar to infinite multijunction cells without the latter's complexity in fabrication (Morti and Luque 2003). The key barrier to implementation has been the absence of highly efficient selective surfaces structured at 10 nm scales (Bitnar et al. 2002; Fleming et al. 2002). With such materials device efficiency should approach 30 %.
- Thermoelectric, a voltage is generated by a temperature difference across a material. The underlying "Seebeck effect" is however small for even application-specific materials (Goldsmid 1960), though research on novel nanomaterials (Chen 2001; Chiritescu et al. 2007).

7.7 Non-convecting Solar Panels

In a non-convecting solar pond, part of the incident insolation is absorbed and converted to heat, which is stored in lower regions of the pond. Solar ponds are both solar energy collectors and heat stores.

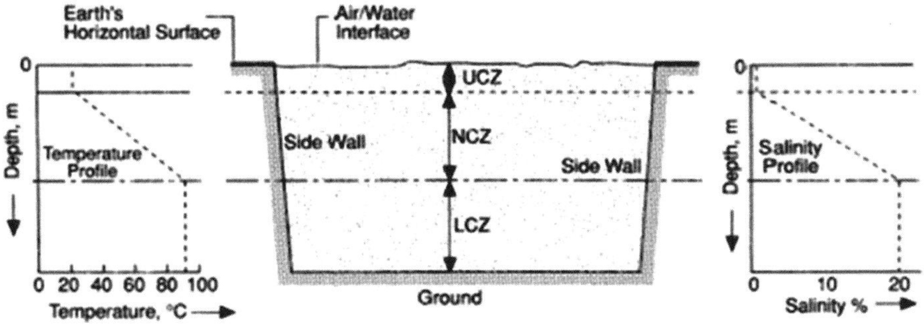

Fig. 7.10 Zones in a non-convecting solar pond

An operating salt-gradient non-convecting solar pond consists of three zones as shown in Fig. 7.10:-

- The upper-convecting zone (UCZ), of almost constant low salinity at close to ambient temperature. The UCZ, typically 0.3 m thick, is the result of evaporation; wind-induced mixing and surface flushing. It is kept as thin as possible by the use of wave suppressing surface meshes and placing wind-breaks near the pond.
- The non-convecting zone (NCZ), in which both salinity and temperature increase with depth. The vertical salt gradient in the NCZ inhibits convection and thus provides the thermal insulation. The temperature gradient is formed due to the absorption of insolation at the pond base.
- The lower-convecting zone (LCZ), of almost constant, relatively high salinity (typically 20 % by weight) salinity at a high temperature. Heat is stored in the LCZ which is sized to supply energy continuously throughout the year. As the depth increases, the thermal capacity increases and annual variations of temperature decreases. However, large depths increase the initial capital outlay and require longer start-up times.

There are Salt gradient lakes that naturally exhibit an increase in temperature with depth. For example Medve Lake in Transylvania (Nielsen 1975) contained a nearly- saturated NaCl solution at a few metres depth, with almost fresh water at its surface. The maximum temperature exceeded 60 °C at a depth of 1.32 m at the end of Summer; the minimum temperature being around 26° during the early Spring. A similar lake near Orville in north-central Washington, USA (Anderson 1958) whose salt is principally magnesium sulphate, exhibited a temperature of around 50 °C in July 1955 at a depth of 2 m, whilst the surface temperature was less than 26 °C. Lake Vanda in the Antarctic is another natural example of the trapping and storing of solar energy as a result of a salt-water density gradient (Wilson and Wellman 1962). The bottom of this lake (−67 m depth) is maintained at −25° despite a mean annual air temperature of about −20 °C. Other natural solar lakes, include the Los Reques lake, Venezuala (Huder and Sonnefeld 1974); Lake Magege in western Uganda (Melack and Kilham 1972); Castle Lake in California, USA

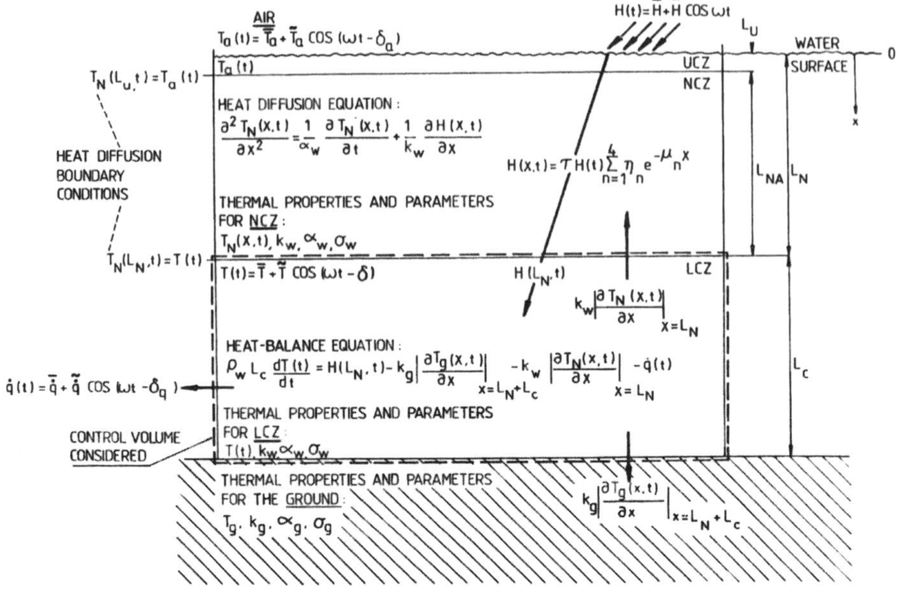

Fig. 7.11 Heat transfer processes in a non-convecting solar pond

(Bachmann and Goldman 1965) and a lake on the shores of the Red Sea, on the east coast of Sinai Peninsula where the temperature rose from 16 °C at its surface to 40° at 1.5 m depth (Por 1968).

The first theoretical analysis of thermophysical behaviours in solar ponds (Weinberger 1964) was a one-dimensional non-convecting model, was employed to predict annual variations of pond temperatures. The key aspects of such analyses are shown in Fig. 7.11. The large thermal capacity of a solar pond dampens its response to diurnal variations in insolation, typically the average temperature of an LCZ varies by less than 1 °C over 24 h (Kooi 1979). However, large annual changes in the daily-average LCZ temperature may occur. Thus an annual sinusoidal variation is an appropriate representation of insolation, (Rabl and Neilsen 1975). Following the approach of Abdel-Salam et al. (1986) varied the heat extraction rate occurs over an annual cycle as a sinusoidal temporal function of phase-lag relative to the insolation. The mean temperature of the UCZ is usually assumed to be that of the ambient air. The LCZ is usually assumed to be homogenous with a uniform salt concentration and fully convective so that its temperature will thus be a function solely of time. The energy balances for a non-convecting solar pond is presumed to have passed through an initial start-up phase and to be operating in a steady-state manner (Abdel-Salam et al. 1986).

From the late 1950s to the late 1980s (with a break in the late 1960/early 1970s) experimental solar ponds have been operating in Israel (Tabor 1981; Tabor and Matz 1964). A salt-gradient solar pond of 2.5 m depth and 200 m^2 effective collector area was constructed in August 1975, at the Ohio State University,

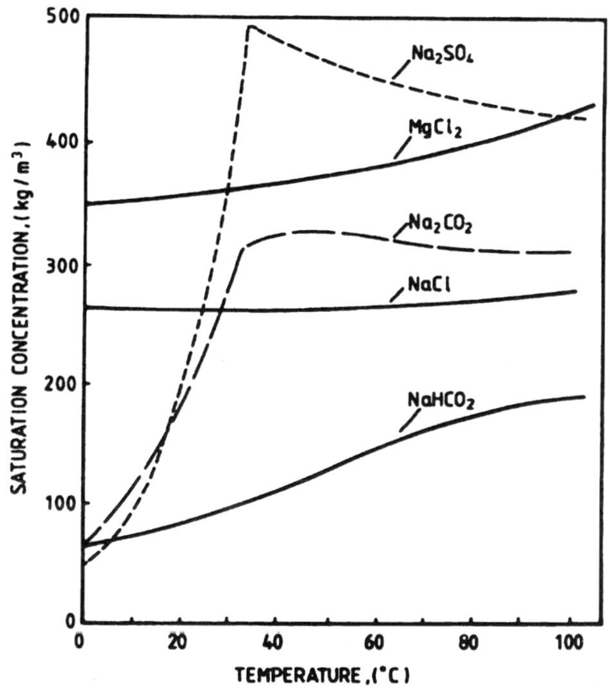

Fig. 7.12 Solarbility of salts in water

USA, for space heating (Nielsen 1976). A solar pond constructed in 1978 has heated an outdoor swimming pool in summer and a recreation building during part of the winter, In Miamisburg, Ohio, USA (Bryant et al. 1979).

The application of solar ponds for electric-power production usually employs an organic vapour Rankine Cycle engine to convert solar-pond heat to mechanical work, and then into electricity. However, to obtain a low cost per generated Watt, solar ponds of several kilometers are required.

Many techniques have been considered in order to suppress natural convection in order to create a solar pond. The most common method used is salt-stratification. Salinity increases with depth in the NCZ until the LCZ is reached: the highest salt concentration occurs uniformly throughout this region. Here the solar radiation will heat the highly saline water, but because of its high relative density (due to its salt content), this hot salt-water will not rise into the lower salinity layers. Thus the heat is stored, yet inhibited from being transferred by convection. Chemically-stable salts, as well as any natural brine can be used to establish a salt-stratified solar pond. A selected salt must be safe to handle; non-toxic; cheap and readily available; not reduce significantly the insolation transmission characteristics of water; and solubility should be temperature dependent. The variation of solubility with temperature for candidate salts is shown in Fig. 7.12.

Sodium and magnesium chlorides though satisfying most criteria (especially sodium chloride in terms of cost) have, solubilities that are modestly temperature dependent. However salts which possess more appropriate variations of solubility

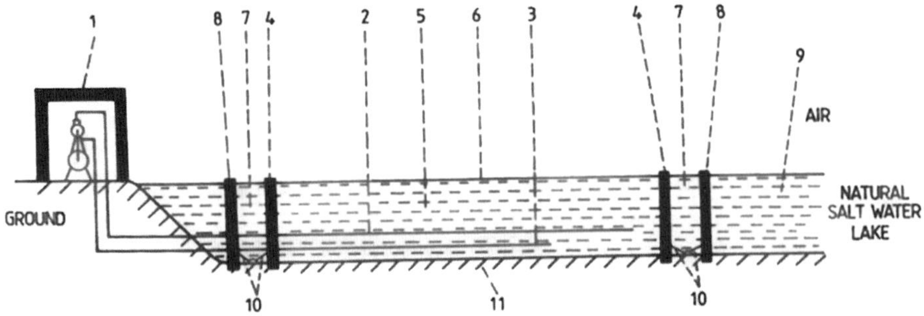

Fig. 7.13 Coffered solar pond

with temperature, do not meet the remaining constraints. Thus sodium chloride remains the most popularly used salt.

Some of the early designs of saline non-convecting solar ponds employed membranes (Rabl and Neilsen 1975) to provide boundaries between the pond zones and prevent upward salt diffusion. However, membranes present major fabrication and maintenance problems, especially for large ponds.

The "Floating" Solar Pond was first suggested in 1976 (Assaf 1976). The Floating Solar Pond is a conventional solar pond "floating" on a large saline lake and enclosed by thermally-insulating walls (Tabor 1981). In the enclosure, the uppermost few metres of the water have an imposed salt-concentration gradient, with an inverse gradient in the lower depths. Two designs have been proposed for floating solar ponds (Crevier and Moshref 1981) which depend on the conventional salt gradients above convecting layers. Intermixing was to be prevented by a horizontal flexible membrane, which were to be held in place by a combination of buoys and weights. Although the Floating Solar Pond avoids the land costs and the need for excavation, the uncertain behaviour of the boundary layer between the LCZ and the lake water beneath it and the need for the use of membranes are major disadvantages. An alternative approach to utilizing existing lakes is the coffered solar pond (Abdel Salam et al. 1986) shown in Fig. 7.13.

In a "saturated solar pond", water at all levels is saturated with salt, and upward salt diffusion is avoided. The pond The chosen salt should possess a solubility, which increases significantly with rising temperatures; the salt gradient would be self-maintained depending upon the local temperature gradient. A saturated solar pond requires far more salt than a conventional non-convecting salt-gradient solar pond. The absence of a salt which has the required properties has prevented the practical construction of a saturated solar pond (Satish and Gurmukh 1980), however aluminium sulphate dodecahydrate has a very temperature dependent solubility and has been proposed (Vitner et al. 1984).

For power production in the multi MW, a solar pond of several-square kilometres surface area is needed. However, this is not feasible economically as excavation and preparation work account for more than 40 % of the total capital cost of the power-generating station (Tabor 1981). So it would appear logical to employ a natural lake, and convert a shallow portion of it into a solar pond. The "coffered solar pond" has been proposed (Abdel-Salam et al. 1986). This is formed by physically isolating a section of a natural lake using walls of thermally-insulating materials.

In a Viscosity-Stabilised Solar Pond, organic thickeners are added to the pond layers thereby creating a static stabilised zone. By increasing the proportion of thickeners, the water becomes more viscous until natural convection is suppressed entirely. Promising prospective thickeners include cellulose methyl ethers, sodium caboxy methyl cellulose and a commercial carboxy vinyl (acrylic) polymer (Shaffer 1978). The problems with thickeners include maintaining the stability of the static zone under the effect of shear stresses and decomposition of the thickeners with time and at temperature above 55 °C.

In a Gel-Stabilised Solar Pond, the stratified non-convecting zone of the traditional salt-gradient solar pond is replaced by a transparent layer of highly viscous or near-solid polymer-gel (Wilkins et al. 1982). As such gels have low thermal conductivity, and are almost solid, they will convect insignificantly. Upward heat transfer will thus be predominantly conductive. Gel ponds overcome the two main problems of conventional solar ponds, maintaining stratification and the inhibition of wind mixing.

The site for a solar pond should be near a cheap source of salt, an adequate source of water, incur low land costs, and experience an all-year solar exposure. The underlying earth structure should be homogeneous and free of stresses and fissures. If not, then increases in temperature may cause differential thermal expansions which could result in earth movements (Tabor 1980). The pond must not pollute aquifiers nor lose heat via underground water streams passing through an aquifier. Any continuous drain of heat will lower the pond's storage capability and effectiveness. Stormy regions should be avoided in order to limit wind surface mixing effect.

Species of fresh-water and salt-water algae grow under the conditions of temperature and salt concentration that exist in a stratified solar pond. Algae growth will inhibit solar transmissivity and insolation (Wittenberg and Harris 1980). Different algae species are introduced by rain water and air-borne dust. To prevent algae formation, copper sulphate has been added at a concentration of about 1.5 mg/l. This has proved to be effective (Poppe and Woomer 1985). The thermal efficiency of a pond depends on the stability of its vertical salt-gradient. The pond will cease to function without the proper maintenance of the stratification. The stability of the salt-gradient is maintained by:

- Controlling the overall salinity difference between the two convecting layers
- Inhibiting internal convection currents if they tend to form in the NCZ; and
- Limiting the growth of the UCZ.

Salt slowly diffuses upwards at an annual average rate of about 20 kgm^{-2} as a result of its concentration gradient. This rate varies and is dependent upon the ambient environment conditions, type of salt and temperature gradient. A combination of surface washing by fresh water and injecting brines of adequate density at the bottom of the pond is usually sufficient to maintain an almost stationary gradient. Several techniques have been developed to achieve this (Nielsen and Rabl 1976; Tabor 1980; Akbarzadeh and MacDonald 1982).

During pond heating, and particularly at higher temperatures, small, unstable, convecting zones may develop within the NCZ. If these zones are left unattended, they will increase in thickness, and this leads to a decrease in the effective thickness of the NCZ. To restore the gradient, brine is injected horizontally from a diffuser placed at the upper boundary of the unstable region. This wave of increased density descends until the lower boundary of the unstable region is reached. The downward velocity of the diffuser and the rate and the density of the brine injected from it are adjusted in such a way that to restore the stability of the gradient (Zangrando 1980). Alternatively external mixed brine of the appropriate concentration may be injected slowly into the unstable region (Nielsen 1979).

Surface flushing is an essential process in maintaining the pond's salt-gradient. Its effect on the UCZ growth is reduced if the velocity of the surface washing water is small. Surface temperature fluctuations will result in heat being transferred upwards through the UCZ by convection, especially at night, and downward by conduction. The thickness of the UCZ varies with the intensity of the incident insolation. In an experimental solar pond at Melbourne University, Australia, the thickness of the UCZ varied between 10 and 15 cm from night time to mid-day (Akbarzadeh et al. 1983).

Evaporation will be caused by insolation and wind-action. The higher the temperature of the UCZ, and the lower the humidity above the pond's surface, the greater will be the evaporation rate. Excessive evaporation results in a downwards growth of the UCZ (Onwubiko 1984). Evaporation can be counter-balanced by surface water-washing, which could compensate for evaporated water as well as reduce the temperature of the pond's surface especially during periods of high insolation. Reducing the wind velocity over the water's surface by using wind breaks will reduce evaporation rates. Evaporation can be the dominant mechanism in surface-layer mixing under light-to-moderate winds. However, under strong winds it becomes of secondary importance. Wind-induced mixing can contribute significantly to the deepening of the UCZ. Winds also induce horizontal currents near the top surface of the pond increasing convection in the UCZ region (Elata and Levien 1966). Wind-mixing has been reduced by floating devices (e.g. plastic pipes, plastic grids and independent rings) and by the use of wind-breaks. UCZ deepening may be suppressed by decreasing the potential energy of the top surface layer of the pond. This may be accomplished by raising the overall salinity of the pond, and then washing this top layer with non-saline water to produce a thin surface sub-layer. More kinetic energy than possessed by surface winds is required to mix this sub-layer with the bulk of the UCZ. Wind energy would thus be dissipated mixing the top layer and little energy would remain to lower the level of the interface between the UCZ and the NCZ (Schladow 1984).

Water is a spectrally selective absorber, only shorter wavelengths reach the bottom of the pond. This has been represented by a sum of four exponential extinction functions, (Rabl and Neilsen 1975) to give an effective absorption coefficient, that is the absorption coefficient divided by the cosine of the angle of refraction. This allows for the increased path length due to refraction of the incident insolation at the water surface.

7.8 Solar Chimney Power Plants

A solar chimney power plant consists of a transparent tubular chimney over 200 m tall rising from a horizontal ground area of over 50,000 m^2 covered with a transparent material (Haaf et al. 1983). At the base of the chimney is located a turbine driven by the natural circulation air flow as shown in Fig. 7.14. In the climate of South Western Algeria such a system can produce between 70 and 43 MW of electricity per month in July/August and December/January respectively (Larbi et al. 2010). Early investigations of solar chimney power plants were undertaken from the 1900s to 1930s (Cabanyes 1903; Günther 1931). Detailed studies by Haaf et al. (1983); Haaf (1984) Schlaich (1995) and Schlaich et al. (2003a, b) have shown that performance is dependent strongly on the incident solar energy but is largely invariant with ambient temperature. Simulation models have been developed to produce optimal designs for particular climates (Larbi et al. 2010; Pasumarthi and Sherif 1998; Bernardes et al. 1999; Maia et al. 2009; Pretorius et al. 2004, 2006a, b). Particular attention has been given to increasing power production and reducing installation cost (Fluri et al. 2009; Chergui et al. 2008).

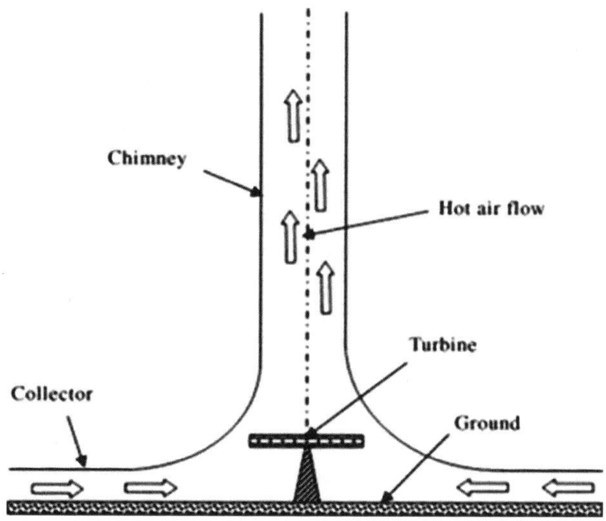

Fig. 7.14 Operation of a solar chimney powerplant

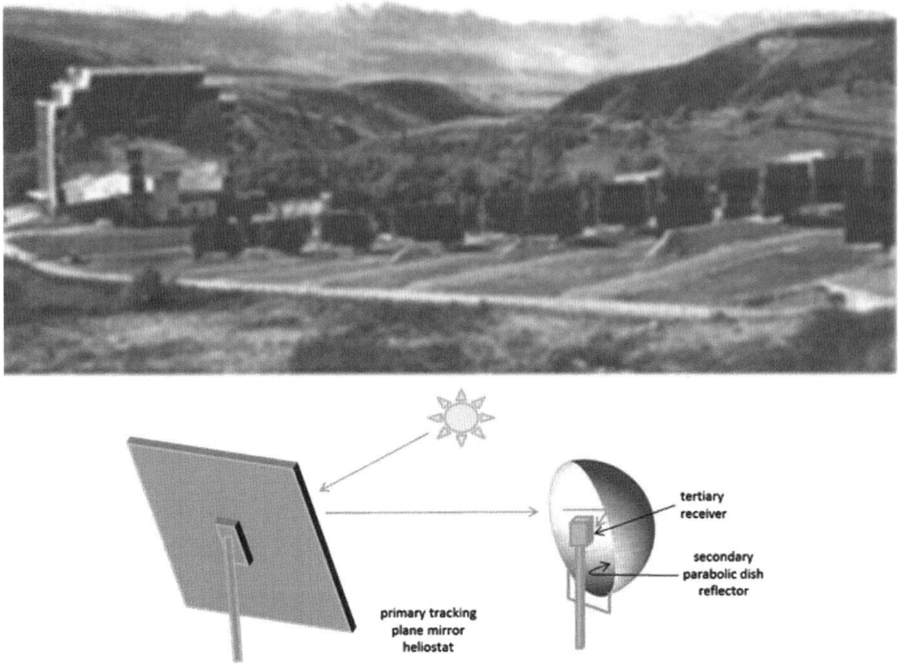

Fig. 7.15 Solar furnace; principle of operation and an installation in Odeillo, France

7.9 Solar Process Heating

7.9.1 Industry

Industrial process heat is an important application of solar energy in medium to high temperature sterilising, pasteurising, drying, hydrolysing, distillation and evaporation, washing and cleaning, and polymerisation (Dincer and Rosen 1998; Schweiger et al. 2000; Kalogirou 2004). However to install solar industrial process heat applications a large initial investment is often needed (Foster et al. 2010).

Industrial process heat at sufficiently low temperatures can be supplied by solar energy. About 40 % of all process heat is in the temperature range from ambient to 180 °C. Process heat requirements below 180 °C can be supplied by flat-plate collectors, solar ponds, evacuated tube collectors or solar concentrators. A comparison of alternative concentrators indicated that the two axes tracking paraboloidal dish is the most promising design for industrial process heat applications in India (Kedare 2005). A design of solar concentrator for medium temperature industrial process heating applications (Chandak and Dubey 2005) that comprised multiple paraboloidal reflector dishes of 12.5 m^2 total aperture area mounted on a structure that rotates parallel to the polar axis tracking the sun delivered heat at 250 °C. Five meter diameter dish concentrators have been used for heat deformation of sheet metal by solar energy (Lytvynenko and Schur 1999). A diagram of a solar furnace is shown in Fig. 7.15.

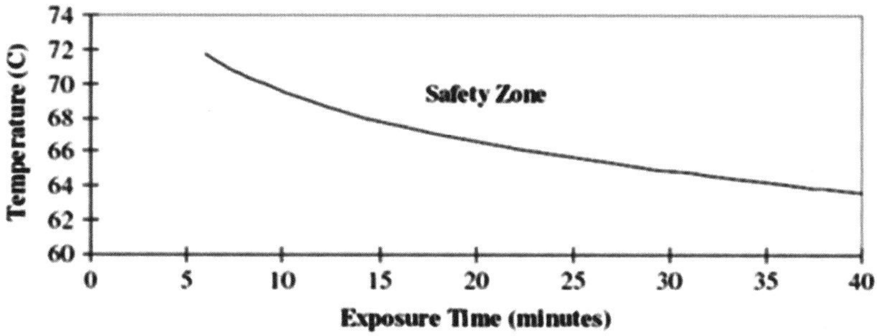

Fig. 7.16 Exposure duration for temperatures between 60 °C and 14 °C required to destroy enteroviruses

7.9.2 Water Treatment Using Solar Energy

Solar energy water treatment systems are a low cost and effective method of reducing microbiological contamination in drinking water (WHO 2007). With no requirement to heat water to boiling, pasteurisation is a promising method for purifying water using solar energy (Duff and Hodgeson 2007). Being unaffected by the turbidity of that is often a feature of many water supplies, gives pasteurisation an advantage over ultraviolet disinfection and filtration; both the latter are negatively influenced by turbidity (Burch and Thomas 1998). Pasteurisation raises, and then maintains, water to an elevated temperature for the duration required to achieve microbiological decontamination. As an example, the duration required to destroy Enteroviruses (Feachem et al. 1983) is shown in Fig. 7.16.

Submicrometer nanoparticles can absorb solar energy across its spectrum. When such nano particles, suspended in water, are exposed to concentrated solar energy a thin layer of steam is formed. This layer both reduces thermal conductance from the nanoparticle to the surrounding water and generates buoyant forces that carries columns of steam bubbles and nanoparticles to the water surface. At the surface the steam is realised and the nanoparticle descends back into the bulk liquid. Eventually the bulk temperature of the water rises until, if the solar energy input insufficient, conventional boiling ensues (Neumann et al. 2013). The phenomena has potential use for solar energy application and in sterilization.

7.9.3 Cooking

A Scheffler concentrator shown in Fig. 7.17 is a paraboloidal reflector with a medium-range concentration ratio used for cooking applications up to 300 °C primarily, to date, in India (Gadhia and Gadhia 2006). In a Scheffler fixed focus solar cooker comprises a primary reflector, a secondary reflector, and clock mechanism powered by clockwork or photovoltaics. The primary reflector produces a converging beam of sunlight aligned with an axis of rotation. The clock mechanism

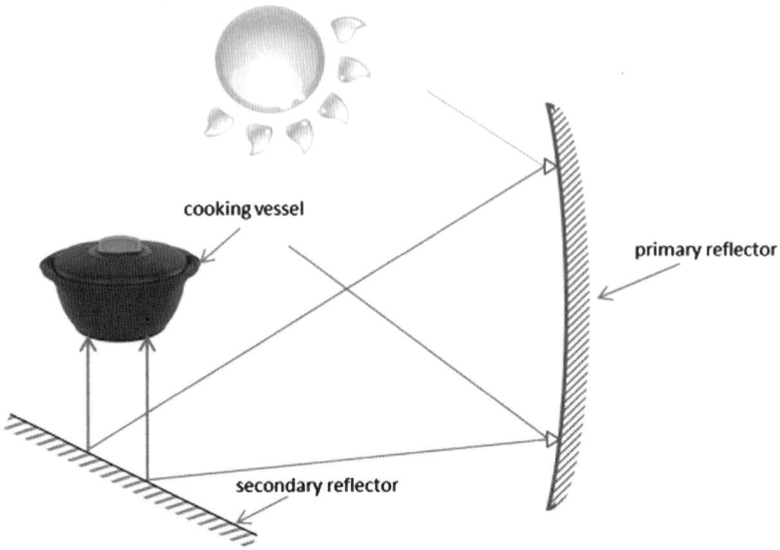

Fig. 7.17 A Scheffler concentrator

Fig. 7.18 Box-type solar cookers

rotates the primary reflector to maintain the alignment of the reflected beam. The fixed secondary reflector reflects the beam from the primary reflector onto a cooking vessel. Each morning, the primary reflector is returned to its starting position and the angle between the axis of rotation and the reflector checked to ensure that seasonal variations in solar azimuth are accommodated. The primary reflector, comprised of flat mirror facets, is a small lateral elliptical section of a much larger paraboloid. Test procedures for such cookers are available (Mullick et al. 1991).

Box type cookers shown in Fig. 7.18 are suitable in appropriate climates for the production of relatively slow-cooked dishes. Rating procedures have been developed for intercomparison of cooking performance (Mullick et al. 1991).

References

Abdel-Salam HEA, Probert SD, Norton B (1986) Predicted performance of coffered solar ponds in the U.K and Egypt. Physio Chemical Hydrodynamics 7(4):217–233

Adams W (1878) Solar heat: a substitute for fuel in tropical countries. Education Society's Press, Bombay

Akbarzadeh A, MacDonald RWG (1982) Introduction of a passive method for salt replenishment in the operation of solar ponds. Solar Energy 29:71–76

Akbarzadeh A, MacDonald RWG, Wang YF (1983) Reduction of surface mixing in solar ponds by floating rings. Solar Energy 31:377–380

Anderson CG (1958) Some limnological features of a shallow saline meromictic lake. Limnol Oceanogr 3:259–269

Assaf G (1976) The Dead Sea: a scheme for a Solar Lake. Solar Energy 1:293–299

Bachmann RW, Goldman CR (1965) Hypolimnetic heating in Castle Lake, California. Limnol Oceanogr 10:233–239

Bernardes MA, dos S, Valle RM, Cortez MFB (1999) Numerical analysis of natural laminar convection in a radial solar heater. Int J Therm Sci 38:42–50

Bitnar B, Dunseh W, Mayor JC, Sigg H, Tschudi HR (2002)

Bryant RS, Bowser RP, Wittenberg LJ (1979) Construction and initial operation of the Miamisburg salt-gradient solar pond, Sun 2. In: Proceedings of the ISES silver jubilee congress, vol 2. Atlanta, pp 1005–1009

Burch J, Thomas K (1998) An overview of water disinfection in developing countries and the potential for solar thermal water pasteurisation. National Renewable Energy Laboratory, Golden

Cabanyes I (1903) Proyecto de Motor solar, La Energia Eléctrica – Revista General de Electricidad y sus Aplicaciones 8:61–65

Chandak A, Dubey D (2005) Innovative solar concentrator for industrial heating applications. International congress on renewable energy, Pune

Chen G (2001) Phonon heat conduction in low dimensional structures. Recent trends in thermoelectric materials research III. Academic, San Diego

Chergui T, Larbi S, Bouhdjar A, Gahgah M (2008) Performances analysis of a solar chimney power plant in South Algeria. In: Proceedings of the world renewable energy congress

Chiritescu C, Cahil DG, Ngayeu M, Johnson D, Bodapati A, Kebhiski P, Zschack P (2007) Ultralow thermal conductivity in disordered WSe_2 crystals layered. Science 315:351–353

Church WC (1907) The life of John Ericsson. C Scribner's Sons, New York

Crevier D, Moshref A (1981) The floating solar pond. ISES-AS conference, Philadelphia, pp 801–807

Dincer I, Rosen MA (1998) A worldwide perspective on energy, environment and sustainable development. Int J Energ Res 18:1305–1321

Duff WS, Hodgeson DA (2007) A simple high efficiency solar water purification system. Solar Energy 79:25–32

Elata C, Levien O (1966) Hydraulics of the solar ponds of the eleventh international congress international association for hydraulic research, Leningrad, pp 1–14

Eneas A (1901) Solar generator. US patent 670,917, 26 Mar 1901

Ericsson J (1868) The use of solar heat as a mechanical motor power. In the report of the centennial celebrations of the University of Lund, Lund

Ericsson J (1884) The sun motor and the sun's temperature. Nature 29:217–218

Feachem RG, Garelick DJ, Mara DD (1983) Sanitation and disease: health aspects of excreta and wastewater management. Wiley, New York

Fleming JC, Liu SY, El-Kady I, Biswal R, Hu KM (2002) Nature 417:52–55

Fluri TP, Pretorius JP, Van Dyk C, Backstrom TW, Kröger DG, Van Zijl GPAC (2009) Cost analysis of solar chimney power plants. Solar Energy 83:246–256

Foster R, Ghassemi M, Alma Cota A (2010) Solar energy: renewable energy and the environment. CRC Press, Boca Raton

Gadhia D, Gadhia S (2006) Parabolic solar concentrators for cooking, food processing and other applications. International solar cooking conference, Granada
Goldsmid HJ (1960) Application of thermo electricity. Wiley, London
Günther H (1931) In Hundert Jahren – Die künftige Energieversorgung der Welt, Gesellschaft der Naturfreunde. Kosmos, Stuttgart
Haaf W (1984) Solar towers: part II: preliminary test results from the Manzanares pilot plant. Solar Energy 2:141–161
Haaf W, Friedrich K, Mayr G, Schlaich J (1983) Solar chimneys, part I: principle and construction of the pilot plant in Manzanares. Solar Energy 2:3–20
Huder PP, Sonnefeld P (1974) Hot brines on Los Roques, Venezuela. Science 185:440–442
Kalogirou S (2004) Solar thermal collectors and applications. Prog Energy Combust Sci 30:231–295
Kedare SB (2005) Solar concentrator for industrial process heat. International congress on renewable energy, Pune
Kennedy CE, Price H (2006) Progress in development of high-temperature solar-selective coating. In: Proceedings of the 2005 international solar energy conference (ISEC2005), 6–12 Aug 2005, Orlando, Paper no ISEC2005-76039, pp 749–755. American Society of Mechanical Engineers (ASME), New York, NREL report: CP-520-36997
Kooi CF (1979) The steady-state salt gradient solar pond. Solar Energy 23(3):7–45
Kraemer D, Poudd B, Feng H-P, Caylor JC, Yu B, Yan YM, Wang X, Wang D, Muto A, McEnaney K, Chiesa M, Ren Z, Chen G (2011) Higher-performance flat-panel solar thermoelectric generators with high concentration. Nat Mater 10:532–538
Larbi S, Bouhdjar A, Chergui T (2010) Performance analysis of a solar chimney power plant in the southwestern region of Algeria. Renew Sustain Energy Rev 14:470–477
Lytvynenko YM, Schur DV (1999) Utilisation of the concentrated solar energy for process of deformation of sheet metal. Renew Energy 16:753–756
Maia CB, Ferreira AG, Valle RM, Cortez MFB (2009) Theoretical evaluation of the influence of geometric parameters and materials on the behaviour of the air flow in a solar chimney. Comput Fluids 38:625–636
Melack JM, Kilham P (1972) Lake Mahega: a mesotropic sulfate chloride lake in Western Uganda. Afr J Trop Hydrobiol Fish 2:141–150
Morti A, Luque A (2003) Next generation photovoltaics; high efficiency through full spectrum utilization. Institute of Physics, Bristol
Mouchout A (1869) Le chaleur solarise et les applications industrielles, 1st edn, Paris
Mullick SC, Kandpal TC, Kumar S (1991) Thermal test procedure for a parabolic concentrator solar cooker. Solar Energy 46:139–144
Neumann O, Urban AS, Day J, Lal S, Nordlander P, Halas NJ (2013) Solar vapor generation enabled by nanoparticles. AesNano 7:42–49
Nielsen CE (1975) Salt-gradient solar ponds for solar energy utilization. Environ Conserv 2:289–292
Nielsen CE (1976) Experience with a prototype solar-pond for space heating, sharing the Sun solar technology in the seventies. Joint conference of the ISES-AS, and the Solar Energy Society of Canada, Winnipeg, pp 169–182
Nielsen CE (1979) Control of gradient zone boundaries, Sun 2. In: Proceedings of the ISES Congress, vol 2. Atlanta, pp 1010–1014
Nielsen CE, Rabl A (1976) Salt requirement and stability of solar ponds, shading the Sun: solar technology in the seventies. Joint conference of the ISES-AS and Solar Energy Society of Canada, Winnipeg, 5 Aug 1976, pp 183–187
Onwubiko C (1984) Effect of evaporation on the characteristic performance of the salt -gradient solar pond. Solar engineering – 1984. In: Proceedings of the ASME Solar Energy Division, sixth annual conference, Las Vegas, pp 6–11
Pasumarthi N, Sherif SA (1998) Experimental and theoretical performance of a demonstration solar chimney model, Part 1: Mathematical model developments. Int J Energy Res 22:277–288
Poppe WI, Woomer NM (1985) Algae management practices for solar salt ponds; practical approaches. In: Progress in solar energy, American Solar Energy Society Meeting, Boulder, 6:439–442

Por FD (1968) Solar lake on the shores of the Red Sea. Nature 218:860–861

Pretorius JP, Kröger DG, Buys JD, Von Backström TW (2004) Solar tower power plant performance characteristics. In: Proceedings of the ISES EuroSun 2004 international sonnenforum

Pretorius JP, Kröger DG (2006a) Solar chimney power plant performance. J Solar Energy 128:302–311

Pretorius JP, Kröger DG (2006b) Critical evaluation of solar chimney power performance. Solar Energy 80:535–544

Pujol R, Moià A, Martinex V (2011) Concentrador solar lineal con reflector estacionario y foco móvil. Era Solar 165:24–30

Rabl A, Neilsen CE (1975) Solar ponds for space heating. Solar Energy 17:1–12

Rodat S, Abanades S, Sans J-L, Flamant G (2009) Hydrogen production from solar thermal dissociation of natural gas: development of a 10 kW solar chemical reactor prototype. Solar Energy 83:1599–1610

Satish CJ, Gurmukh DM (1980) Laboratory demonstration of self-creation, self-maintenance and self-correction of saturated solar ponds. IECEC'80 energy to the 21st century. Proceedings of the15th inter-society energy conversion engineering conference, vol 2. Seattle, pp 1448–1452

Schladow SG (1984) The upper mixed zone of a salt-gradient solar pond: its dynamics, prediction and control. Solar Energy 33:417–426

Schlaich J (1995) The solar chimney: electricity from the Sun, Axel Menges edition, Stuttgart

Schlaich J, Bergermann R, Schiel W, Weinrebe G (2003a) Design of commercial solar tower systems-utilization of solar induced convective flows for power generation. In: Proceedings of the international solar energy conference

Schlaich J, Bergermann R, Schiel W, Weinrebe G (2003b) Sustainable electricity generation with solar updraft towers. Struct Eng Int 3:222–229

Schweiger H et al (2000) The potential of solar heat in industrial processes: a state of the art review for Spain and Portugal. In: Proceedings of the Eurosun, Copenhagen

Shaffer HL (1978) Viscosity-stabilized solar ponds mankind's future source of energy. Proceedings congress, New Delhi

Shuman F (1911) Power from the Sun; a pioneer solar power plant. Scientific American, September

Shuman F, Boys CY (1917) Solar boiler. US patent 1,240,890, 25 Sept 1917

Seinfeld A (2005) Solar thermochemical production of hydrogen – a review. Solar Energy 78:603–615

Tabor H (1980) Non-convecting solar ponds. Solar energy. Proc Roy Soc A295: 422–433

Tabor H (1981) Review article: solar ponds. Solar Energy 27(3):181–194

Tabor H, Matz R (1964) A status Re ort on a solar-pond project. Solar Energy 9:177–182

Tellier C (1889) Elevation des eaux par la chaleur atmospherique, Paris

Vitner A, Reisfeld R, Sarig S (1984) Self-generation of a laboratory-scale saturated solar pond. Solar Energy 32:671–675

Weinberger H (1964) The physics of the solar pond. Solar Energy 8:45–56

WHO (2007) Combating waterborne disease at the household level. World Health Organisation, Geneva

Wilkins E, El-Genk M, El-Husseini K, Thakur D (1982) An evaluation of the gel pond performance. ASME, 82-WA/Sol-29

Wilson AT, Wellman HW (1962) Lake Vanda: an Antarctic lake. Nature 196:1171–1173

Wittenberg LJ, Harris MJ (1980) Management of a large operational solar pond. IECEC'80, Energy to the 21st century. Proceeding of the 15th inter-society energy conversion engineering conference, vol 2. Seattle, pp 1435–1437

Zangrando F (1980) A simple method to establish salt-gradient solar ponds. Solar Energy 25:467–470

Chapter 8
Solar Water Heating and Combisystems

The Sun, the hearth of affection and life, pours burning love on a delighted earth

Rimbaud (1962)

8.1 Overview

Solar energy water heaters can be categorized as either active or passive. An active system is defined as requiring a pump to drive the collecting medium through the system, whereas passive systems require no external power. Distributed systems comprise a solar collector, hot water store and connecting pipework; they may be either active or passive. In the former, temperature sensors, a control circuit and a pump are required to convey the fluid from the collector to the store. Generic system types are shown schematically in Figs. 8.1, 8.2, 8.3, 8.4, 8.5, 8.6, 8.7, and 8.8.

Taxonomies such as that shown in Table 8.1 have been developed to classify the diverse different forms of solar energy water heaters (Norton 2011). The annual variations of space and/or water heating loads that each system should accommodate are quite different. The principle types available are;

- Thermosyphon solar water heaters (Norton and Probert 1986) in which the heat transfer fluid is conveyed around a circuit comprised of the collector, hot water store and intervening pipe work by natural convection, can be direct or indirect and can use either flat-plate or evacuated tube collectors
- Pumped (or forced) circulation solar water heaters (Reddy 1987; Duffie and Beckman 1991) are usually indirect and have generally larger collector area than thermosyphon systems, they also can use either flat plate or evacuated tube collectors.
- Combisystems (Weiss 2003) (a contraction of "combined systems") provide both space heating and hot water. These, invariably indirect, systems require

B. Norton, *Harnessing Solar Heat*, Lecture Notes in Energy 18, DOI 10.1007/978-94-007-7275-5_8,

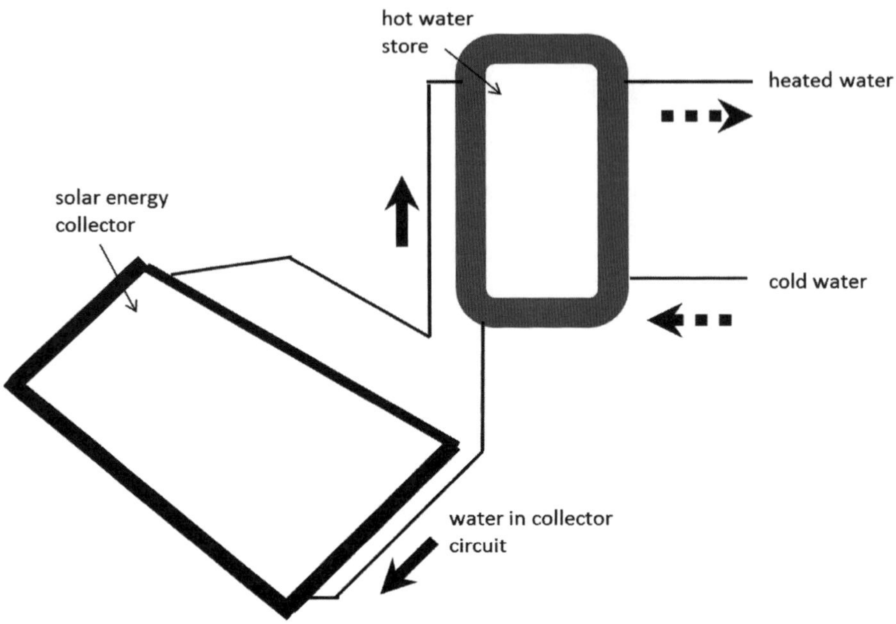

Fig. 8.1 Once-through flow from a collector or integral passive solar water heater

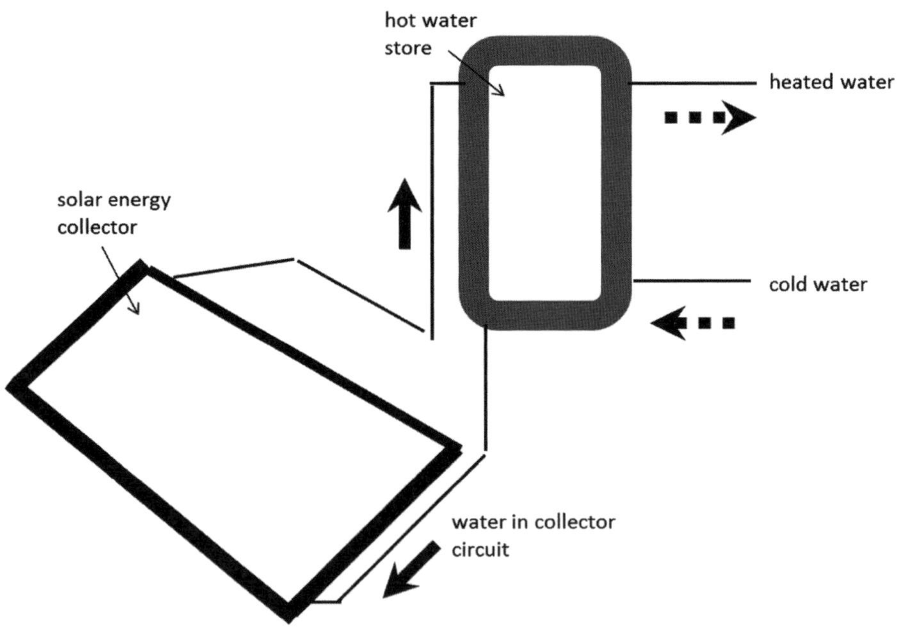

Fig. 8.2 Direct pressurised system employing thermosyphon flow

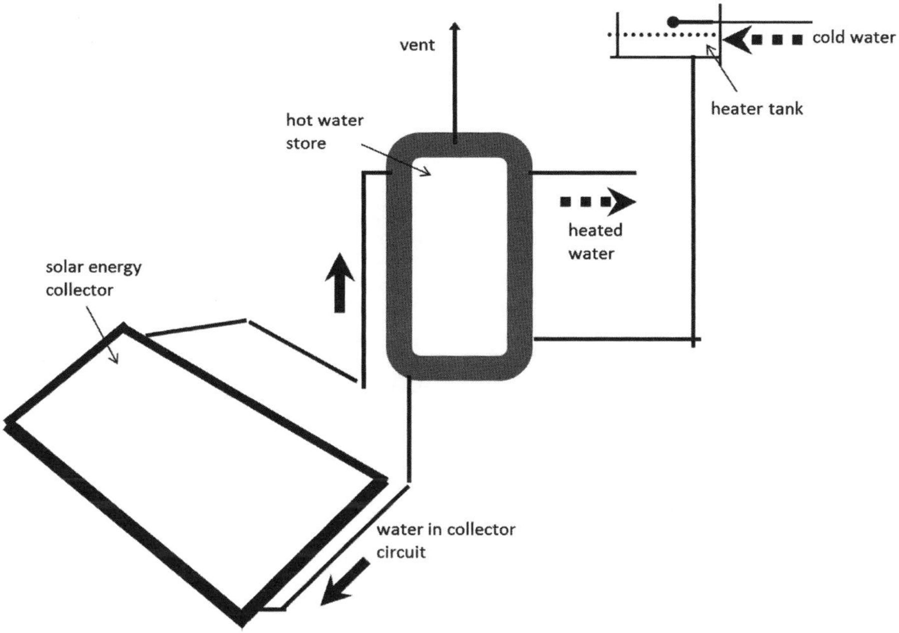

Fig. 8.3 Direct unpressurised system employing thermosyphon flow

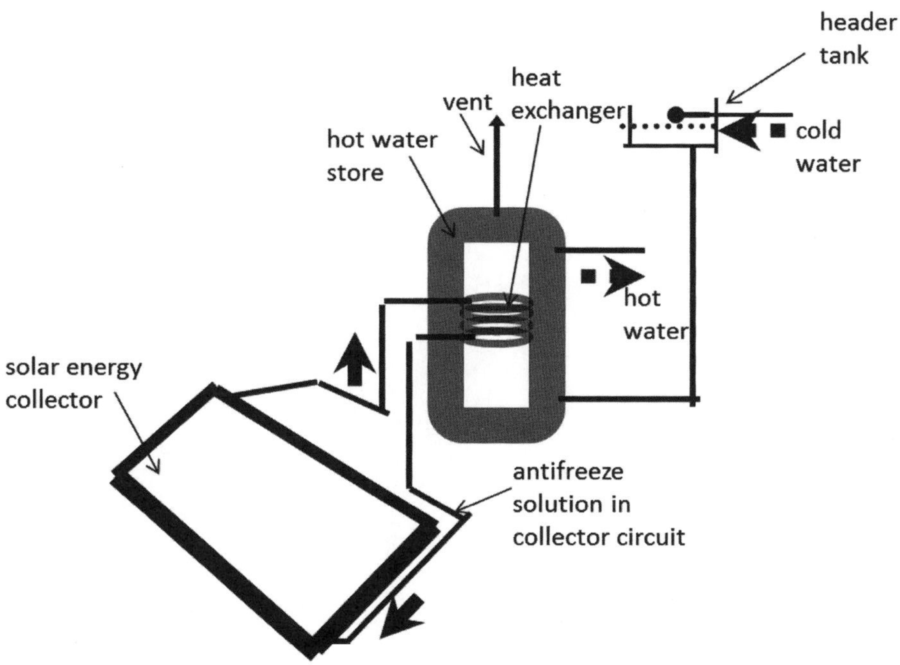

Fig. 8.4 Indirect impressional system employing thermosyphon flow

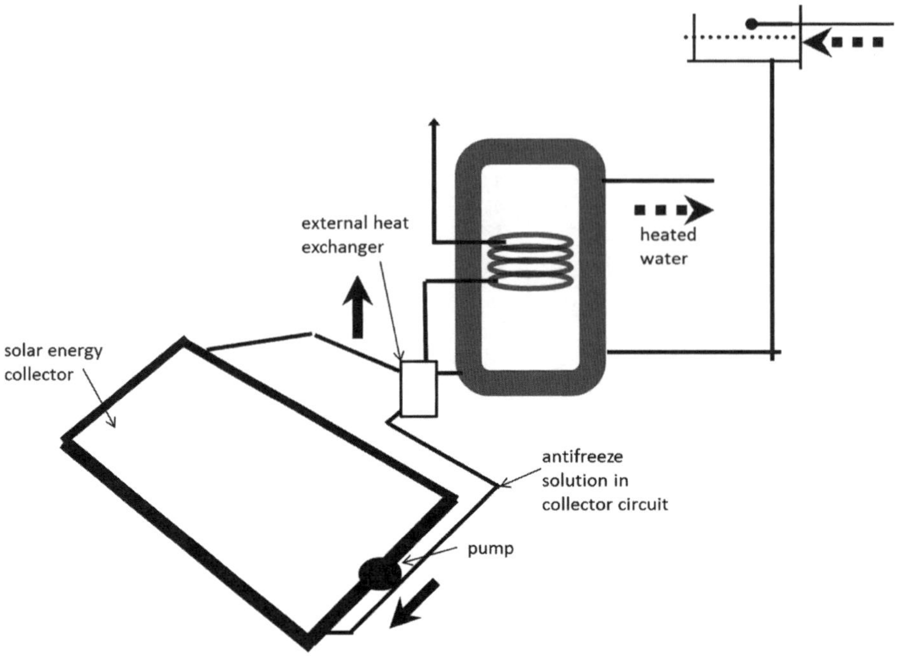

Fig. 8.5 Indirect system with pumped-circulation employed in the collector circuit and thermosyphon flow employed recording circuit

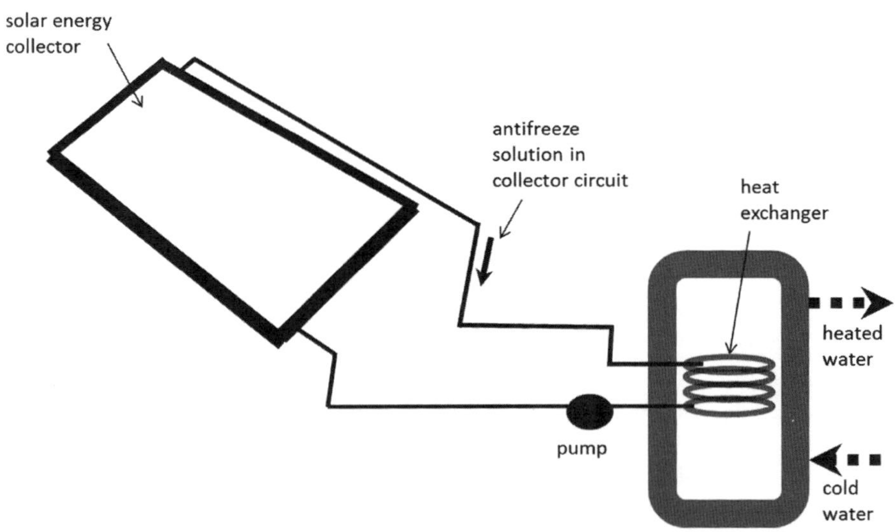

Fig. 8.6 Pressurised indirect forced circulation system with heat exchanger and collector in the same circuit

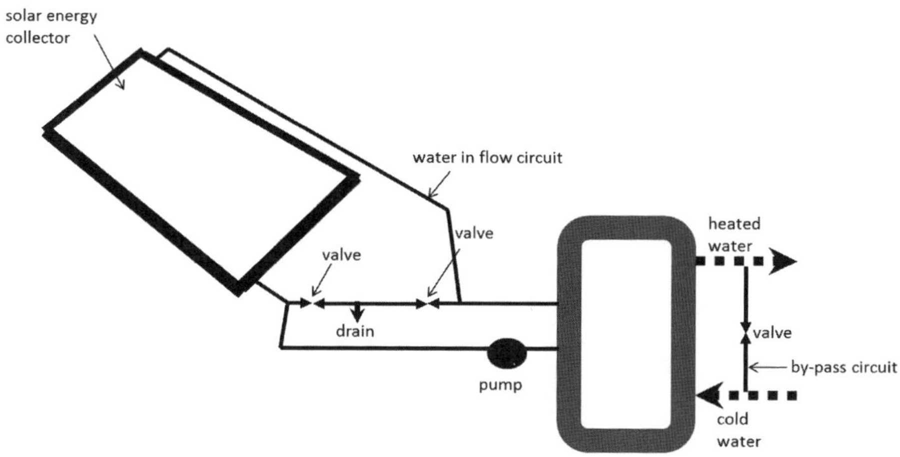

Fig. 8.7 Direct forced circulation system with drain-down for water freeze or pump power failure protection

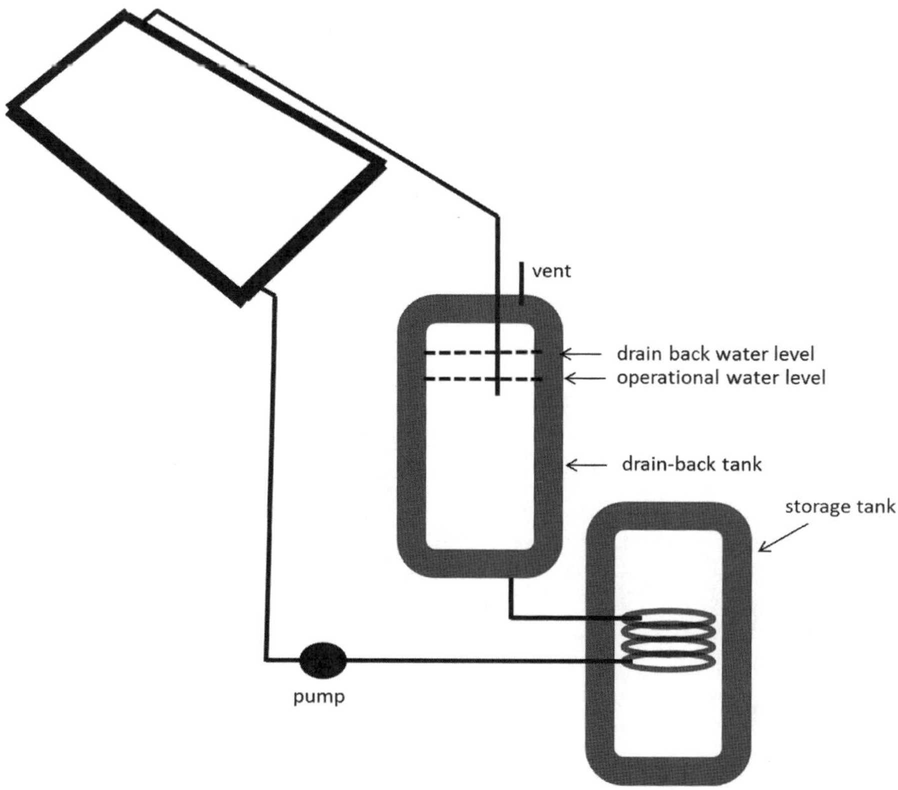

Fig. 8.8 Drain-back system in which when the pump stop the collector fluid drains by gravity to the drain-back tank thus providing water freeze and power failure protection

Table 8.1 Taxonomy of solar water heaters

					Prevention of		Position of store			
				Hot-water store	nocturnal reverse thermosyphon	Collector Freeze protection	Relative to collector	Relative to building	Installation	Collector type
Hydronic	**Integral collector storage**			Single or multiple cylindrical usually horizontal	Hot water withdrawal	Heat retained in thermal mass of the store	Same item	On roof exterior	Installed as a complete factory – built unit	Integral
				Plastic pillow or bags	Double glazing. Movable aperture installation					
	Thermosyphon	Close-coupled	Single phase	Horizontal cylindrical	Height of collector above store. Pipework arrangement	Indirect	Directly above			Evacuated tube
			Two phase		Cessation of fluid evaporation	Drained				Flat-plate
		Distributed		Vertical cylindrical	Height difference between collector and store	Indirect	Above	Inside loftspace	Assembled from components during installation	Photovoltaic/Thermal
	Pumped	Solely hot water		Vertical cylindrical	Non-return valve	Indirect	Below	Variety of possible locations inside building		
		Combisystem: space and water heating		Vertical tank-in-tanks	Flow resistance of pipework and stationary pump					
				Bikini tanks						
		Solar-assisted heat pump (Combi $^{+}$)		Vertical cylindrical						

		Swimming pool heating	Swimming pool		Outdoor pool collectors drained in freezing weather	Pool usually located below	Solar heated pools are indoor and outdoor		Unglazed Glazed flat plate
Aeolic	Solar heated air		An air-to-water heating exchanger	Unnecessary		Various			Air-heating
Electric	Photovoltaic		Vertical cylindrical			Any	Interior	Water heating is an optional use of installed photovoltaic electricity	Photovoltaic

careful design of heat exchanger arrangements and control strategies from the optimal solar savings fraction is to be provided (Yazdanshenas and Furbo 2007; Letz et al. 2009; Yazdanshenas et al. 2008).

- Integral collector storage (Smyth and Norton 2006) in which the heat water store is also the solar energy collector.
- Combi+; are solar assisted heat pumps (Morrison 1984; Troi et al. 2008). To obtain higher temperatures than provided by a solar collector either additional auxiliary heating or thermodynamic conversion is necessary. As an example of the latter, heat provided by solar collectors may be used to evaporate the working fluid in the evaporator of a heat pump that transfers heat from a colder reservoir to a warmer reservoir. During compression the temperature of the heat pump working fluid increases to well above the temperature provided by the solar collector. During condensation, heat is rejected at a higher temperature to hot water store (Charters et al. 1980; Morrison 1984; Huang and Lue 2003). Such systems are sometimes referred to as Combi^{+} (Troi et al. 2008). For a solar assisted heat pump hot water system the key determinants of performance are collector area, speed of the heat pump compressor and insolation (Hawlader et al. 2001).
- Large-scale interseasonal energy storage systems (Schmidt et al. 2004). The majority of large-scale interseasonal storage systems serve housing via distinct heating networks. Table 8.2 summarises such systems with an output >4MWth in operation in 2010 (Dalenbäck 2010).
- Swimming pool heating (Ruiz and Martinez 2010); employing flat plate collectors often unglazed and occasionally with low cost plastic pipe absorbers.
- Photovoltaic solar water heaters (Fanney and Daugherty 1997) where a electrical heating element immersed the water to be treated is powered by a photovoltaic array.
- In a two phase thermosyphon system the natural circulation, circuit uses a fluid with a low boiling point as the working fluid (e.g. methanol or freon). The liquid absorbs heat when passing through the collector and boils. The gas rises to a heat exchanger, where it gives up its latent heat to the storage medium and returns to the liquid state to begin another cycle. This is somewhat similar to a heat pipe. Passive indirect closed two-phase transfer of heat from the solar collector to the store avoids scaling, fouling and the need for freeze protection, with suitable fluids, corrosion is far more limited than with aqueous systems. Working fluid options include acetone, petroleum ether (Sion et al. 1979), trichlorofluoromethane (Schreyer 1981), R134a (Ong and Hiader-E-Alahi 2003; Esen and Esen 2005), R407C and R410A (Esen and Esen 2005). Acetone, R134a and R410A are all low cost and available readily. Though acetone obviates the use of high pressures, it is flammable. The use of R134a is advocated as it requires a lower operating pressure than R410A whilst providing similar thermal performance (Ordaz-Flores et al. 2012). In novel integral collector storage solar water heater with an outer absorbing vessel and an inner storage vessels, the space between the two vessels contains a small pool of water, which when heated evaporates and consequently condenses on the surface of the colder surface of the inner vessel, thus transferring thermal energy to the store (De Beijer 1998).

Table 8.2 Large-scale interseasonal solar heating systems with output >4MWth (Adapted from Dalenbäck 2010)

Collector area (m^2)	Nominal power output MW_{th}	In operation from	Location	Country
18,300	12.8	1996	Marstal	Denmark
10,700	7.5	2009	Broager	Denmark
10,073	7.0	2009	Gram	Denmark
10,000	7.0	2000	Kungälv	Sweden
8,012	5.6	2007	Braedstrup	Denmark
8,012	5.6	2008	Strandby	Denmark
7,300	8.1	2003	Grailsheim	Germany
7,284	5.1	2009	Torring	Denmark
6,000	4.2	2008	Soenderberg	Denmark
5,670	4.0	1997	Neckarsulm	Germany

In both pumped circulation and thermosyphon units vertical tanks tend to be employed in climates where, for a significant part of the year, not all the water in the store will be heated to the desired temperature. As vertical tanks provide a greater vertical aspect ratio for buoyant warmer water to form thermally stratified layers usually with a thermocline as shown in Fig. 8.14. Such thermal stratification ensures that the temperature of withdrawn hot water is closer to collector outlet temperature. This segregation of hot water from colder layers means the temperature of this water withdrawn is higher than the mean store temperature. There is thus less likelihood of supplementary heating being required to further warm to the desired demand temperature. Larger solar fraction thus ensures than would from a fully mixed store as withdrawal would be at the, lower, mean solar temperature.

In high insolation climates where often all the water in the store is usually at the demand temperature, enhancing thermal stratification usually will not provide a larger solar fraction. Conversely, it could require the addition of cooler water to achieve a temperature suitable for baths or showers. In addition there may be a heightened probability of high temperature damage, particularly when no withdrawal of heated water takes place. Thus as the stratification-enhancing attributes of a vertical tank are not required horizontal tanks are thus more often used in high insolation climates.

8.2 Integral Passive Solar Water Heaters (IPSWH)

The Integral Passive Solar Water Heater (IPSWH), called commonly the breadbox or batch water-heater, consists of an integral collector and storage unit. Its simplest form consists simply of a tank, painted black to absorb insolation. Variations are comprised of one or more tanks, painted black or applied with a selective absorber surface, within a well insulated box, possibly with reflectors and covered with single, double or even triple glazing material of glass or plastic or a combination of the two.

The first IPSWH's were just bare tanks of water left out to warm in the sun. These were used on a few farms and ranches in the southwest of the USA in the late

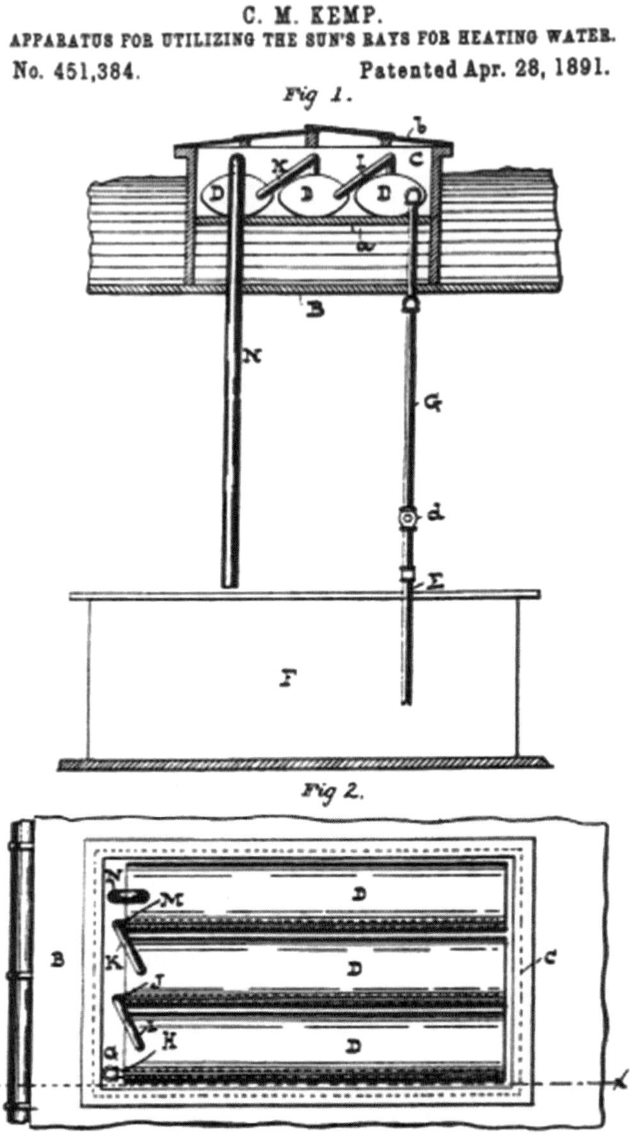

Fig. 8.9 Kemp's patent for an integral passive solar water heater

1800s, and were reportedly quite capable of producing water hot enough for showering by the late afternoon on clear days (Butti and Perlin 1977). The first commercially-manufactured solar water heater was an IPSWH, patented in 1891 by Kemp (1891). This was sold in Maryland, in the USA under the trade name "The Climax Solar-Water Heater". A reproduction of Kemp's original patent is shown in Fig. 8.9. During the early to mid 1970s, there were many different IPSWH designs

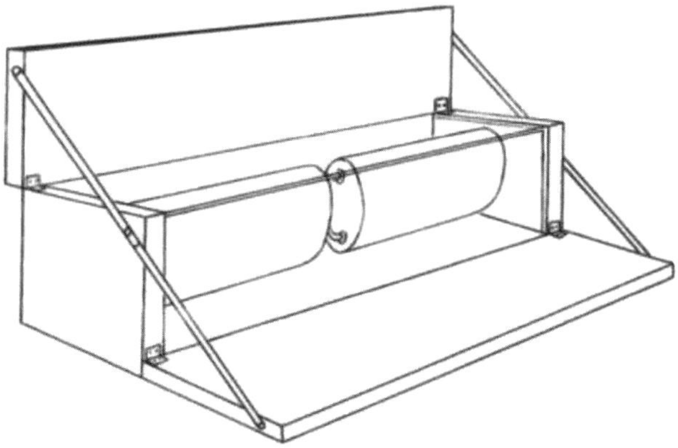

Fig. 8.10 Breadbox solar water heater

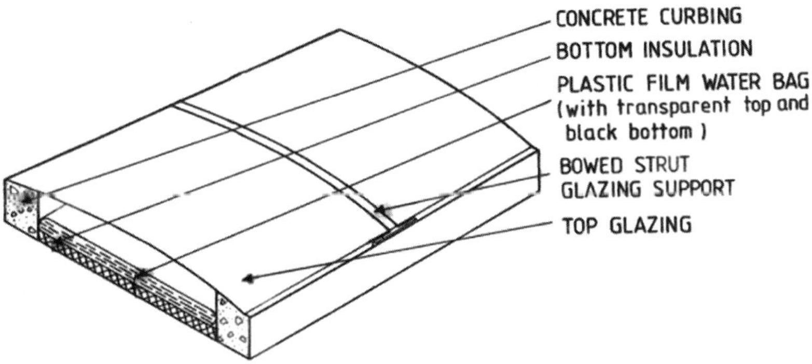

Fig. 8.11 The shallow solar pond integral passive solar water heater

built and tested. One of the most innovative designs of this period was the "Breadbox" water heater of Baer (1975). As shown in Fig. 8.10 this IPSWH comprised two cylindrical tanks mounted horizontally. Moveable insulated lids were used to enhance collection during the day when open, and to reduce night-time heat loss when closed. Several other enthusiasts from the 1970s who built and tested their own collectors, based their own designs on Baer's water heater (Bainbridge 1981a, b). A ground-mounted system referred to as a "shallow solar pond" (not to be inadvertently confused with a non-convecting solar pond) was developed as a low cost solar water heater for high insolation conditions. A diagram of a shallow solar pond is shown in Fig. 8.11.

Studies of the economic viability of solar energy water heaters have found for warm climates IPSWH and thermosyphons to be most viable when compared with pumped systems (Lange 1981; Czarnecki and Read 1978; Fanney and Klein 1983).

Simplified single-node models in which an IPSWHH is a fully mixed store, can be used to predict the mean temperature and the solar savings fraction. However, in a stratified hot water storage tank, the water drawn off will be the hottest layer of water in the topmost part of the tank, at temperature and not at the mean temperature, thus the 'true' solar savings fraction, will be that given by the uppermost temperature. Solar savings fractions predicted by simplified single-node simulation models will consistently be lower than the true value. This under-prediction could be significant when determining the economic viability of a particular system, as an under-prediction of the solar savings fraction would indicate correspondingly lower financial savings implying, a longer payback time. The key problem when using 'one-node' analysis is how to relate the mean store temperature to the outlet temperature and incorporate this into a model in order to predict the solar savings fraction more accurately.

In the first computer-based simulation model to predict the performance of an integral passive solar water heater (Garg et al. 1972), the collector was considered as a single-node system, and with appropriate input parameters was employed as a collector design tool to predict the water temperature at any time (with no draw off) for various collector designs. A sizing procedure for integral passive solar water heaters using data available from testing methods, monthly average climatic data and load size has been developed (Zollner et al. 1985) to predict the long-term performance of integral collector systems. Monthly performance predictions where in the form of monthly solar saving fractions, defined as the fraction of the load met by solar (neglecting any auxiliary jacket losses). Such "f-chart" correlations (Klein et al. 1976) for IPSWH were also developed by Tully (1983). IPSWHs design methods have been proposed (Burns et al. 1985; Panico and Clark 1984; Proctor 1984; Yeh and Chen 1986) where assumptions about IPSWH behaviour had been made. The effect of changing operating variables was examined on the top losses and the heat storage delivery efficiency, a latter term was used to describe the effect of stratification on the temperature of the outlet water during a draw-off. A simplified model of the heat loss, where all heat losses (radiative, convective and conductive) are lumped together in one term has been found to be adequate in the analysis of multi-tank IPSWH units (Weller et al. 1985). Testing methodologies, have been developed for IPSWHs (Huggins and Block 1984), to predict the annual performance.

8.3 Distributed Water Heaters

A distributed system circuit is comprised of the solar collector, hot-water store (or heat-exchanger located therein) and the associated pipework. In a forced circulation system flow is induced by a pump. In a thermosyphon fluid flows due to buoyancy forces occurring in the closed loop. These forces arise because of the difference in densities of the water in the collector, which is heated by the sun, and that in the cooler hot-water store. In a forced circulation solar energy water heater, the heat transfer fluid is recirculated through the collector at a constant flow rate by the action of a pump. However, for a thermosyphon system, the rate and the direction of the flow are dependent on prevailing weather conditions and on the geometry of the pipework. All types can be constructed as direct or indirect systems, in the latter

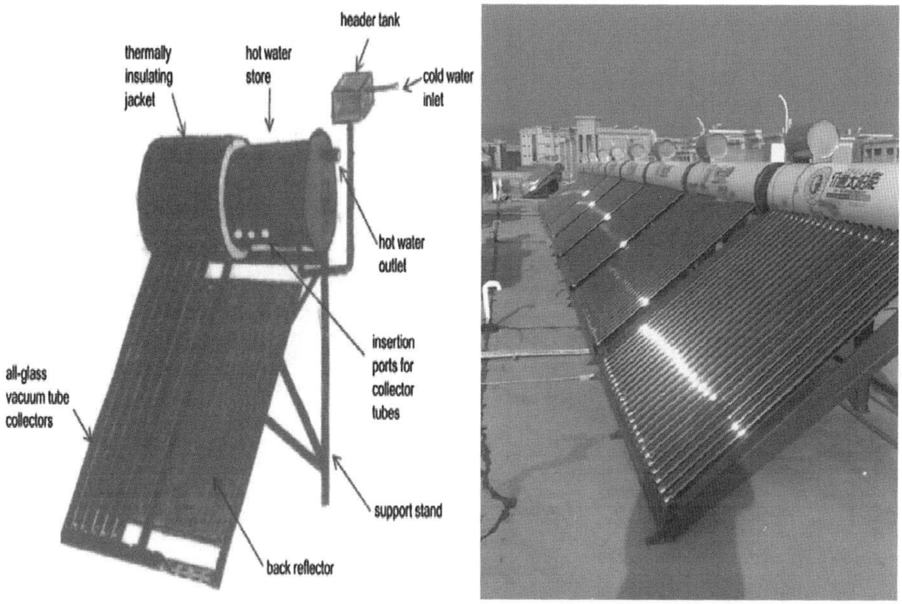

Fig. 8.12 Examples of a close-coupled evacuated tube thermosyphon solar water heater

case usually an aqueous anti-freeze solution flows through a heat-exchanger immersed in the hot water store. Because they are often designed individually for each application, distributed systems exhibit a wider variety of specification. In close-coupled units, a horizontal hot water store and collector are juxtapositioned to form a single unit as shown in Figs. 8.12 and 8.5; fluid circulation in close-coupled units is invariably thermosyphonic.

The first patent for a natural-circulation solar water heater was granted to Bailey who lived near Los Angeles, California (Bailey 1910) shown in Fig. 8.13. Although probably not the first to develop such a system (Butti and Perlin 1980), his firm was the earliest commercial manufacturer and installer of a standard unit. It was successfully marketed in southern California as a solar heater which would store heated water for use at night. This attribute represented a significant advance over contemporary IPSWHs. Due to the competition provided by cheap natural gas in the early 1920s, from the then abundant reserves in the Los Angeles Basin, sales of solar water heaters declined. The firm founded by Bailey became primarily manufacturers of gas-burning water-heaters, producing only 40 solar heaters in 1930 and ceasing production of them altogether by 1941.

However in 1923 Bailey had already sold his patent to H.M. Carruthers of Miami Beach. With the boom in house building in Florida at the time, thermosyphon water heaters found a ready market there. More importantly, the unit cost of electricity – the most common alternative fuel, was such that these water heaters had pay-back periods only slightly longer than 2 years (Butti and Perlin 1980). Their relative popularity thus soon spread to other southern states of the USA (Carnes 1932; Alt

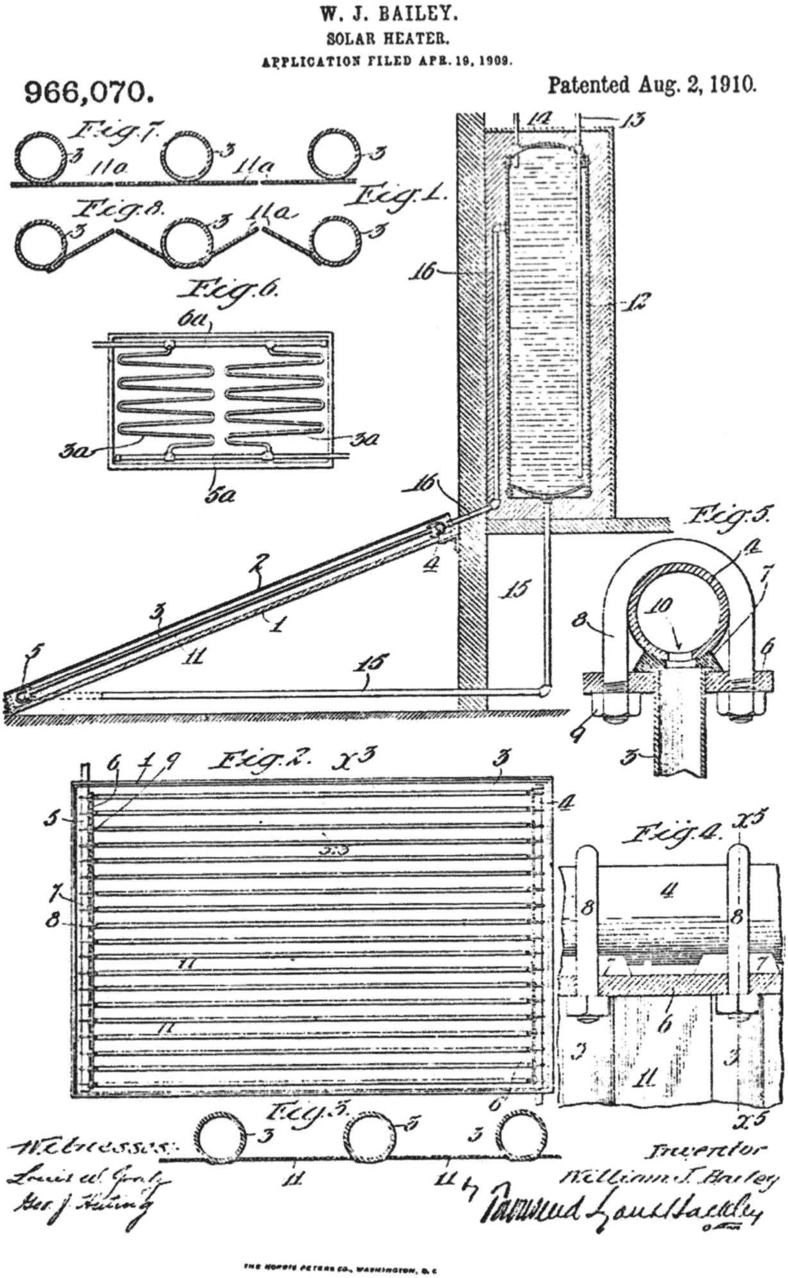

Fig. 8.13 Bailey's patent

1935; Merle 1940). During the 20 year period following the end of the First World War, as many as 60,000 thermosyphon solar water heaters were installed in the Miami area alone (Scott 1976), and 85 % of small houses in Florida had them by 1946 (Anon 1946).

With the increasing availability of cheap fuel oils and piped gas in the 1950s, solar energy became relatively unattractive financially, even in Florida, and so its rate of utilisation declined. A study of the early solar water heater industry in Florida identified additional factors which contributed to this decline (Scott 1976); a rapid growth occurred in the number of developers of large housing traits. Who to competed with one another to offer the lowest initial price for a new house, tending to install initially, cheaper, fossil fuelled hot-water heaters. As well as effectively depriving many house buyers of a choice concerning the type of hot-water system which they wished to have installed, this also caused a contraction of the "solar-industry". There was thus a reduction in the availability of maintenance services for existing installed units. Even those were usually too small to satisfy a significant fraction of the post-war increases in demand per household for hot water. These problems, together with that of corrosion, gave rise to a poor opinion of solar water heaters among those householders who still had a choice of which system to install. Such effects reinforced one another and led to the reduction by 1960 in the number of thermosyphon solar water-heating units still in use in Florida to about 25,000 (Andrassy 1961).

While thermosyphon solar-heated water declined in popularity in the southern USA, an increasing interest was shown in its use in Israel (Sobotka 1961). Australia (Morse 1955a), South Africa (Whillier 1995), the Indian Subcontinent (Mathur et al. 1959) and the Mediterranean countries, which had in common, ample year-round insolations, the necessary engineering expertise, and high fuel costs particularly in remote locations. In Australia and South Africa, this interest was encouraged by the publication of design and construction manuals and by field trials, (Morse 1955b; Chinnery 1967) respectively. In 1957, a committee of the Australian government recommended that thermosyphon solar water heaters should be installed in government residences in northern tropical areas (Anon 1957). The rapid fuel-price inflation that ensued after November 1973 initiated a renewed interest in harnessing solar energy. Commercial manufacturers of thermosyphon units are now prevalent throughout the world from Cyprus to China. In Isreal, since 1980, natural-circulation solar water heaters have satisfied one-third of the annual national energy requirements for hot water (Shitzer et al. 1979). In Austria one of the strongest markets for pumped solar water heaters developed from self-built solar water heater initiatives. In China since 2007 the installation of solar water heaters has been mandatory in some areas. The policy is enforced through the construction approval (Runqing et al. 2012). In 2010, 98 % of the 145 million square metres of solar collectors were vacuum tubes almost always in thermosphyon systems (Zhentao and Zhichen 2010).

As shown in Fig. 8.14, in a thermosyphon solar water heater the buoyant pressure arises from differences in water temperature as shown in Fig. 8.14. The collector flow rate is determined by the extent that buoyant force overcomes the

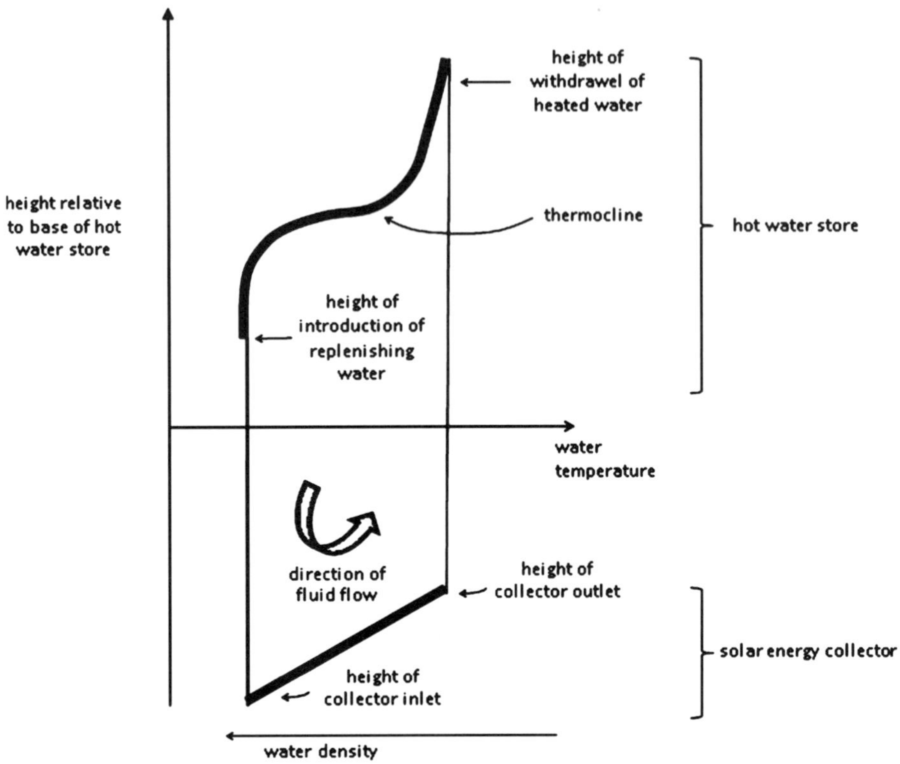

Fig. 8.14 Features of the flow circuit in a thermosyphon solar water heater

frictional losses in the natural circulation circuit. A height difference is usually employed to prevent nocturnal reverse circulation. Two other techniques employed less frequently; one is the use of novel pipework arrangements that create an intermediate task limb enabling absorber and store to be at the same height as shown in Fig. 8.15. The other technique is the use of low flow resistance one-way check valves (Buckley 1979) shown in Fig. 8.16.

To analytically predict the performance of a natural-circulation solar- energy water heater, three alternative broad approaches can be adopted, these being:

- Simplified models (Morrison and Tran 1984; Huang and Hsieh 1985).
- Correlation of performance characteristics from either the simulation or monitoring of generic systems (Song and Zhang 1982; Morrison and Sapsford 1983; Hobson et al. 1987; Malkin et al. 1986).
- Rigorous simulation via the simultaneous solution of the coupled energy and momentum equations (Close 1962a, b; Ong 1974, 1976; Young 1980; Huang 1980; Mertol et al. 1981; Riddle 1985; Hobson and Norton 1988).

The first two approaches are intended for the estimation of the long- term performance of a system and for the determination of the system size that achieves

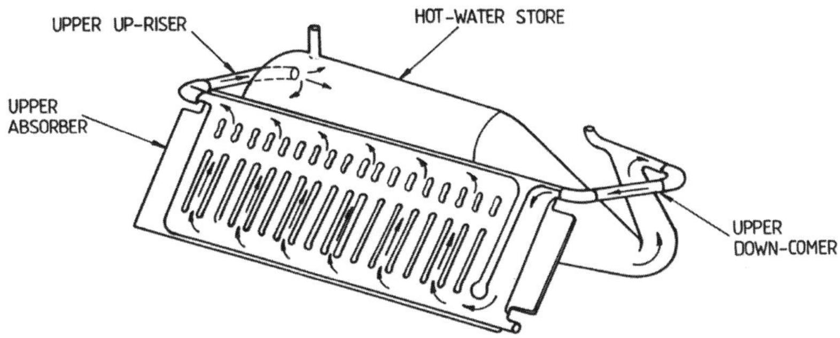

Fig. 8.15 Use of a tank-limb arrangement to prevent nocturnal reverse circulation in a thermosyphon solar water heater

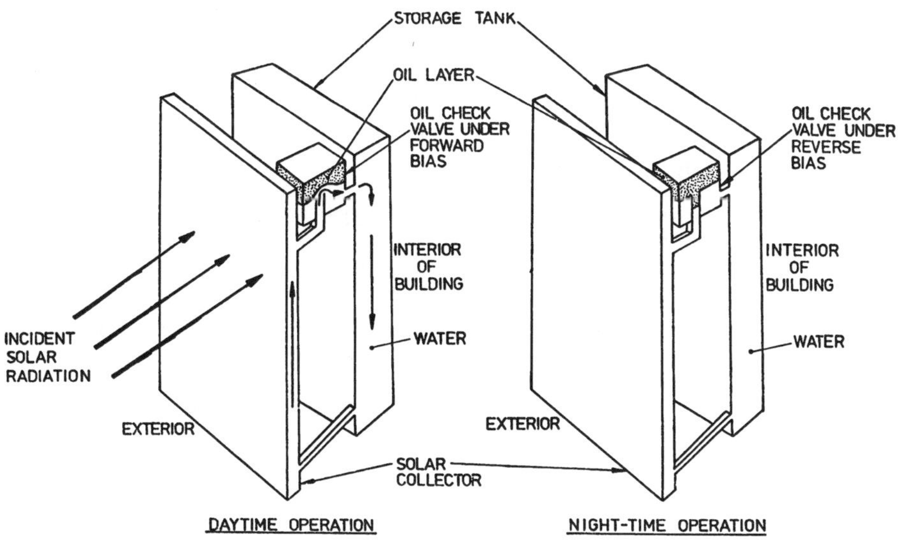

Fig. 8.16 Use of an oil check valve to prevent nocturnal reverse circulation

the optimum solar fraction. Because of the simplifications inherent in the first approach, such models are limited by the range of operating conditions and system configurations over which the simplifying assumptions are valid. Models referred to in this category often require experimentally-determined information which is only obtainable once the system has been constructed. The second approach cannot be applied reliably to those systems for which dimensions and climatic conditions a correlation has not been determined.

The third approach of a rigorous simulation can be used for:

- The engineering optimisation of systems from short term performance simulations,

- To establish the long term correlation of system behaviour over a wide range of system types, and operating conditions,
- To determine the useful prediction limits both spatially and to smaller time intervals of simplified models.

That in practice, the mean collector temperature was only slightly higher than the mean storage tank temperature (Close 1962), enables a simple analytical model to predict day-time mean tank temperature for no water draw off and sinusoidal variations with time in both isolation and ambient temperature.

Such analysis are simple and accurate when applied to specific systems under known conditions, however they are unable to provide information on performance as a function of system parameters or environmental conditions. Fourier series expansion may be used to model variations in the ambient temperature. In validating this model experimentally, a close correlation was obtained during the day-time. The observed night-time system loss was almost half way between the loss rates predicted for the two extreme cases of (i) perfect coupling, high reverse flow and (ii) perfect decoupling, no reverse flow between the collector and storage tank.

The difference method to predict the system temperature distribution and thermosyphonic flow rates

- Equal mean temperatures in the collector and storage tank, and
- The assumption that the variation in temperature was linear in the direction of flow.

There was some agreement between predicted temperatures and those measured in one particular system near the middle of the insolation period, but large discrepancies were noted at other times.

These arose mainly from:

- The neglect of the thermal capacities of the flat plate collector and connecting pipes, and
- The use of linear dependencies for the density and viscosity of the circulating fluid.

In the "Detailed Loop Model" (Mertol et al. 1981), the coupled energy and momentum equation were solved assuming unsteady flow with provision for both direct and indirect heat transfer to the storage water. An empirically determined one dimensional vertical temperature profile was assumed for the hot water storage tank and an unnecessarily simplified approach used to model withdrawal of water from the tank.

For a comprehensive thermosyphon water heater simulation: a finite difference, transient heat transfer analysis should be applied to the circulating liquid in all the components (i.e. collector, upriser, storage tank and downcomer) of the thermosyphon loop. For transient accuracy, collector plate and cover capacitance terms should be included in the simulation. The density, specific heat, viscosity, conductivity and Prandtl number of the circulating liquid are best represented as second-order polynomial functions of temperature. All heat transfer coefficients must be temperature dependent and based on the ambient and mean component temperatures. These coefficients should be updated at each time step in the numerical solution. A simulation of buoyancy induced mixing between stratified layers, (which occurs for a warmer fluid introduction below a cooler layer) should be included in the storage tank model. A transmission coefficient of the glass collector cover should be employed that is a function of the sun hour angle. In determining the mass flow rate, a form of the momentum equation which includes transient terms should be used. Friction factors should be calculated using correlations appropriate to both non-isothermal thermally destabilised low Reynolds numbers flow and isothermal developing laminar flow in the straight sections of the ducting and empirically determined laminar loss coefficients employed for pipework bends.

In a simple mixing model (Morrison and Tran 1987), if a layer of warm fluid exists below a cooler layer, complete mixing is assumed to occur and the two adjacent nodes take on a single temperature. This process is repeated throughout the tank until the thermocline is restored.

Pressure terms due to body forces acting on the fluid arise from the total vertical hydrostatic head integrated around the thermosyphon loop.

Using a finite difference technique, the equations representing the main energy and momentum transfer processes are solved simultaneously with their associated boundary conditions. First order forward difference should be used for transient terms, first order windward difference for convective terms and second order central difference for conductive terms. Unless small time steps and small node dimensions are used, considerable errors can be introduced by the windward differencing scheme due to artificial diffusion (Young 1980). Replacing the convective terms in the main equation by a second order, central differencing rather than a first order accurate windward differencing scheme would at first sight appear advantageous. However such a scheme does not prevent artificial diffusion and indeed introduces oscillations in the solution if the velocity/thermal diffusivity is greater than 100. For a typical thermosyphon solar energy water heater, this ratio is -150 and consequently, oscillations could be expected. Artificial diffusion can be eliminated by employing an explicit scheme which is accurate to second order in time and space. However, this would only produce a stable solution when the cell Reynolds number is less than or equal to 2. This particular restraint may limit the range of system geometries and operating conditions that may be considered. When solving the simultaneous equations using a Gauss-Seidel iterative method in order to find variation of the temperature for each node with time, the solution is unconditionally stable and the only limit on the size of the time step employed is the required accuracy of the solution.

The identification of generalised dimensionless grouped parameters for both pumped (Klein et al. 1976; Liu and Hill 1979) and buoyancy-driven (Morrison and Tran 1987; Song and Zhang 1982) solar-energy water heaters, has provided a practical approach by which their long-term performance can be predicted readily. One of the major inherent draw-backs of correlating parameters derived previously has been that they are based on steady-state analyses and therefore relate to

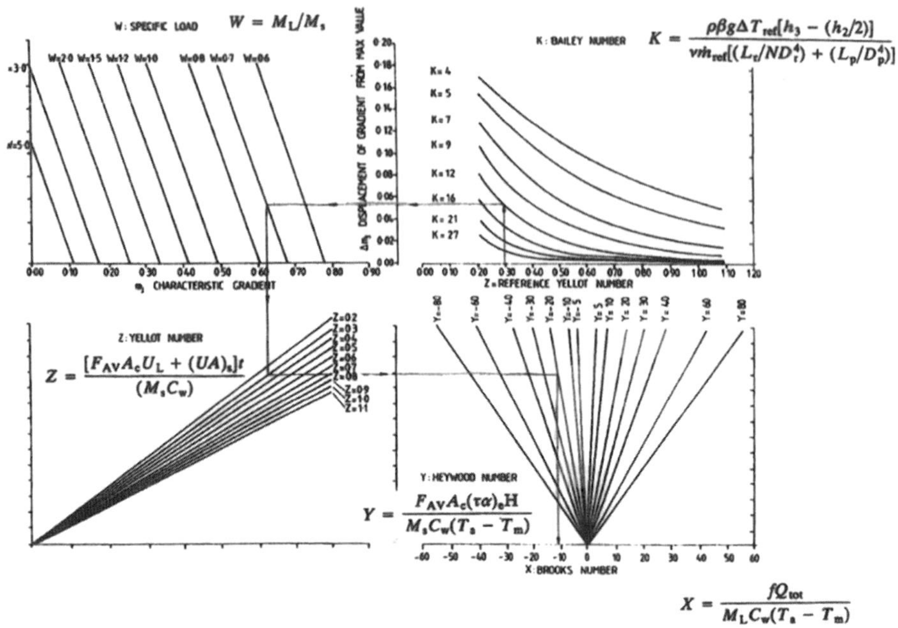

Fig. 8.17 A design nomogram for direct thermosyphon solar energy water heaters

performance over relatively-long periods of operation. A method of determining a characteristic correlation curve for an individual natural-circulation solar-energy water-heater based on a transient analysis which relate to diurnal performance, has been developed (Hobson et al. 1987).

From a transient heat balance on a generic directly-heated thermosyphon solar-energy water-heater, the following dimensionless parameters Y, Z and X, designated the Heywood, Yellot and Brooks numbers respectively, have been identified (Hobson et al. 1987) as shown in Fig. 8.17 based on experimental data collected via the monitoring of systems operating in England (Hobson et al. 1987) and Portugal, (Norton et al. 1988). A dimensionless parameter, the Bailey number, K, represents the system parameters effecting flow within the system. The correlation technique outlined was derived using thermal performance data generated by a validated numerical simulation model using Kew (London, UK) weather data (Hobson and Norton 1988). The relationships between the Yellot, Z, Bailey, K, Heywood, Y, and Brooks, X, numbers and the Specific Load, W, may be summarised as a nomogram as shown in Fig. 8.17, (Hobson and Norton 1988). The Heywood, Y, and Yellot, Z, numbers and the Specific Load, w, are functions of the applied conditions whereas the Bailey number, K, is a function essentially of the system design. However, all these dimensionless groups include information available readily to a designer who, using the nomogram in Fig. 8.17 can thus determine the Brooks number, X, and thus the solar fractions.

Pumped circulation solar water heaters do not generally either provide more hot water or heat water more efficiently than a comparable thermosyphon system. However pumped systems have a significant advantage of layout flexibility over thermosyphons which means, for example, that only the collectors need be located external to the building. The roof does not therefore need to sustain externally the weight of a hot water store; this, particularly for larger installations has the advantage of rendering installation easier and in certain cases is the only feasible means. In forced circulation systems, to avoid cooling water from the hot water store, the pump must cease operation at night and during periods when the solar heat gained by the collector is less than the heat lost from it. To achieve this, temperature sensors generally at the collector's inlet and outlet feed signals a controller that activates the pump. Various control algorithms have been developed (Wuestling et al. 1985; Prud'homme and Gillet 2001) though in most practical systems the pump is activated simply by an unmodified temperature difference across the collector (Deutsche Gesellschaft für Sonnenenergie 2005). Systems using a photovoltaic-powered circulation pump (Parker 1976; Stefanizzi et al. 2003) operate only when the solar energy intensity is sufficient to meet the power requirement of an optimally-selected (Al-Ibrahim et al. 1996) pump. For most pumped-circulation systems, the collectors are roof mounted through novel wall-mounted systems have been proposed (Ji et al. 2006) and very-large scale solar district heating applications have used free-standing ground-level collector arrays (Schmidt et al. 2004; Dalenbäck 2010).

Solar combisystems satisfy both part of a building's space heating demand and part of the domestic hot water consumption. High energy savings can ensue when a well-insulated thermally-stratified small capacity auxiliary heat store operates at a low set-point temperature (Weiss 2003). The auxiliary heating system should also operate with high efficiency. Combisystems have been installed in a wide variety of configurations depending on plumber's preferences and experience, manufacturers' specifications and, for retrofitted installations, the arrangement of pre-existing tanks and their associated pipework. This presents particular challenges when seeking to develop broadly applicable system rating characterisation methods (Letz et al. 2009) In tank-in-tank solar combisystems, a domestic hot water tank is integrated into the space heating hot water store: solar heat is transferred by an internal spiral heat exchanger located in the lower part of the tank. Domestic hot water is withdrawn directly from domestic hot water tank in the store. One alternative design uses a "bikini" tank with two separate mantles encircling circumferentially the hot water tank at upper and lower levels, these circulate domestic hot water and solar heated water respectively both transferring heat to the space heating water. The domestic hot water is withdrawn directly from the tank. Theoretical investigations of bikini solar combisystems (Yazdanshenas and Furbo 2007) have shown that bikini systems are suitable for low-energy buildings because they operate most effectively at the low auxiliary volume set point temperatures that then prevail (Yazdanshenas et al. 2008).

As a high water temperature is not sought, unglazed flat plate collectors are used to heat swimming pools with simple arrangements of black plastic pipes often being employed. Swimming pool heating is currently the main solar water heater type in the USA, Canada and Australia (Weiss and Mauthnen 2010). Systems almost always use pumped circulation and together with the pool itself collectors are drained in winter in locations at higher latitudes. Parametric analyses have been conducted to optimize solar heated swimming pool configurations using hour-by-hour simulation (Ruiz and Martinez 2010).

8.4 Photovoltaic Solar Water Heaters

Photovoltaic solar water heating ensues when electricity from a photovoltaic array is dissipated as heat from a resistive heater immersed in the hot water store (Fanney and Daugherty 1997; Fanney et al. 1997). The maximum power point of the photovoltaic array's current-voltage characteristic is maintained by varying optimally the electrical load imposed by the heating element. Solar energy conversion efficiencies of photovoltaic systems are generally lower (typically a factor of two or greater) that of solar thermal collectors. For a photovoltaic solar water heater the collection area is thus over twice that for a solar thermal water heater. The cost per unit area of photovoltaic cells is also higher currently than for solar thermal collectors. As large areas of a more expensive solar energy collection component are therefore necessary, photovoltaic solar water heaters are not competitive economically with solar thermal systems. However as only a cable connects the heating element in the store to the photovoltaic array, the installation process is simple and the use of electricity to transfer energy obviates any need for mains water pressure reduction and/or freeze protection. Many building integrated photovoltaic systems (Norton et al. 2011) act as embedded generators that supply a grid (Mondol et al. 2009). Where photovoltaic-to-grid feed–in tariffs for electricity sold to the grid are not in place or do not provide a sufficient return on investment, it may be that the best use of any building integrated photovoltaic energy in excess of the contemporaneous building electrical load is often to heat water. Indeed this circumstance can prevail almost inadvertently where instantaneous electric water heaters are employed in a building supplied with photovoltaic electricity.

8.5 Freeze Protection

Water freezing in the collector can lead to the need to replace the collector and where joints located internally burst to significant water damage to a building. Both of these incur costs that adversely affect economic viability and consumer confidence. Evacuated collectors (Rabl 1985) are used in the majority of solar water heaters globally mainly because of their ubiquity in China. The evacuated envelope, solar-selective absorber surface low long-ware transmittance and, where included, a heat pipe's inherent duration of freeze damage pressures of glass combine to protect the water in the collector from freezing. Multiple glazing layers (Duffie and Beckman 1991) and transparent insulation materials can provide collector freeze protection. Similar measures can be applied to integral passive solar water heaters, though performance will be reduced by optical losses when too many multiple glazed layers are used such as a six cover superinsulated system (Bishop 1983). Thermal mass provides significant freeze protection for integral pressure solar water heaters (Smyth et al. 2001a, b). Systems of all types may also be drained in winter (Duffie and Beckman 1991; Deutsche Gesellschaft für Sonnenenergie 2005).

The most common inherent protection against potential damage in cold weather arising from water freezing within the collector pipework is to have an aqueous glycol solution flowing in a closed loop comprised of the collector, a heat exchanger, and the associated pipework. This system has the advantage of requiring neither automated nor manually-operated drain valves. In some climates even in very cold conditions, utizable insolation can follow swiftly after cloudy cold weather. Under such conditions a glycol-filled indirect system can produce warm water unlike a drained down system. This advantage is counteracted when operations in warmer conditions as a lower thermal efficiency is shown by an indirect system when compared with a direct one. For thermosyphon solar water heaters, the sources of an indirect system's relative inefficiency are (Norton and Edmonds 1991); higher viscosity of aqueous glycol solutions (compared with water) reducing the natural-circulation flow rate, additional (compared with a direct system) flow resistance introduced by the heat exchanger, heat transfer resistance in the heat exchanger, and the lower specific heat capacity of the heat transfer fluid compared with water.

The performance reduction in warm collectors arising from the use of an aperture antifreeze solution can be outweighed by the enabled extended annual operation. An optimal aperture antifreeze solution concentration will exist for a particular combination of climate, solar water heater and patterns of lost water withdrawal.

For the freezing points of commercially available propylene-glycol solutions in concentrations of up to 50 % and ambient temperature distributions for each of the months when frost is recorded in London, England, critical concentrations for system survival are shown in Fig. 8.18. For a particular system and conditions (Norton and Edmonds 1991), the optimum aqueous propylene-glycol concentration is shown in Fig. 8.19 to be 25 %, which is the critical concentration for survival throughout the year. The output from an indirect system with the optimum solution operated throughout the year is approximately 12 % greater than that from a direct system used from March to October only and drained-down in winter.

An alternative to preventing freezing damage via the use of indirect anti-freeze-conveying collector-store circuits in to drain the system. This can be accomplished in two different ways;

- Drawn-down; where all water a direct system is removed in winter until weather conditions are no longer likely to cause frost.

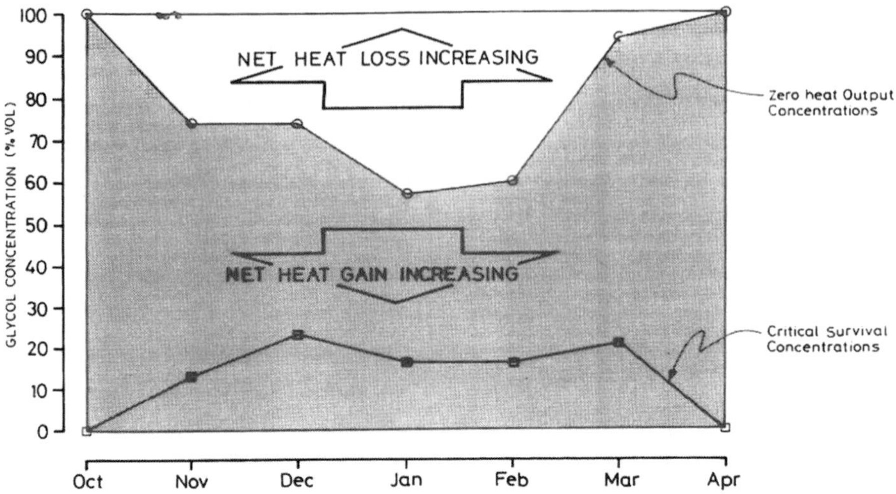

Fig. 8.18 Critical glycol concentrations for use in an indirect thermosyphon solar water heater

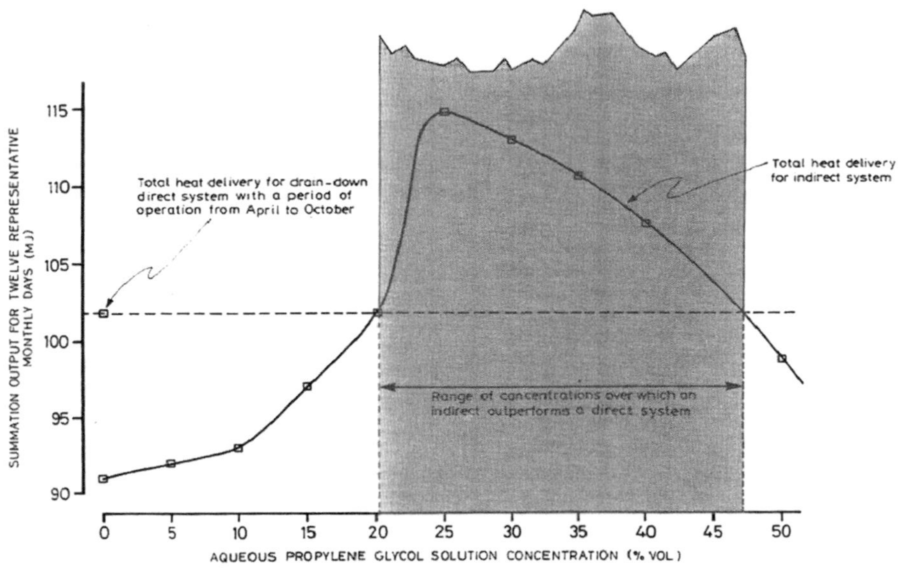

Fig. 8.19 Hot water production for different concentrations of aqueous proplylene glycol (Norton and Edmonds 1991)

- Drain-back where water only flows through in the collectors when the pump is operating. When the pump, at a lower height than the collectors, ceases to operate, the water falls back to a predetermined level in pipework within the building. The advantage over anti-freeze indirect systems of such direct drain-back systems is the efficiency gain from obviating the need for heat exchangers.

The advantage over drain-down systems is the extended duration of operation. The only disadvantages in the larger pump and additional pumping energy required. These systems have found particular use in France (Siré 2011) and the USA. In one example of the latter a drainback system has been combined with a PV-powered pump (Dontje 2007).

8.6 System Testing Methods

A performance test for solar collectors and systems, as in many other manufacturing industries, is necessary to allow the purchaser to compare different types of collectors from different manufacturers using easy-to-understand performance indices. Testing is, of course, also advantageous and necessary for the industry as a whole acting as an independent quality control mechanism for the manufacturer and providing information (test data) to allow development of their products. An acceptable testing method must be reliable and repeatable and must generate the required thermal performance data by which collectors can be assessed both technologically and economically and to allow systems to be compared. During tests, ambient and water temperatures, insolation and windspeed are recorded. Multiple or single draw-off representative profiles selected, and the corresponding efficiencies are calculated.

The first widely adopted American industry standard, was ASHRAE Standard-95 (Anon, ASHRAE 1981) applied to all solar domestic hot water systems. The water heater is exposed to controlled conditions of a specified irradiation, ambient temperature and draw-off profile and after about 3 days, the collector attains a state of equilibrium, whereby the energy withdrawn is constant. However, the test did not provide a prediction of long-term performance, nor are system parameters obtained with which a long-term performance prediction can be made. Thomas (1985) extended the ASHRAE Standard-95 to include the testing of integral passive solar water heaters using the alternative in-line heat source test. Treating the integral passive solar water heaters as a conventional collector with a very long time constant, the thermal efficiency and incident-angle response characteristics were measured. Using this data, an in-line heat source was programmed to simulate insolation according to the Standard-95 alternate test procedure.

Four distinct approaches towards testing solar water heater characteristics can be identified

- Long-term outdoor tests (Morrison and Sapsford 1983); (Fanney and Klein 1983); (Western et al. 1980).
- Short-term outdoor tests (Cooper and Lacey 1981; McLean 1978; Chinneru 1971; Reichmuth and Robison 1982)
- Indoor tests using a solar simulator (Jarnes and Proctor 1983).
- Indoor tests using electrical heating (Fanney 1984).

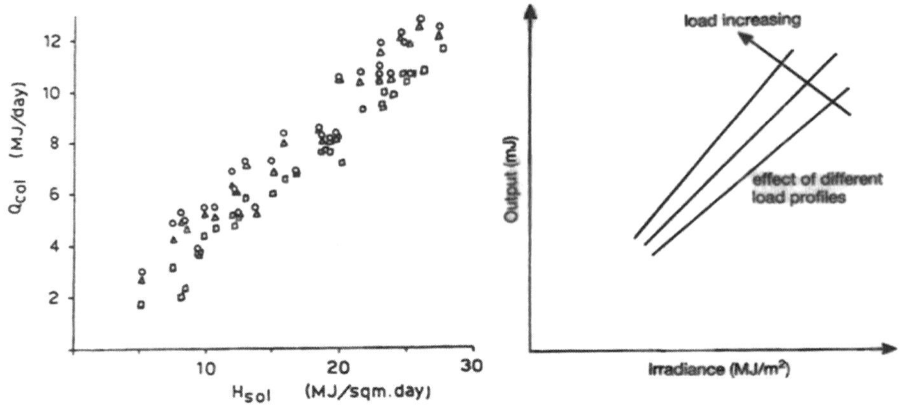

Fig. 8.20 Input/output rating method, on the left is shown a typical example of the daily results obtained and on the right is shown the effect of different hot water withdrawal loads

There are three main generic techniques for extending short test data to give longer-term overall system performance (Gilliaert and Tebaldi 1987).

- Input/Output methods – where the system is considered as a complete unit, as a "black-box" and only the inputs and outputs are relevant to the analysis. The daily irradiation incident on the system is taken as the input and the hot water delivered is the output. The results can be plotted on a simple solar input to heated water diagram as shown in Fig. 8.20, or on a diagram where the input parameter is a modified daily insolation that characterises insolation utilizability
- Correlation-based methods such as the collector loop efficiency method – where initially the efficiency of the primary loop is determined in a short system test (5 h). This parameter is then used in conjunction with two collector parameters and store parameters to predict the long-term performance.
- Simulation methods where component tests and short term system data provide parameters for a computer simulation model (Haberl et al. 2009). The simulation is then run using relevant hour-by-hour weather data to give long-term performance. This approach requires that the simulation model employed includes appropriate algorithms for components (as example particular concentrators and heat exchanges) and phenomena (for example natural circulation flow rate in a thermosyphon solar water heater).

Computer simulation requires measurement of particular parameters such as

- Thermal capacity of the collector
- Optical efficiency
- Heat loss coefficient
- Heat exchanger coefficient

These are determined by short-term field tests or tests under a solar simulator and together with data for ambient temperature, input water temperature, insolation and load demand are entered into the simulation model to obtain a long-term performance prediction (Boussemare 1986; Boussemare and Bougard 1988)

Currently systems are rated to standards EN12976 for factory-made solar water heating units and ISOIDIS 9459-5 (de Jalón et al. 2012). These tests serve as the basis for product certifications such as the "Solar Keymask" scheme in Europe (Nielsen 2007).

References

Al-Ibrahim AM, Klein SA, Beckman WA, Mitchell JW (1996) An investigation of photovoltaic powered pumps in direct solar domestic hot water systems. In: Proceedings of the ASES Solar '96, Asheville

Alt HL (1935) Sun effect and the design of solar heaters. Heating, Piping and Air Conditioning, 111–118

Andrassy S (1961) Solar water heaters. In: Proceedings of the United Nations conference on new sources of energy, Rome, pp. 20–22

Anon (1946) Putting sunbeams on the roof to work. Plumb Heat J 117:44–46

Anon (1957) Report and recommendations, inter- departmental committee on equipment in tropical staff houses and community amenities: appendix D. Government of the Commonwealth of Australia

Anon (1981) Method of testing to determine the thermal performance of solar domestic water heating systems. ANSI/ASHRAE Standard, 95–1981, ASHRAE, 1791, Atlanta

Baer S (1975) "Breadbox" water heater plans. Zomeworks, Albuquerque

Bailey WJ (1910) Solar heater. US patent no. 966, 070

Bainbridge DA (1981a) Integral passive solar water heater performance. In: Proceedings of the 6th national passive solar conference, Portland, pp 163–167

Bainbridge DA (1981b) The integral passive solar water heater book. The Passive Solar Institute, Davis

Bishop RC (1983) Superinsulated batch heaters for freezing climates. In: Proceedings of the 8th national passive solar conference, Sante Fe, pp 807–810

Boussemare C (1986) Computer simulation of the CSTG experimental test procedure for ICS DHW systems. In: Proceedings of the 4th meeting of the solar collector and systems testing group, Lisbon, pp 285–298

Boussemare C, Bougard J (1988) Contribution to the development of test procedures for integral collector storage DHW systems. In: Proceedings of the 3rd meeting, solar collector and systems testing group, Lyon, pp 285–312

Buckley S (1979) Sun up to Sun down. McGraw-Hill, New York

Burns S, Zeenni E, Guven HM (1985) Stratification and performance characteristics of a solar breadbox water heater. In: Proceedings of the 10th national passive solar conference, Raleigh, pp 430–435

Butti K, Perlin J (1977) Solar water heaters in California, 1891–1930. The CoEvolution Quarterly, 4–13

Butti K, Perlin J (1980) A golden thread. Van Nostrand Reinhold, New York

Carnes A (1932) Heating water by solar energy. Agri Eng 6(13):156–159

Charters WWS, de Forest L, Dixon CWS, Taylor LE (1980) Design and performance of solar boosted heat pumps. In: Proceedings of the Australian and New Zealand Solar Energy Society, annual conference, Melbourne

Chinnery DNW (1967) Solar water heating in South Africa. SCIR Research Report, 284

Close DJ (1962a) The performance of solar water heaters with natural circulation. Solar Energy 6:33–40

Close DJ (1962b) CSIRO Engineering Section Report ED 7, Melbourne

Cooper PI, Lacey JC (1981) Evaluation of a household solar water heating system rating procedure using a reference system for performance comparison. Solar Energy 26:213–222
Czarnecki JT, Read WRW (1978) Advances in solar water heating for domestic use in Australia. Solar Energy 20:75–80
Dalenbäck J-O (2010) Take off for solar direct heating in Europe. Polska Energetyka Sloneczna 1–4:9–15
De Beijer HA (1998) Product development in solar water heating. Renew Energy 15:201–204
De Jalon AG, Mateu E, Sallaberry F, Fritzzsche U (2012) EN-12976 guide of procedures for reliability tests of factory made systems. CENER report, CENER, Sarriguren
Deutsche Gesellschaft für Sonnenenergie (2005) Planning and installing solar thermal systems: a guide for installers. Earthscan, London
Dontje JH (2007) Designing a pv-powered drainbank solar hot water system. Homepower Am Eng 120:96–100
Duffie JA, Beckman WA (1991) Solar engineering of thermal processes. Wiley, New York
Esen M, Esen H (2005) Experimental investigation of a two phase closed thermosyphon solar water heater. Solar Energy 79:459–468
Fanney AH (1984) An experimental technique for testing thermosyphon solar hot water systems. ASME J Solar Energy Eng 106:457
Fanney AH, Daugherty BP (1997) A PV solar water heating system. ASME J Solar Energy Eng 199:126–133
Fanney AH, Dougherty BP, Kramp KP (1997) Field performance of photovoltaic solar water heating systems. ASME J Solar Energy Eng 119:265–272
Fanney AH, Klein SA (1983) Performance of solar domestic hot water systems at the National Bureau of Standards – measurements and predictions. ASME J Solar Energy Eng 105:311–321
Garg HP, Ganguli R, Puri JS, Pal RS (1972) Design and performance prediction of low cost solar water heater. Res Ind 17:125–129
Gilliaert D, Tebaldi P (1987) Development of standard test procedures for the determination of the long term performance of solar domestic hot water systems, a concerted action of the Commission of the European Communities, European Solar Collector and Systems Testing Group. In: Proceedings of the 7th meeting, Catania
Haberl R, Frank E, Vogelsanger P (2009) Holistic system testing. In: Proceedings of the ISES solar world congress, Johannesburg
Hawlader MNA, Chou SK, Ullah MZ (2001) The performance of a solar-assisted heat pump water heating system. Appl Therm Eng 21:1049–1065
Hobson PA, Lo SNG, Norton B, Probert SD (1987) Correlating the daily performance of indirect thermosyphon solar energy water heaters. In: Proceedings of the ISES solar world congress, Hamburg
Hobson PA, Norton B (1988) A design nomogram for direct thermosyphon solar-energy water heaters. Solar Energy 43:85–93
Huang BJ (1980) Similarity theory of a solar water heater with natural circulation. Solar Energy 25:191–198
Huang BJ, Hsieh CT (1985) A simulation method for solar thermosyphon collector. Solar Energy 35:31–43
Huang BJ, Lee CP (2003) Long-term performance of solar-assisted heat pump water heater. Renew Energy 29:633–639
Huggins JC, Block DL (1984) Development of solar domestic hot water systems performance prediction methodology based on system test results. Contract report FSEC-CR-100-84, Florida Solar Energy Center, Cape Canaveral
Jarnes SR, Proctor D (1983) Evolution of a standard for evaluating the thermal performance of a domestic solar hot water system. In: Proceedings of the international solar energy society solar world congress, Perth
Ji J, Hun J, Chow T-T, Yi H, Lu J, He W, Sun W (2006) Effect of fluid flow and packing factor on energy performance of wall-mounted hybrid photovoltaic/water heating collector system. Energy Build 38:1380–1387

Klein SA, Beckman WA, Duffie JA (1976) A design procedure for solar heating systems. Solar Energy 18:113–127

Lange FS (1981) And the winner is, New Shelter, May/June: 32–39

Letz T, Bates C, Perers B (2009) A new concept for combisystems characterisation, The FBC method. Solar Energy 83:1540–1549

Liu ST, Hill JE (1979) A proposed technique for correlating the performance of solar domestic water-heating systems. ASHRAE Trans 85, part (i): 96–109

Malkin MP, Klein SA, Duffie JA (1986) A design method for thermosyphon solar domestic hot water systems, Solar engineering – 1986. In: Proceedings of the ASME solar energy conference, Anaheim

Mathur KM, Khanna HL, Davey TM, Suri SP (1959) Domestic solar water heater. J Scient Ind Res 18A:51–58

McLean DI (1978) Performance testing of solar equipment. Solar energy conference. Energy Research Institute, University of Cape Town, South Africa, 18–20 Sept 1978

Merle A (1940) Solar heaters for service water. Heat Ventilation 37:22–26

Mertol A, Place W, Webster T, Greif R (1981) Detailed loop model Davis analysis of liquid thermosyphons with heat exchangers. Solar Energy 27:367–386

Mondol JD, Yohanis YG, Norton B (2009) Optimising the economic viability of grid-connected photovoltaic systems. Appl Energy 86:985–999

Morrison G (1984) Simulation of packaged solar heat pump water heaters. Solar Energy 53:249–257

Morrison GL, Sapsford CM (1983) Long-term performance of thermosyphon solar water-heaters. Solar Energy 30:341–350

Morrison GL, Tran HN (1984) Simulation of the long term performance of thermosyphon solar water heaters. Solar Energy 33:515–526

Morrison GL, Tran NH (1987) Correlation of solar water heater test data. Solar Energy 39:135–142

Morse RN (1955a) Solar water heaters. In: Proceedings of the world symposium on applied solar energy, Phoenix, pp 919–200

Morse RN (1955b) The design and construction of solar water heaters. Report E.D.l, central experimental workshops, Commonwealth Scientific and Industrial Research Organisation, Melbourne

Nielsen JE (2007) Introduction to solar keywork – the story and process in short. Solar keymark workshops, Vienna

Norton B (2011) Solar water heaters: a review of systems research and design innovation, Green. Int J Sustain Energy Convers Storage 1:189–207

Norton B, Eames PC, Yadav YP (1991) Symmetric and asymmetric linear compound parabolic concentrating solar energy collectors. The state-of-the-art in optical and thermophysical analysis. Int J Ambient Energy 12(4):171–190

Norton B, Eames PC, Mallick TK, Huang MJ, McCormack SJ, Mondol JD, Yohanis YG (2011) Enhancing the performance of building integrated photovoltaics. Solar Energy 85:1629–1664

Norton B, Edmonds JEJ (1991) Aqueous propylene glycol concentrations for the freeze protection of thermosyphon solar energy water heaters. Solar Energy 47(5):375–382

Norton B, Edmonds JEJ, Kovolos E (1992) Dynamic simulation of indirect thermosyphon solar energy water heaters. Renew Energy 2:283–297

Norton B, Probert SD (1986) Thermosyphon solar energy water heater's. Adv Solar Energy 3:125–170

Norton B, Reis A, Lo SNG, Hobson PA, Inverno AC (1988) Correlation of the performance of passive solar-energy water-heaters for the Portuguese climate. In: Proceedings of the P.L.E.A. 88 passive and low energy architecture, Porto, pp 687–692

Ong KS (1974) A finite difference method to evaluate the performance of solar energy water heater. Solar Energy 16:131–147

Ong KS (1976) An improved computer program for the thermal performance of a solar water heater. Solar Energy 18:183–191

Ong KS, Md H-E-A (2003) Performance of a R134a-filled thermosyphon. Appl Thermal Eng 23:2373–2381
Ordaz-Flores A, Garcia-Valladares O, Gomez VH (2012) Findings to improve the performance of a two-phase flat plate solar system, using acetone and methanol as working fluids. Solar Energy 86:1089–1098
Panico D, Clark G (1984) General design method of integral passive solar water heaters In: Proceedings of the 9th national solar conference, Columbus, pp 81–86
Parker GJ (1976) A forced circulation system for solar water heating. Solar Energy 18:475–479
Proctor D (1984) A generalised method for testing all classes of solar collector. Solar Energy 32 (3):387–393
Prud'homme T, Gillet D (2001) Advanced control strategy of a solar domestic hot water system with a segmented auxiliary heater. Energy Buildings 33:463–475
Rabl A (1985) Active solar collectors and their applications. Oxford University Press, Oxford
Reddy SJ (1987) The estimation of global solar radiation and evaporation through precipitation. Solar Energy 38:97–104
Reichmuth H, Robinson D (1982) An analytical model and associated test procedure for predicting the performance of batch type DHW systems. In: Proceedings of the 7th national solar conference, Knoxville, pp 963–968
Riddle DE (1985) Modelling the transient performance of natural circulation solar hot water systems. Num Meth Thermal Problems 4:1029–1039
Ruiz E, Martinez PJ (2010) Analysis of an open air swimming pool solar heating system using an experimentally validated TRNSYS model. Solar Energy 84:116–123
Runqing H, Peijun S, Zhongying W (2012) An overview of the development of solar water heater industry in China. Energy Policy 5:46–51
Schmidt T, Mangold D, Müller-Steinhagen H (2004) Central solar heating plants with seasonal storage in Germany. Solar Energy 76:165–174
Schreyer JM (1981) Resiential application of refrigerant-charge solar collectors. Solar Energy 26:307–312
Scott JE (1976) The solar water heater industry in South Florida: history and projections. Solar Energy 18:387–393
Shitzer A, Kalmanoviz D, Zvirin Y, Grossman G (1979) Experiments with a flat-plate solar water heating system in thermosyphonic flow. Solar Energy 22:27–36
Siré R (2011) Drainback systems use an efficient and safe solution for large solar domestic hot water installations; practical experience. In: Proceedings of the 5th European solar thermal energy conference, Marseille
Smyth M, Eames PC, Norton B (2001a) Evaluation of a freeze resistant integrated collector/ storage solar water heater (ICSSWH) for Northern Europe. Appl Energy 68:265–274
Smyth M, Eames PC, Norton B (2001b) Annual performance of heat retaining integrated collector/ storage solar water heaters in a northern maritime climate. Solar Energy 70(5):391–401
Smyth M, Norton B (2006) Integral collector storage solar water heaters. Renew Sustain Energy Review 10:503–538
Sobotka R (1961) Solar water heaters. In: Proceedings of the United Nations conference on new sources of energy, Rome, pp 96–99
Song ZP, Zhang HJ (1982) Prediction of system performance of solar water heaters for specified locality. Solar Energy 28:433–441
Stefanizzi P, Cardinale N, Piccininni P (2003) Economic optimisation of loss-flow domestic hot water plants. Renew Energy 28:1899–1914
Thomas WC (1985) Investigation of standard test procedures for integral storage solar domestic hot water systems. Report No VPI-E-85.5, Virginia Polytechnic Institute and State University
Troi A, Vangiouklakis Y, Korma E, Jähnig D, Wiemken E, Franchini G, Mugnier D, Egilegor B, Melograno P, Sparber W (2008) Combi+ ; identification of most promising markets and promotion of standardised system configuration for small scale solar space heating and cooling applications. In: Proceedings of the EuroSun, Lisbon

Tully N (1983) Performance prediction of integral solar water heaters. In: Proceedings of the ISES solar world congress, Perth, pp 254–258
Weiss W (2003) Solar heating for houses – a design handbook for solar combisystems. James and James, London
Weiss W, Mauthner F (2010) Solar heat worldwide. International Energy Agency, Solar Heating and Cooling Programme
Weller P, Clark G, Collins W (1985) Indoor testing of an integral passive solar water heater. In: Proceedings of the l0th national passive solar conference, Raleigh
Western BE, Benseman RF, Robinson WH (1980) The saving from domestic solar water heaters. New Zeal Energy J 53:90–93
Whillier A (1995) PhD thesis, MIT, Cambridge
Wuestling MD, Klien SA, Duffie JA (1985) Promising control alternatives for solar water heating systems. ASME J Solar Energy Eng 107:215–221
Yazdanshenas E, Furbo S (2007) Theoretical comparison between solar water/space-heating combisystems and stratified design options. ASME J Solar Energy Eng 129:438–448
Yazdanshenas E, Furbo S, Bates C (2008) Theoretical comparison between solar combisystems based on bikini tanks and tank in-tank solar combisystems. In: Proceedings of the EuroSun 2008, Lisbon
Yeh H, Chen L (1986) Analysis of a built-in storage type of solar water heater. Energy 11 (7):727–732
Young MF (1980) Solar domestic hot water systems: a comparative study and storage tank investigation. Ph.D. thesis, University of California, Davis
Zhentao L, Zhichen H (2010) The development and future of solar thermal application industry in China. Convection on China's solar thermal application industry, Hefei
Zollner A, Klein SA, Beckman WA (1985) A performance prediction methodology for integral collection-storage solar domestic hot water systems. ASME J Solar Energy Eng 107:265–272

Chapter 9
Solar Drying

> *... after our stores of oil and coal are exhausted the human race can receive power from the rays of the sun*
>
> Shuman

Inadequate preservation of food products result in 30 % loss due to spoilage. Drying is heat and mass transfer between the wet product and the heated air flow. Ideally a solar dryer should be designed and operated to dry the product uniformly and efficiently from an initial as-harvested moisture-content to a final safe storage moisture content (Fig. 9.1).

Traditional open sun-drying shown in Fig. 9.2 is neither efficient nor clean. Solar drying being more efficient, clean and controllable, shortens drying time, maintains a consistent final moisture content and reduces bacterial decay of the product.

Various types of dryers are available as shown in Fig. 9.3. The food solar-dryers commonly used in the tropics are:

- The natural convection solar dryers are constructed mostly as box type dryers of limited mass capacity.
- The forced convection type solar dryer is more efficient and can be used in the sub tropics.

The tunnel dryer construction is recommended (Esper and Mühlbauer 1993). The traditional method for fruits and vegetables drying in rural areas is to spread the products on the ground with exposure to the sun in the open air. Sun-drying method may be efficient and cheap process but has disadvantages such as contamination by dirt, insects and bacteria and loss due to wetting by rain squalls. These are usually accepted as an inherent part of the method of processing. In order to protect the products from above mentioned disadvantages and also to accelerate the time for drying the products, control the final moisture and reduce wastage through bacterial action, solar dryers can be used. The drying process preserves foods by removing enough moisture from food to prevent decay and spoilage. Water content of properly dried food varies from 5 % to 25 % depending on the type of food. Solar

B. Norton, *Harnessing Solar Heat*, Lecture Notes in Energy 18,

DOI 10.1007/978-94-007-7275-5_9,

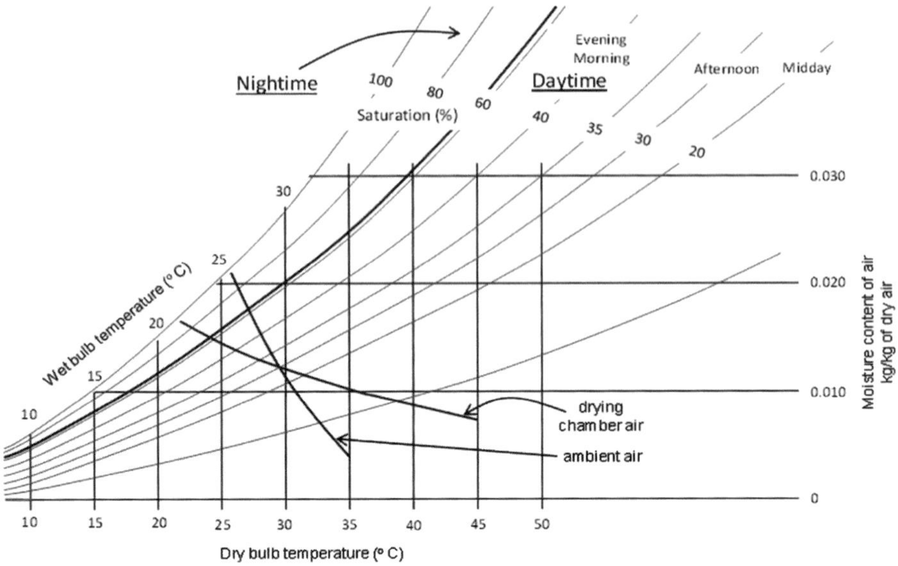

Fig. 9.1 Psychrometric representation of the drying and ambient air for a natural circulation solar dryer in Nigeria

Fig. 9.2 Open sun drying of grain

Fig. 9.3 Taxonomy of solar dryer types

drying can be considered as an elaboration of sun drying and is an efficient system of utilizing solar energy. The tropics and subtropics have abundant solar radiation. Successful drying depends upon:

- Enough heat to draw out moisture without cooking the food
- Dry air to absorb the released moisture
- Adequate air circulation to carry off the moisture

Natural convection solar dryers do not require power from the electrical grid or fossil fuels. Hence the obvious option for drying would be the natural convection solar dryers. Many studies on natural convection solar drying of agricultural products have been reported. The several designs available include

- Cabinet type solar drier suitable for drying fruits and vegetables (Sharma et al. 1995), shown in Fig. 9.4
- Mixed mode drier for drying paddy (Exell and Kornsakoo 1978; Oosthuizen 1995), shown in Fig. 9.5
- Natural circulation greenhouse dome type shown in Fig. 9.6
- Natural circulation greenhouse chimney type dryer shown in Fig. 9.7
- Glass-roof natural circulation solar dryer shown in Fig. 9.8
- Polythene tent solar dryer shown in Fig. 9.9

These dryers have been widely tested in the tropical and subtropical countries. Considerable studies on simulation and optimization have also been reported. With indirect natural convection solar dryers, care in both design and operation is essential if satisfactory drying rates are to be achieved. Typical box type solar dryer is suitable for drying of 10–15 kg of fruits and vegetables (Sharma et al. 1995).

Forced convection solar dryers include:

- Indirect forced convection solar dryer (Oosthuizen 1996), as shown in Fig. 9.10
- Greenhouse type solar dryer (Esper and Mühlbauer 1993; Janjai 2004), an example with interior drums is shown in Fig. 9.11

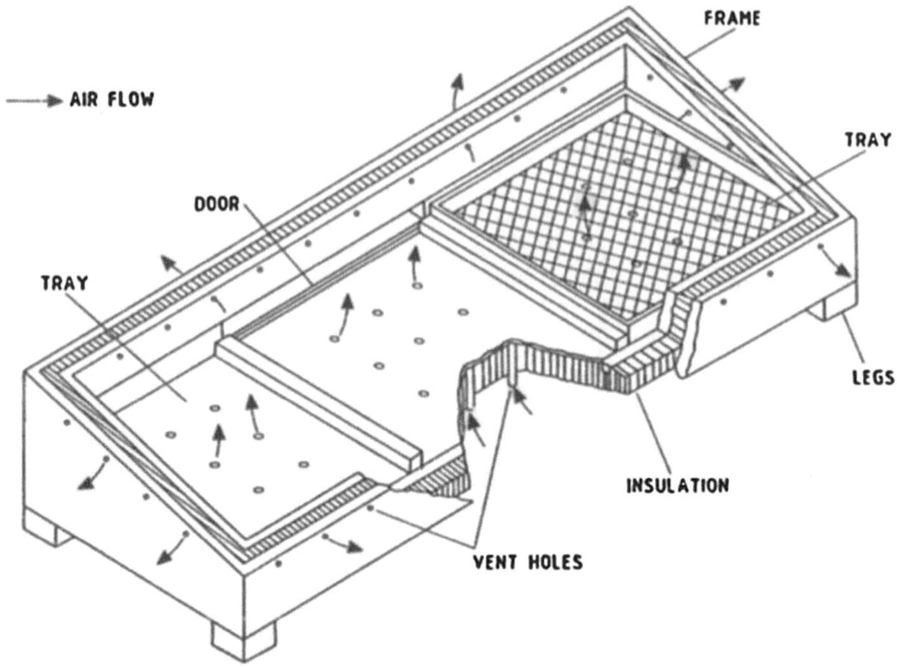

Fig. 9.4 Natural-circulation solar-energy cabinet dryer

- Roof integrated solar drier (Janjai et al. 2006), shown in Fig. 9.12
- Transparent roof type shown in Fig. 9.13

Numerous tests in the different regions of the tropics and subtropics have shown that fruits, vegetables, cereals, grain, legumes, oil seeds, spices, fish and even meat can be dried properly in the solar tunnel dryer. Solar dryers can though be cost effective because relatively low skills are required to construct, operate and maintain the dryers using low cost readily available local materials for construction. However, care, experience and/or informed advice is required to design a dryer and select appropriate materials.

In both forced convection and natural convection indirect solar grain drying systems (Simate 2003; Bala and Janjai 2009) grain at the bottom of the bed dries rapidly while that at the top is still wet due to condensation. Optimisation of indirect natural convection systems has resulted in long systems with thin grain beds.

Before the decision for a dryer is formed the climatic, technical and economic environment in which the dryer will be operated has to be analysed. Not all dryers fit for all produce. Fruits are in general dried for export, in such cases driers which can handle up to 100 kg daily and more are required in order to develop a sustainable business situation. Dried vegetables are mainly suited for local markets; in such cases smaller dryers with a daily drying capacity up to 10 kg are sufficient. Fresh fruits and vegetable finished cut and sliced for drying deteriorate rapidly in

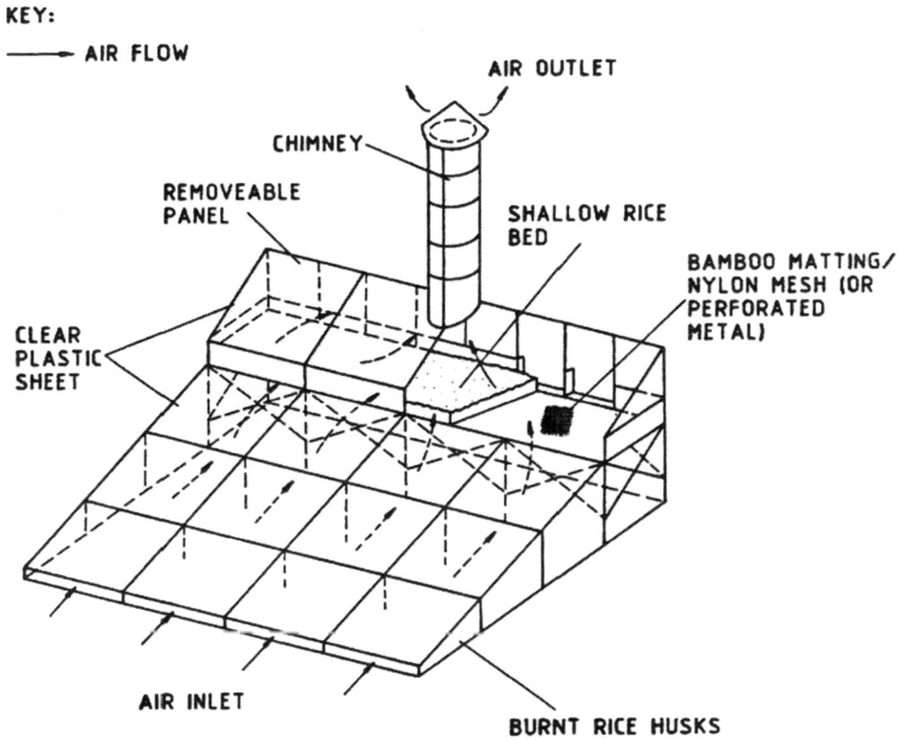

Fig. 9.5 Mixed mode solar rice dryer by natural circulation, Exell et al.

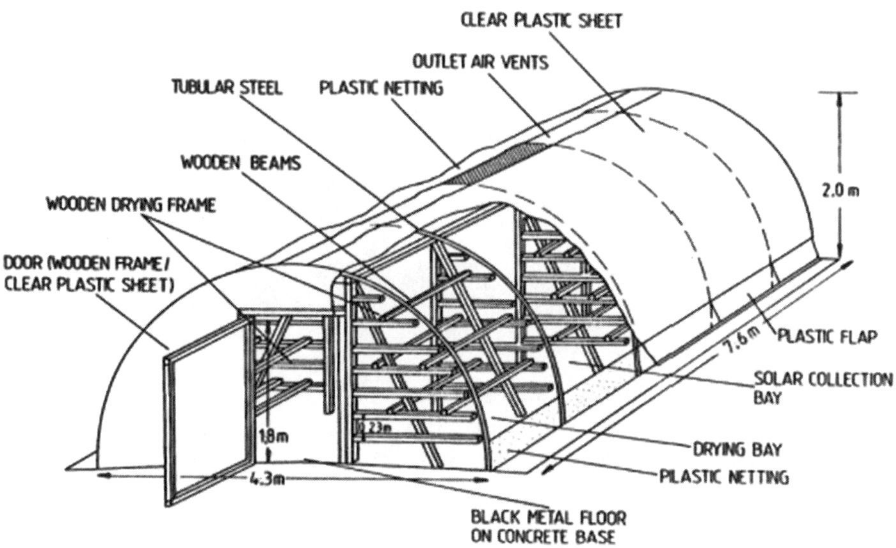

Fig. 9.6 Natural circulation dome type dryer

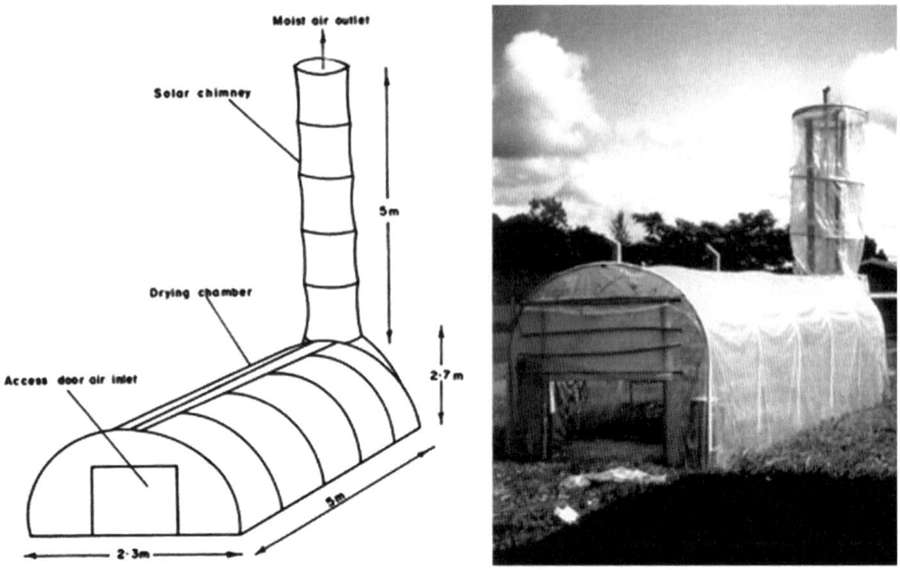

Fig. 9.7 Natural circulation chimney-type solar dryer

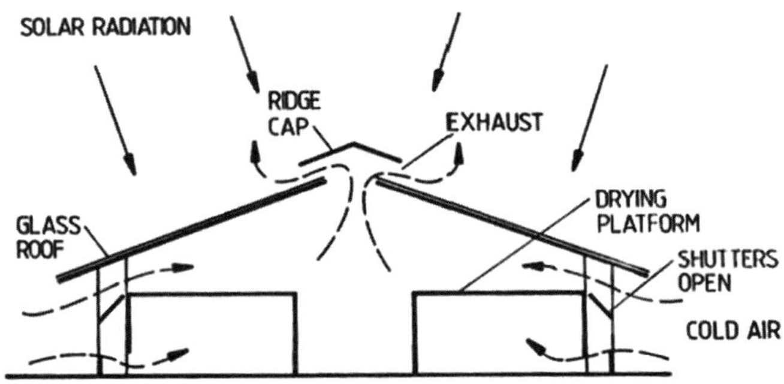

Fig. 9.8 Schematic illustration of a natural-circulation glass-roof solar-energy dryer

hot climates. Drying should reduce returned water content to a safe storage level to avoid spoilage during nightimes.

Successful solar drying requires:

- A suitable, and suitably prepared product;
- Are the insolation and ambient humidity conditions during the harvesting period favourable for drying (sunshine duration around 6–8 h)
- A dryer capacity and product sources market supply demands
- The ability to observe international focus quality standards
- Packaging and storage facilities suitable for intermediate storage of dried products

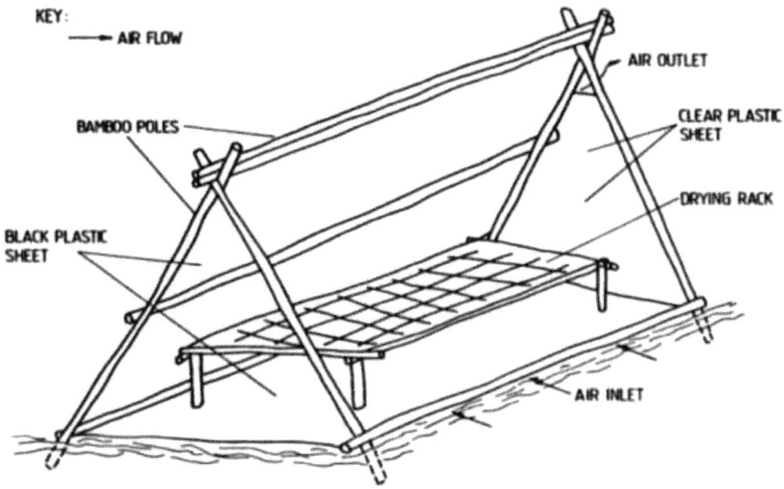

Fig. 9.9 A natural-circulation integral-type polythene tent dryer

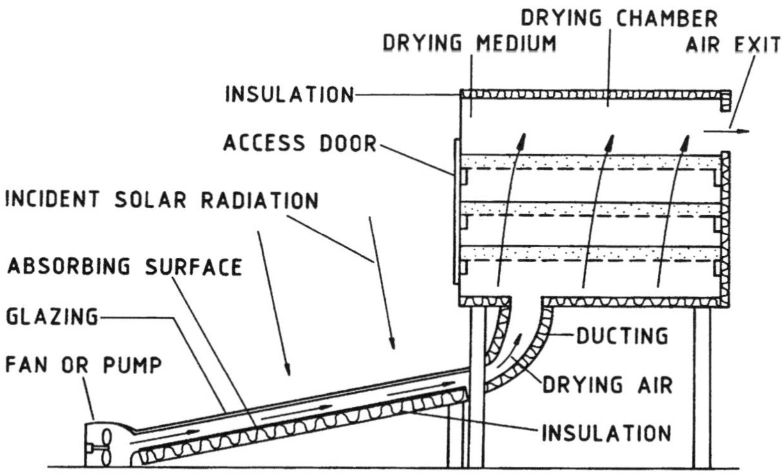

Fig. 9.10 A distributed-type active solar-energy dryer

The dryer should be operated in optimal technical conditions, it should always be positioned in full sunlight, air inlets and outlets should be free of any obstructions. Air inlets should be protected from dust and any fumes. Trays on which the product is positioned for drying and the whole interior of the dryer should always be clean and free from any residues. In order to assure optimal hygienic conditions measures have to be taken to keep especially insects and rodents out of the dryer. Special care has to be given to the transparent cover it should be clean, free of dirt and dust and tightly stretched over the dryer frame also when the dryer is not in operation. It is not advisable to operate dryers when the environmental relative humidity very high

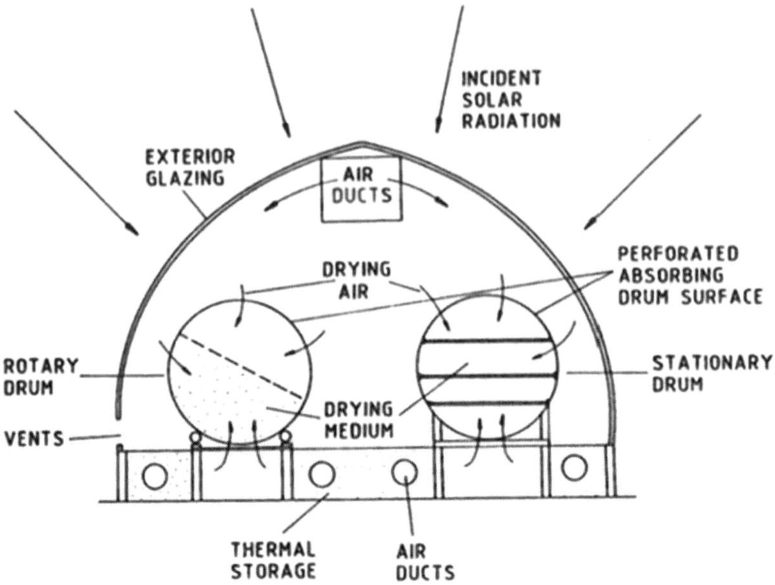

Fig. 9.11 An active interior-drum-absorber greenhouse solar dryer

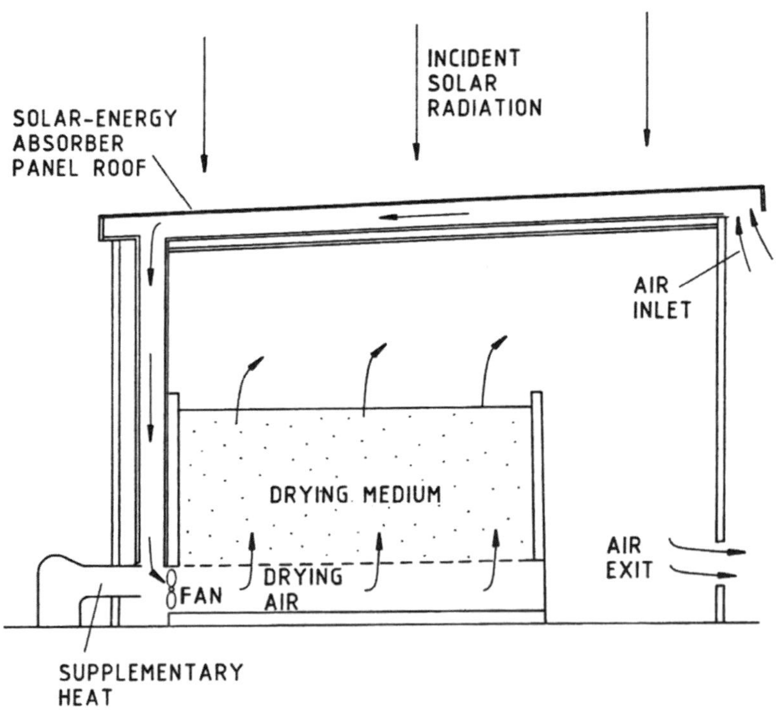

Fig. 9.12 An active collector-roof solar-energy storage dryer

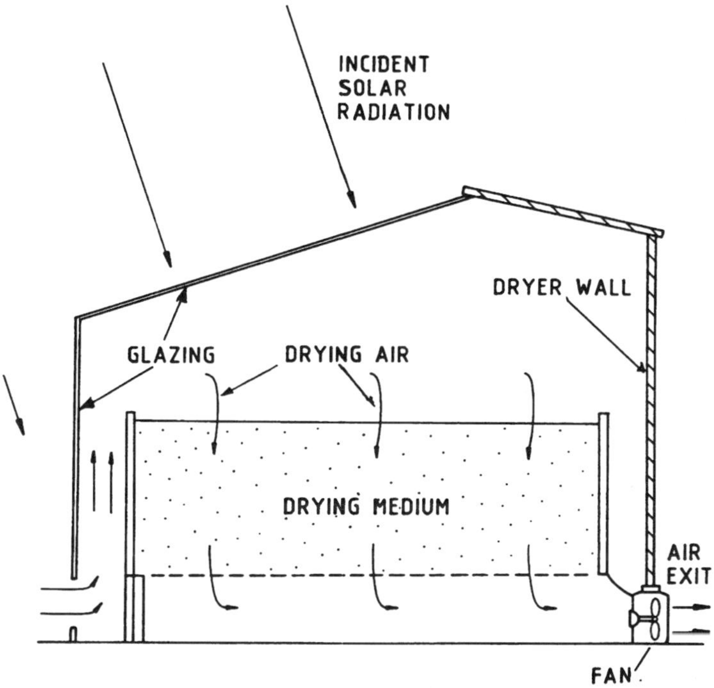

Fig. 9.13 A forced-convection transparent-roof solar barn

since the relative humidity of the air used for drying has a pronounced effect on the water removal capacity of the air. The linear velocity of the air within the drying chamber is an important consideration for sweeping away the saturated stagnant boundary layer around the material being dried. Without sufficient air movement, drying will be slow which may allow time for spoilage to occur. The temperature of the drying air should optimally be in the range of 50–55 °C for most fruits and vegetables. For drying herbs, 45 °C is a reasonable maximum. Temperatures that are too high (above 60 °C) may cause case hardening which will reduce the effectiveness of the dryer and lead to future problems with the dried material. The dryer trays and other food contact surfaces should be made of materials appropriate for this purpose e.g. wood or stainless steel. When loading fresh product on the trays it is recommended not to exceed a thickness of about 0.5–0.6 cm since it allows for good removal of the water and diffusion to the surface while giving a good thickness of finished product. In case of products with a waxy cuticle e.g. berries it is in addition helpful to slice the products or disrupt the outer surface otherwise. The product should be monitored frequently during the drying process. It should be turned at regular intervals and racks may need to be rotated and their positions changed to enhance drying uniformity. The direction of airflow should be across the surfaces of the material being dried rather than from a bottom up direction through the dryer. If this cannot be avoided, then a solar-powered fan

Table 9.1 Food crops suitable for solar drying

Fruits	Apples, apricots, banana, dates, figs, lemon, mango, oranges, papaya, peaches, pears, pineapple, plums
Vegetables	Broccoli, cabbage, carrots, celery, chives, corn, garlic, green beans, mushrooms, leeks, okra, onions, peas, peppers, potatoes, shallots, spinach, sweet potatoes, tomatoes

inside the dryer may create horizontal flow patterns (which will also address the stagnant boundary layer problems). In cases where the material cannot be dried in a single day, to store the partially dried product overnight, either the product should (i) remain in the dryer with inlets and outlets closed to prevent the ingress of humid air during the cooler night-time hours or (ii) be removed for storage in a dedicated facility overnight. The latter may be required especially when drying products with high moisture contents (such as tomatoes with ca. 95 % wet basis moisture) that thus have long drying times.

Products with high moisture content have a lower yield (kg of dried product per 100 kg of fresh product) compared with products of moderate moisture content; For example 100 kg of tomatoes at 95 % moisture will give about 6 kg of finished dried product whereas 100 kg of fresh apples at 84 % moisture will deliver over 17 kg of dried product at 10 % moisture. Fruits and vegetables suitable for drying are summarized in Table 9.1.

In developing for solar dryers model that it is assumed:

(i) There is uniformly mixed air inside the dryer.
(ii) Crop drying behaviour can be represented by thin layer drying correlations.
(iii) Specific heat capacities of air, cover, ground and product are constant.
(iv) Fraction of solar radiation lost through the north wall is negligible and absorptivity of air is negligible.
(v) A time interval is employed in the numerical solution of the system of equations that ensures constant air conditions prevail.

The rate at which energy is stored in the cover is equal to the convective heat energy transfer rate between the air inside the dryer and the cover plus the rate of radiation heat transfer between the sky and the cover plus the thermal convecture heat transfer rate between the cover and ambient air plus the radiation heat transfer rate between the crop and the cover plus the rate of solar radiation absorbed by the cover. This energy balances in a greenhouse dryer cover (Bala and Janjai 2009) are shown in Figs. 9.14 and 9.15.

Where C_{pc} is the specific heat capacity of the cover ($Jkg^{-1}K^{-1}$), M_c is the mass of the cover, T_{a}, T_{am}, T_c, T_p and T_s are the temperatures (K) of the internal air, ambient cover, product and sky respectively, A_c and A_p the areas (m^2) of the cover and product respectively, h are the relevant heat transfer co-efficients ($Wm^{-2}K^{-1}$) (with radiative heat transfer co-efficients, h_r calculated by iteration for the applicable temperature range). I_t is the insolation (Wm^{-2}) and α_c is the absorptance of this cover. The energy balance of the air within the layer is equal to the rates of convective heat transfer between the crop and air and floor and air, plus the sensible

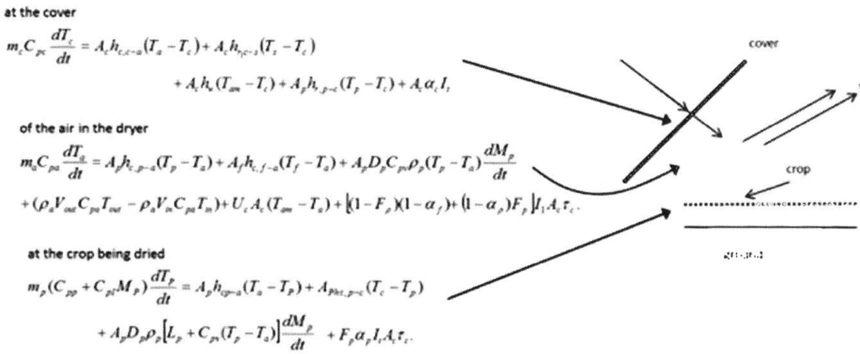

Fig. 9.14 Energy balances in a greenhouse-type solar dryer

Where L is the latest heat of vapourisation of moisture from the product(Jkg^{-1}). The conductive heat flow into the floor equals the rate of solar radiation absorption on the floor plus the rate of convecture heat transfer between the air and the floor. Where k_f is the thermal conductivity of the floor$(Wm^{-1}K^{-1})$.

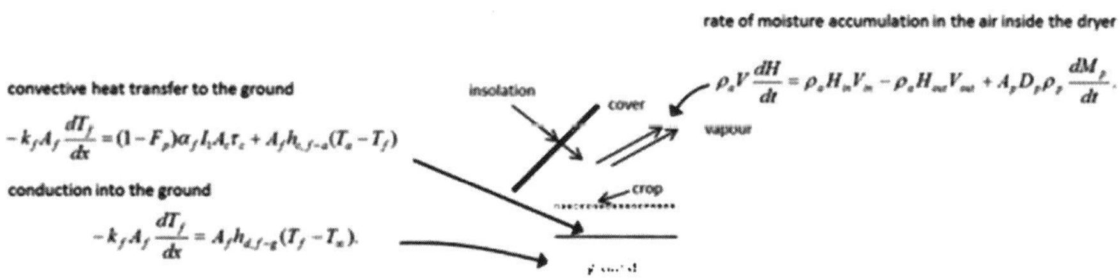

Fig. 9.15 Heat stored in the ground and moisture accumulated in the air of a greenhouse-type solar dryer

heat transfer from the crop to air plus the heat associated with flow of air in and out of the dryer taking account of heat loss from air in the layer to ambient and solar energy collected where M_a and C_{pa} are the mass and specific heat of air in kg and $Jkg^{-1}K^{-1}$ respectively, $M_\rho \rho, A_\rho, D_\rho \, and \, T_\rho$ are the mass, density, area, depth, temperature and specific heat capacity of the product in kg, kgm^{-3} m^2, m, K and $Jkg^{-1}K^{-1}$ respectively, α_f, A_f and T_f apply to the floor, V_{in} and V_{out} are the inlet and outlet flow rates (m^3s^{-1}) respectively and T_{in} and T_{out} the corresponding temperatures (K). F_p is fraction of insolation incident on the product.

The rate at which thermal energy is stored in the crop equals the rate of thermal energy convective heat transferred to the crop plus the rate of thermal energy received from cover by the product due to radiation plus the rate of thermal energy lost from the crop due to sensible and latent heat loss from the crop plus the rate of thermal energy absorbed by the crop. In Fig. 9.14 L is the latest heat of vapourisation of moisture from the product (Jkg^{-1}), The conductive heat flow into the floor equals the rate of solar radiation absorption on the floor plus the rate of convecture heat transfer between the air and the floor. Where k_f is the thermal conductivity of the floor $(Wm^{-1}K^{-1})$.

The rate of thermal energy flow into the floor due to conduction assumes a temperature T_∞ at a depth is interseasonally invariant. The rate of moisture accumulation in the air inside dryer is equal to the rate of moisture inflow into the dryer due to entry of ambient air minus the rate of moisture outflow from the dryer due to exit of air from the dryer plus rate of moisture removed from the crop inside the dryer H is the humidity ratio with suffixes "in" and "out" referring to the dryer inlet and outlet respectively. The radiative heat transfer coefficient from the cover to the sky $(h_{r,c-s})$ is given by

$$h_{r,c-s} = \varepsilon_c \sigma (T_c^2 + T_s^2)(T_c + T_s).$$

where ε_c is the emittance of the cover, σ is the Stefan-Bolzmann constant $(Wm^{-2}k^{-4})$. Radiative heat transfer coefficient between the crop and the cover $(h_{r,p-c})$ is given by

$$h_{r,p-c} = \varepsilon_p \sigma (T_p^2 + T_c^2)(T_p + T_c).$$

where ε_p is the emittance of the crop. As $h_{r,c-s}$ and $h_{r,p-c}$ are functions of temperatures, these are computed iteratively at each time during a simulation. The sky temperature (T_s) is

$$T_s = 0.552 T_{am}^{1.5},$$

Convective heat transfer coefficient from the cover to ambient due to wind h_w is

$$h_w = 5.7 + 3.8 V_w.$$

Convective heat transfer coefficient inside the solar greenhouse dryer for either the cover or product and floor (h_c) is computed from

$$h_{c,f-a} = h_{c,c-a} = h_{c,p-a} = h_c = \frac{Nu k_a}{D_h}$$

where D_h is given by: $D_h = \frac{4WD}{2(W+D)}$

where W and D are the width and height of the dryer (m) respectively and the Nusselt number is $Nu = 0.0158\mathrm{Re}^{0.8}$, Re is the Reynolds number $\mathrm{Re} = \frac{D_h V_a}{\nu_a}$

Where V_a is air speed in the dryer and ν_a is kinematic viscosity of air. The overall heat loss coefficient from the greenhouse dryer cover (U_c) is computed from

$$U_c = \frac{k_c}{\delta_c}$$

Where k_c and δ_c are the thermal conductivity $(W_m{}^{-1}K^{-1})$ and thickness (m) of the cover respectively.

Thin layer drying correlations are obtained for particular crops by determining experimentally the best fit to an equation of the from

$$\frac{M(t) - M_e}{M_o - M_l} = X \exp(-Yt^z)$$

Where $M(t), M_o$ and M_e are moisture contents (as percentage of dry bulb) at time t, originally and at equilibrium respectively. X, Y and Z are constants. Different values for X, Y and Z are found for different crops and often for different methods of crop preparation before drying.

References

Bala BK, Janjai S (2009) Solar drying of fruits, vegetables, spaces, medicinal plants and fish, developments and potentials. In: International solar food processing conference, pp 1–24

Esper A, Mühlbauer W (1993) Development and dissemination of solar tunnel. Dr. Solar World Congress, Budapest, 22

Exell RHB, Kornsakoo S (1978) A low cost solar rice dryer. Appropriate Technol 5:23–24

Janjai S (2004) Personal communication. Silpakorn University, Nakhon Pathom

Janjai S, Srisittipokakun N, Bala BK (2006) Experimental and modelling performances of a roof integrated solar drying system for drying herbs and spices. Energy 33:91–103

Oosthuizen PH (1995) The design of indirect solar rice dryers. J Eng Int Dev 2(1):20–27

Oosthuizen PH (1996) An experimental study of simulated indirect solar rice dryer fitted with a small fan. J Eng Int Dev 3(1):22–29

Sharma VK, Colangelo A, Spagna G (1995) Experimental investigation of different solar driers suitable for fruits and vegetable drying. Renew Energy 6(4):413–424

Simate IN (2003) Optimization of mixed-mode and indirect-mode natural convection solar dryers. Renew Energy 28:435–453

Chapter 10
Solar Cooling, Refrigeration and Desalination

In spite of the seeming paradox [it is] possible to utilize the rays of the sun to make ice

Augustin Mouchot (1879)

10.1 Introduction

There are many options available to us to integrate solar energy into the process of "cold" production. The different routes via which the scorching heat of the sun may be converted into a desirable chill are delineated in Fig. 10.1.

One important application of solar refrigeration is to cool vaccine stores. The need for such systems is greatest in peripheral health centres in rural communities of developing countries. In the absence of main-grid electricity, the vaccine cold chain can be extended to these areas through the use of autonomous solar-energy operated vaccine stores in rural health centres there where electricity is either not available or its supply is too unreliable to operate a vaccine store. Four litres of vaccine immunises 150 infants and their mothers. A 30 l capacity vaccine store will be adequate for storing the vaccine and in addition some life-saving medicines. The solar operation of widely-used refrigerators, working on a compression cycle as shown in Fig. 10.2, use solar energy into electricity using photovoltaic electricity converted to a.c. electricity.

High temperature thermal energy produced by concentrating collectors can be transformed into mechanical energy via a heat engine to drive a refrigerator compressor. Unfortunately the high temperature concentrating collectors most suited to this application will need daily tracking of the sun, rendering small scale systems complex and expensive. In addition, high temperature heat storage is essential if the unit is to continue to run in the absence of sunlight.

B. Norton, *Harnessing Solar Heat*, Lecture Notes in Energy 18,

DOI 10.1007/978-94-007-7275-5_10,

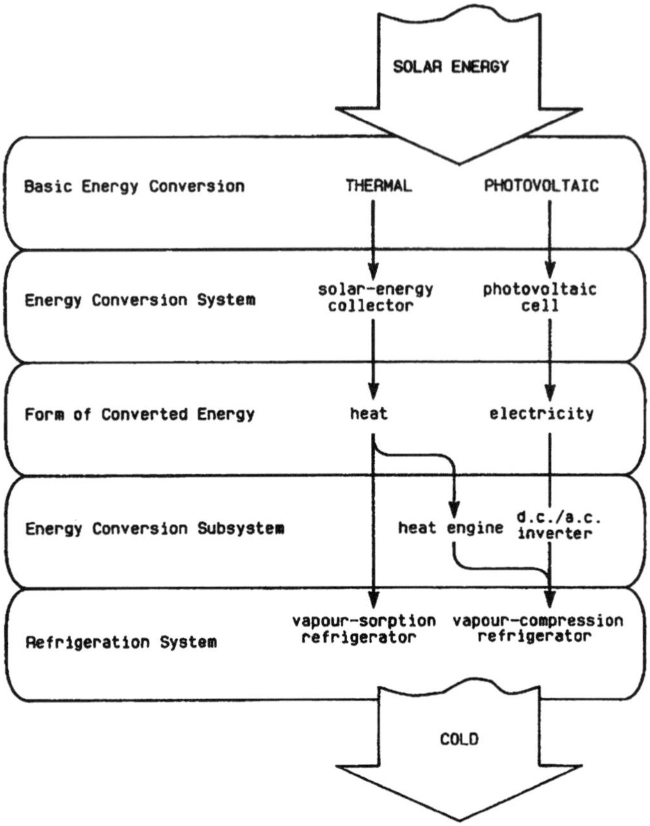

Fig. 10.1 Routes via which cold can be produced from solar energy

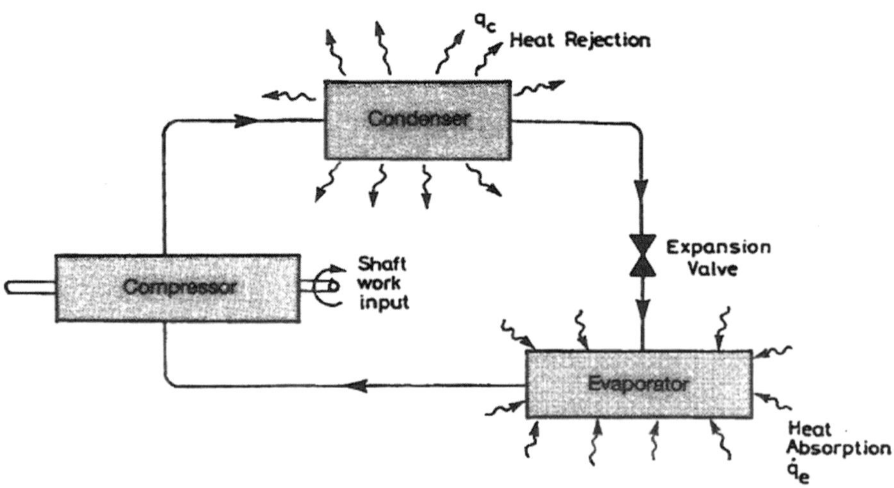

Fig. 10.2 Schematic diagram of a vapour compression refrigerator

10.2 Adsorption Units

Porous solids, termed adsorbents, can physically and reversibly adsorb large volumes of a vapour, termed the adsorbate. Though this phenomenon, called solid adsorption, was recognised in the nineteenth century (McBain 1932) its practical application in the field of refrigeration is relatively recent. The concentration of adsorbate vapours in a solid adsorbent is a function of the temperature of the "working pair" (i.e. mixture of adsorbent and adsorbate) and the vapour pressure of the adsorbate. The dependence of adsorbate concentration on temperature, under constant pressure conditions, makes it possible to adsorb or desorb the adsorbate by varying the temperature of the mixture. This forms the basis of the application of this phenomenon in the solar-powered intermittent vapour adsorption refrigeration cycle as shown in Figs. 10.3 and 10.4.

Worsoe-Schmidt (1985a, b) developed a unit that worked on the principle of solid absorption using calcium chloride as the absorbent and ammonia as the refrigerant. A reversible chemical reaction takes place when the refrigerant is absorbed by the solid absorbent. This results in physical changes in the mixture. When ammonia is absorbed into calcium chloride, swelling of the mass up to 400 % takes place (Bechtoft-Neilsen and Worsoe-Schmidt 1977). To overcome this a small quantity of another salt was added to calcium chloride and then ammonia was mixed to prepare a paste, which was subsequently heated in a controlled manner to produce a new granulated absorbent. The desorption of ammonia (from a mixture of

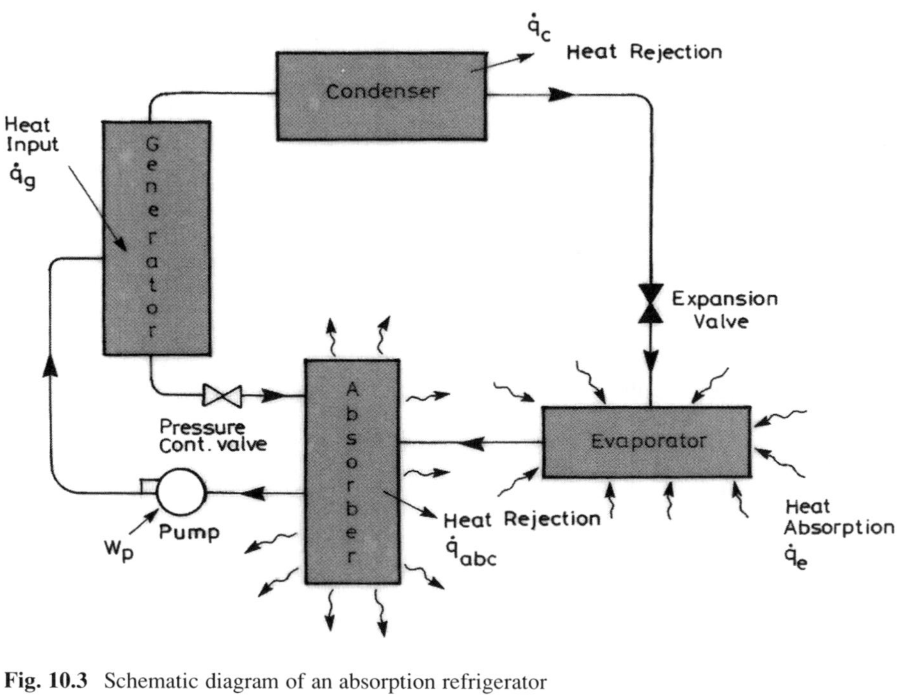

Fig. 10.3 Schematic diagram of an absorption refrigerator

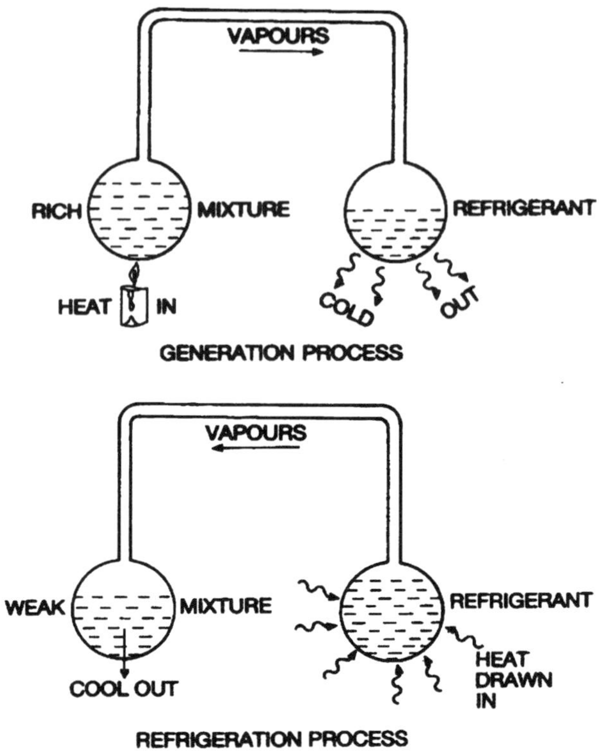

Fig. 10.4 The intermittent vapour sorption refrigeration cycle and removing water vapour from ammonia during generation

$CaCl_2.8NH_3$), as shown in Fig. 10.5, takes place at selected temperatures. For the design condensation temperature of 40 °C two molecules are desorbed at 88 °C. Four of the remaining six molecules cannot start desorbing before the temperature reaches 106 °C. One of the remaining two ammonia molecules needs the mixture to be heated to 250 °C to be released. This means that if the maximum temperature of a particular day cannot reach 106 °C only a third of the possible total amount of ammonia will be generated. The heat of adsorption and desorption for the working pair is high; almost twice the latent heat of evaporation of ammonia. The consequence is that a large combined collector/absorber area is required, thus increasing the initial capital cost and size of the unit.

Ice made during the night in a separate unit is used to refrigerate the vaccine cabinet (Worsoe-Schmidt 1985a, b). Thus the storage cabinet is inconveniently indirectly cooled with additional capital cost is incurred.

As ammonia systems inherently work at high pressures, heavy gauge metal is required for fabrication of components. Ammonia being very corrosive only permits the use of steel in the system.

When different solid adsorption working pairs (i.e. zeolite-water, zeolite-methanol and activated carbon-methanol) are compared the activated-carbon-methanol working pair performed the best as shown in Fig. 10.5. A reported COP of a system designed to produce ice of 0.14, with a calcium chloride-ammonia working pair (Delgardo

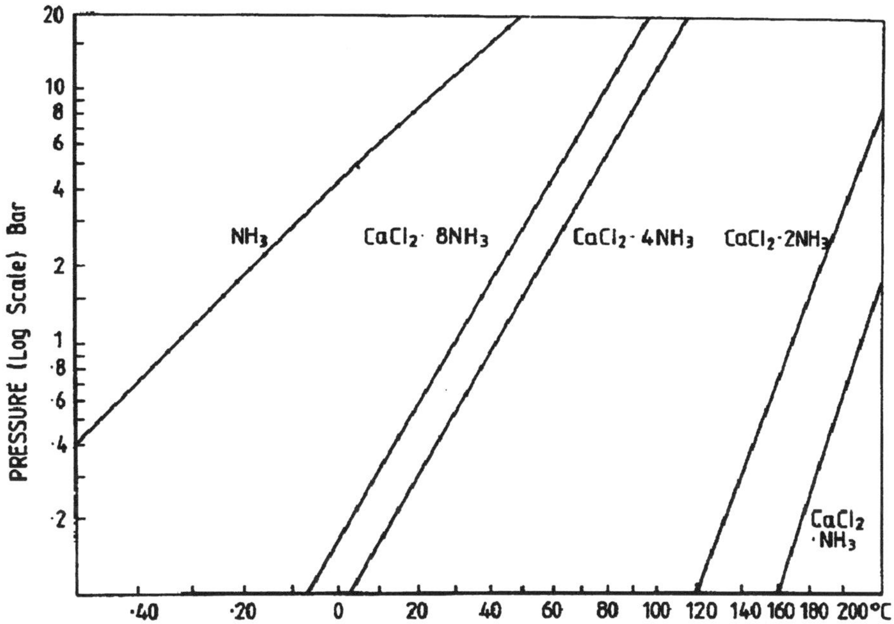

Fig. 10.5 Isosteres for ammonia and calcium chloride

et al. 1983) better than the 0.11 achieved by Worsoe-Schmidt's unit. A silica gel-water adsorption pair can also be used (Hildbrand et al. 2004; Luo et al. 2010).

An adsorbent-refrigerant working pair for a solar refrigerator requires the following characteristics: (1) a refrigerant with a large latent heat of evaporation, (2) a working pair with high thermodynamic efficiency, (3) a small heat of desorption under the envisaged operating pressure and temperature conditions and (4) a low thermal capacity (Fig. 10.6).

10.3 Absorption Units

A refrigerant (e.g. ammonia) which has a great affinity to water is employed. The ammonia, dissolved in water, is pumped to the generator where it is solar-heated until this "strong" solution reaches the required operating pressure. The refrigerant vapour is driven from the strong solution when the strong solution (in the generator) attains the generating temperature, leaving a lower concentration solution in the generator. The condensed ammonia vapour is then stored as liquid refrigerant at the condenser pressure. Via an expansion valve the liquid refrigerant is throttled to the evaporator pressure. This produces the cooling required. The ammonia refrigerant passes to the vapour state in the evaporator and upon reaching the absorber, is absorbed by the weak solution arriving from the generator. The refrigerant is redissolved, forming a concentrated solution which is pumped back to the generator for the cycle to repeat.

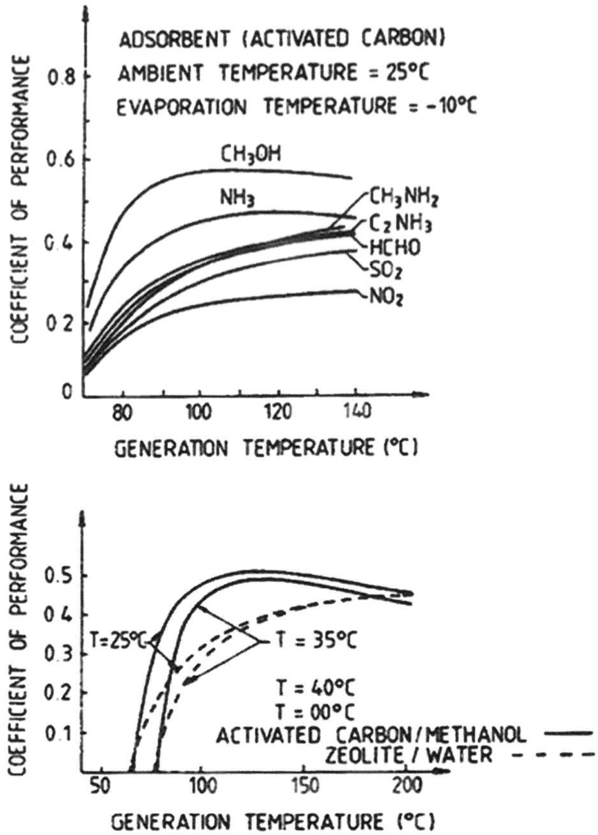

Fig. 10.6 Variation of coefficient of performance with generation temperature for a range of desorbent pairs

By circulating cooling water to absorb the heat of solution in the absorber vessel, the rate of ammonia vapour absorption equals the rate of evaporation, to maintain a constant system pressure. If such cooling water is at 30 °C, then typically the aqueous ammonia solution in the absorber would be maintained at 35 °C. The evaporation of ammonia in the evaporator then occurs at a temperature of 12 °C and at a saturation pressure of 6.58 bar.

Solar-operated Plattons-Muns cycle water-ammonia refrigerators have been developed (Chinnappa 1961; Farber 1970; Swartman and Swaminathan 1970; Exell and Kornsakoo 1981; Exell et al. 1984, 1986; Uppal et al. 1986a, b). The efficiency of such systems is limited by the condensing temperature.

10.4 Solar Desalination

In solar desalination solar energy is used either to produce heat directly to evaporate water or is converted to electricity to fulfil a membrane process as shown in Fig. 10.7 (Zheng et al. 2006). Solar desalination is becoming a mature technology (Abdelrassoul 1998).

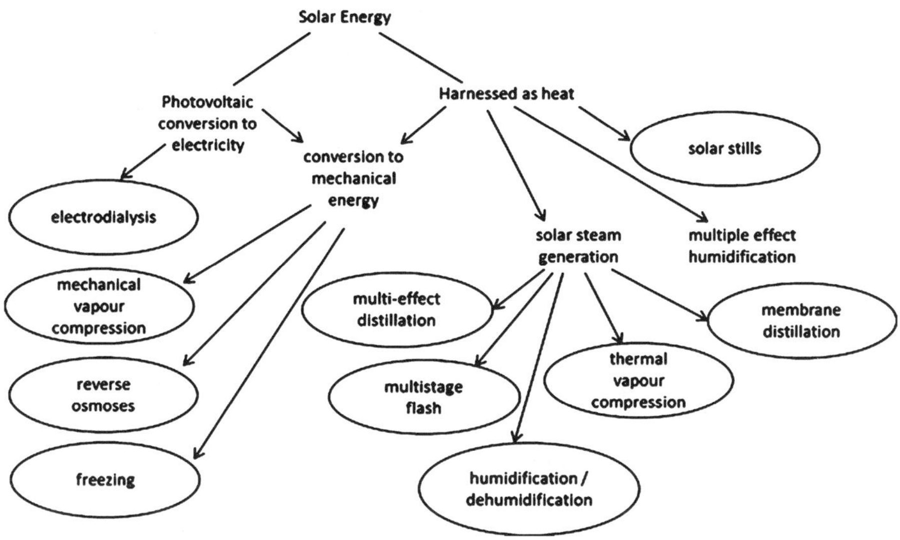

Fig. 10.7 Taxonomy of solar desalination systems

Table 10.1 Total disused salts in water sources

Type	Total dissolved solids
Freshwater	Up to 1,500
Brackish water	1,500–10,000
Salt water	>10,000
Seawater	10,000–45,000
Standard seawater	35,000

Freshwater differs substantially from seawater by the amount of salts as shown in Table 10.1. Solar desalination units used for freshwater production based on the evaporation of water by solar heating or the use of polymer membranes that allow the selective passage of water via semi-permeable membranes or ions via ionic exchange membranes. Most evaporative processes need thermal energy together with mechanical and electrical energy to power circulation pumps and auxiliary systems (Delyannis 2003).

In multi-stage flash desalination seawater evaporates in flashing chambers in which the pressure is maintained below the feed vapour pressure. The production of vapour cools the remaining brine. The flashing stages are arranged in series by decreasing pressure. The large amounts of vapour produced are condensed on the external surface of a tube bundle and the latent heat of condensation heat released to the condensing bundle is used for preheating the seawater feed stream. The pre-heated seawater is further heated to the required temperature by low-pressure steam from provided either directly or via an intermediate heat transfer oil, from a concentrating solar collector.

In multi-effect desalination evaporation occurs on the external surface of a tube bundle which is heated by solar-generated steam condensing inside its tubes. The vapour produced in one effect is then used as the steam in each following effect, each operating at lower pressures and, therefore, at a lower boiling temperatures. Up to 10–12 kg of distilled water is produced per kg of steam fed into the first effect.

To enhance overall energy use effectiveness, multi-effect desalination units can be coupled to Thermal Vapour Compression (TVC) and Mechanical Vapour Compression (MVC) vapour recovery systems. In TVC system, part of the vapour from the last effect is compressed by a steam ejector powered by solar generated mid-pressure steam. This is used as motive steam for the first effect increasing performance by up to16 kg of distilled water per kg of steam fed into the ejector. In a MVC system all the vapour exiting from the last effect is compressed mechanically to provide the steam for the first effect. Humidification-dehumidification and membrane distillation processes are also used for desalination.

As the latent heat of fusion to produce ice from water is much lower than water's heat of vaporisation, the production of frozen desalted water, via the removal of heat from a salt solution, is likely to be less energy intensive than most current desalination techniques. This approach has not been researched extensively to-date. Freshwater can be extracted from salt water without the water undergoing a phase change. For example in pressure-driven membrane separation processes of which the Reverse Osmosis (RO) has been adopted most widely adopted. In RO a semi-permeable membrane allows the passage of water whilst concomitantly preventing the passage of salts into the permeate stream. Salts can also be removed from brackish salty waters by using devices able to capture the salt ions either by ionic exchange membranes in an electrodialyses process in which salt ions are forced to pass through the membranes and thus separated from the main water stream, or ionic exchange resins, where salt ions are captured from water within the solid matrix of the resin.

Concentrating solar power and seawater desalination can be combined. Multi-Stage Flash distillation can be operated with steam extracted from turbines or supplied by boilers. Low-temperature Multi-Effect Distillation using steam extracted from a turbine as shown in Fig. 10.8. These systems have the largest scope for performance improvement.

- Reverse osmosis (RO) supplied with electricity from steam power plant or combined gas/steam power cycle

The cooling required for condensation of exhaust steam from the turbine can be supplied by water from the desalination system replacing the conventional cooling system. Concentrating solar power plants used for desalination of seawater must, obviously, be close to the sea. Unfortunately this can affect adversely both initial and operating costs. Land close to the sea is more expensive than the arid interior locations often selected for solar power plants. As coastal climates may be subject to sea fog, curtailing the prevalence high direct insolation, a shorter duration of productive operation ensues. Continuous operation of solar desalination plants requires hybridisation of power plant with usually gas or biofuel and/or diurnal high temperature energy storage.

A passive solar still is a sealed glazed unit at atmospheric pressure in which solar heated water vaporates from a basin (Dunkle 1961; Malik et al. 1982). The water vapour condenses on the glass cover to be collected by internal gutters. A variety of designs are available as illustrated in Fig. 10.9. As the glass cover cools rapidly at night, distillation of the warm feed water continues after sunset (Malik and Tran 1973; Onyegegbu 1986).

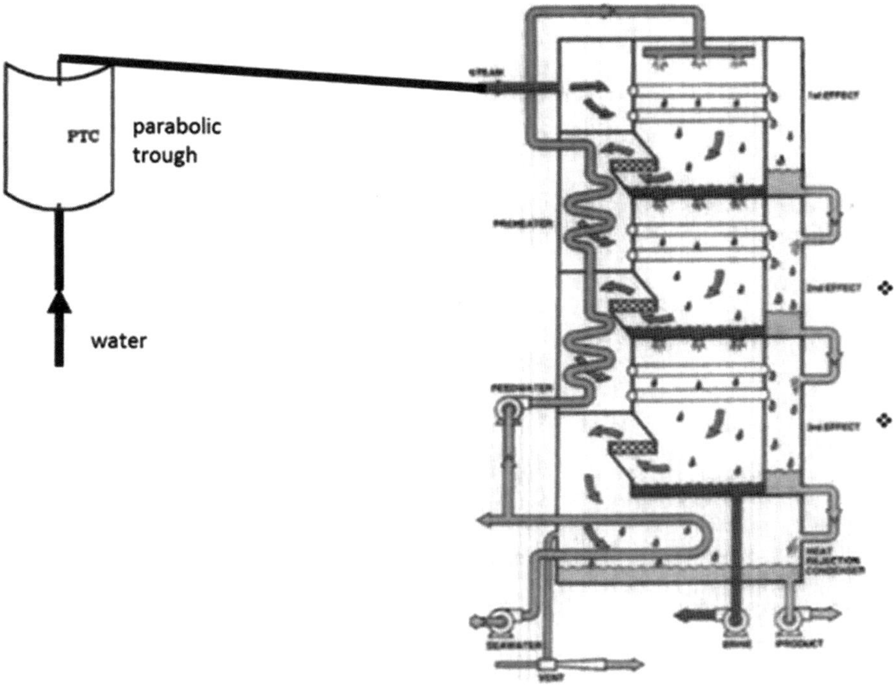

Fig. 10.8 A solar steam generated multiple effect stack evaporation

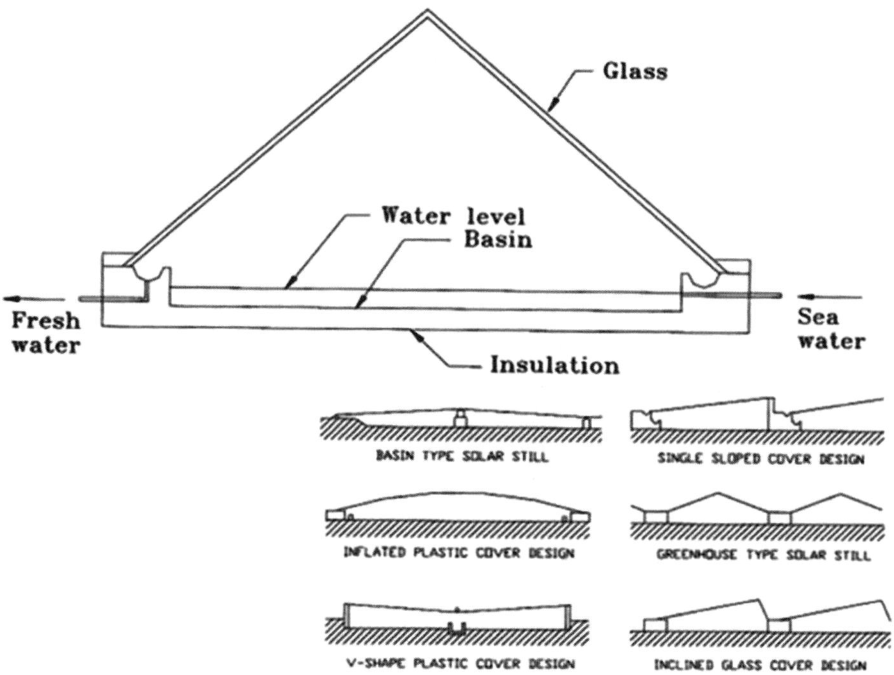

Fig. 10.9 Passive solar stills

References

Abdelrassoul RA (1998) Solar water desalination potential for economic solar desalination in the Middle East. Renew Energy 14:345–349

Chinnappa JCV (1961) Experimental study of the intermittent vapour absorption cycle employing the refrigerant – absorbent systems of ammonia – water and ammonia – lithium nitrate. Solar Energy 5:1–35

Delgardo R, Choiser A, Grenier Ph, Ismail I, Meunier F, Pons M (1983) Etude du cycle intermittent charbon actif/methanol en vue de la Realisation d'une machine a fabriquer de 1a glace fonctionnant a l'Energie solaire. Report Commissions E1-E2, International Institute of Refrigeration

Delyannis E (2003) Historic background of desalination and renewable energies. Solar Energy 75:357–366

Dunkle RV (1961) Solar water distillation; the roof type still and a multiple effect diffusion still. In: Internation developments in heat transfer, ASME proceedings in international heat transfer conference. Part V. Boulder, pp 895–902

Exell RHB, Kornsakoo S (1981) Design and testing of a solar powered refrigerator, Research report 126. Asian Institute of Technology, Bangkok

Exell RHB, Kornsakoo S, Oeapipatanakul S, Chanchaona S (1984) A village-sized solar refrigerator. Research report 172, Asian Institute of Technology, Bangkok

Exell RHB, Bhattacharya SC, Upadhaya YR (1986) Solar cooling for cold-storage applications using solid dessicants and adsorbents. Presented at the workshop on physics of solar energy, New Delhi

Farber EA (1970) Design and performance of a compact solar refrigeration system, Paper no. 6/58, ISES conference, Melbourne

Hildbrand C, Dind P, Pons M, Buchter F (2004) A new solar powered adsorption refrigerator with high performance. Solar Energy 77:311–318

Luo H, Wang RL, Dui Y (2010) The effects of operation parameter on the performance of a solar powered adsorption chiller. Appl Energy 87:3018–3022

Malik MAS, Tran VV (1973) A simplified mathematical model for predicting the nocturnal output of a solar still. Solar Energy 14:371–385

Malik MAA, Tiwari GN, Kumar A, Sodha MS (1982) Solar distillation. Pergamon Press, Oxford

McBain JW (1932) The sorption of gases and vapours by solids. Routledge, London

Onyegegbu SO (1986) Nocturnal distillation in basin-type solar stills. Appl Energy 24:29–42

Swartman RK, Swaminathan C (1970) Further studies on solar powered intermittent absorption refrigerators. Paper No 6/114, ISES conference, Melbourne

Uppal AH, Norton B, Probert SD (1986a) Solar energy stimulated absorption refrigerator for vaccine storage. Appl Energy 25:167–174

Uppal AH, Norton B, Probert SD (1986b) Solar absorption-cycle refrigerator for vaccine storage. In: Proceedings of the UK-ISES conference, solar energy for developing countries, power for villages, London, pp 28–36

Worsoe-Schmidt P (1985a) Solar refrigeration at the village level based on a solid-absorption cycle. In: Proceedings of the solar energy society of India, National Solar Energy Convention, 1984, Bhopal

Worsoe-Schmidt P (1985b) Some results from the development of a solid-absorption refrigerating system, Paper D-18. Proceedings of the commission of the European communities, absorption heat pumps congress, Paris

Zheng H, He K, Yang Y, Chen Z, Li H (2006) Study on a multi-effects regeneration and integral-type solar desalination unit with falling film evaporation and condensation processes. Solar Energy 80:1189–1198

Chapter 11
Greenhouses

"Just living is not enough, said the butterfly
One must have sunshine, freedom
and a little flower for a companion"

The Butterfly, Hans Christian Anderson (1861)

11.1 Overview

The driving process in plant growth is photosynthesis; the transformation of co2 and Ho, using solar energy, to carbohydrates and oxygen. Metabolism involves the conversion of these carbohydrates into energy which maintains the life processes of the plant. Photosynthesis is highly sensitive to environmental factors. The basic function of a greenhouse is to provide environmental conditions which either make possible, or accelerate, the process of photosynthesis. The interior microclimate of a greenhouse can be characterised by its temperature, the illumination therein and the constituents of the enclosed atmosphere including water vapour, carbon dioxide and pollutants, such as oxides of nitrogen and sulphur. The supply of nutrients is a further factor affecting the rate of plant growth, and should be taken into account when designing to achieve optimal conditions. These requirements vary according to the particular plant species and its stage of growth. The method and economic viability of creating a specified environment depend on the prevailing ambient conditions and the value of the crop eventually to be harvested. A given greenhouse in a particular climate can produce an environment suitable for a certain botanical species yet the same greenhouse, in another location or at a different time of year, may be unsuitable for that same plant. So, to some extent, the plant varieties to be grown should be chosen to suit the artificial environment which can be achieved economically within the greenhouse.

The protected cultivation of crops may be traced to long ago as the Babylonians and ancient Chinese (van den Muijzenberg 1980). In Europe, the evolution of

B. Norton, *Harnessing Solar Heat*, Lecture Notes in Energy 18,
DOI 10.1007/978-94-007-7275-5_11,

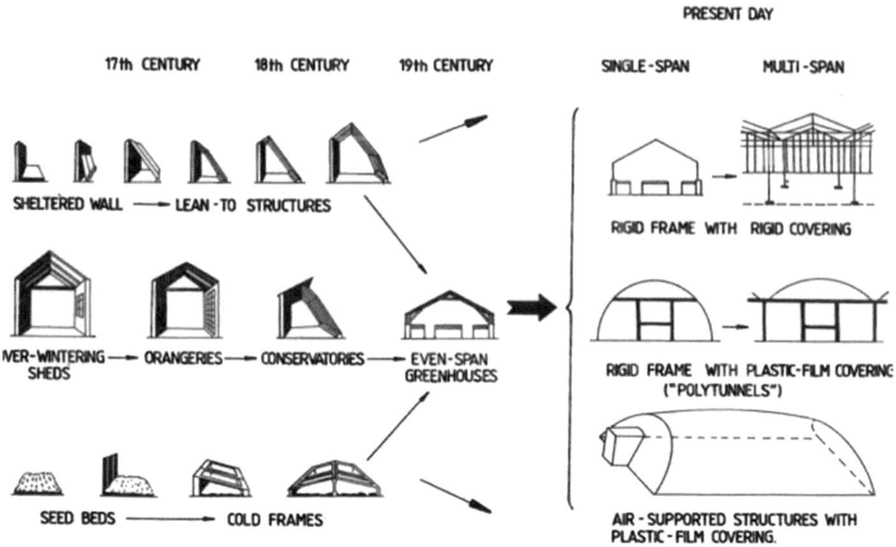

Fig. 11.1 Evolution of the greenhouse

greenhouses can be traced from the sixteenth century along three lines of development (i) hot-beds with glazed covers for growing semi-tropical fruits, such as melons; (ii) lean-to structures built onto south-facing walls in order to provide added protection for grapevines; and (iii) over-wintering sheds for sub-tropical plants, such as small orange trees as shown in Fig. 11.1. The latter developed into orangeries and conservatories, culminating in great cast-iron and glass structures such as the Crystal Palace in London during the Victorian era. The first even-span greenhouse, a symmetrical structure fully-glazed on both sides of its longitudinal axis, was built in the late eighteenth century.

The need for frost protection was the main impetus for the development of covered growing areas and over-wintering sheds. Heat was derived from solar radiation, and additionally, in larger buildings, open-fires and stoves were sometimes used. Furnaces with flues to distribute the heat came into common use during the eighteenth century, followed by steam, hot-water and warm-air systems in the nineteenth century. Root-zone heating, using decomposing animal manure and bark-pits, dates back to pre-Renaissance times and persisted, on a commercial scale, until early this century. Underground heating by (i) sub-surface flues was first tried in the eighteenth century; (ii) steam and hot-water pipes in the nineteenth century; and (iii) electrical-resistance coils in the twentieth century. Over-wintering sheds often incorporated a high degree of thermal insulation and shutters – the latter were closed at night and can be regarded as the antecedents of modern thermal screens, whose purpose is to reduce the rate of heat loss from greenhouses, especially during the night. Both internal and external screens or shades were recorded as being used for greenhouses during the eighteenth century (van den Muijzenberg 1980). Electricity also enabled air-conditioning and controlled ventilation, which were introduced during the 1930s. The influence of carbon dioxide gas concentration in the artificial environment was investigated as early as the 1920s, but it was not until the 1960s that flue gases from methane-burning heaters were used as a source of supplementary carbondioxide for the internal environment. Humidification and evaporative cooling of greenhouses were introduced in the 1950s (van den Muijzenberg 1980).

11.2 Greenhouse Materials

As an alternative cover material to glass, oiled paper was tried in the Netherlands during the late eighteenth century, and it was in common use in Japan well into the twentieth century. Cellulose-based film was also tried, without much success, early in this century. After the second World War, plastic (i.e. polymeric) materials became more readily available. Techniques for using rigid plastic sheets were developed, and PVC and polyethylene films are now often attached internally to the greenhouse framework to create an insulating air-space between the outer cover and the protected artificial environment. Cheap 'polytunnels' and the less common air-supported greenhouses have been developed. These utilise a lightweight transparent plastic film as the cover material.

The internal environment of a greenhouse heats up primarily due to the short-wave solar insolation being transmitted through the cover and absorbed by the internal surfaces of the greenhouse. These surfaces re-emit the radiation at far longer wavelengths, to which the cover, traditionally of glass, is relatively opaque. This theory was questioned long ago as 1909 (Wood 1909). The heat trapped by this radiative effect alone would cause a temperature rise inside a typical greenhouse of only about 4 °C (Businger 1963). Thus radiation rectification is a significant effect, but the *suppression of convection losses* by the presence of the enclosure is the major cause of the temperature rise inside a greenhouse. Generally the long-wave radiation rectification mechanism amounts to only 10 + 25 % of the total, so-called, 'greenhouse effect'.

The thermal performances of greenhouses covered with plastic films have been compared with those covered with glass (Bowman 1962). The rate of heat loss from a polyethylene-covered greenhouse can be 10 + 15 % higher than from a similar glass greenhouse: when the cover is dry, this difference can be attributed largely to the transparency of polyethylene to long-wave radiation. Apart from cost considerations, the most important property of a cover material must be its light transmittance. In a typical commercial glasshouse, up to 20 % of the incident sunlight can be obscured by opaque structural elements (van den Kieboom et al. 1981). The non-perfect transmittance of the glass further reduces the illumination at ground level inside the greenhouse. The proportion of incident sunlight which is transmitted into a greenhouse varies both with the season of the year and with the time of day. For typical commercial glasshouses, estimates of the overall transmissions range from 65 % to 80 % (Whittle and Lawrence 1959; Morris et al. 1972;

Harnett et al. 1971; van den Kieboom et al. 1981). The remainder is obscured by structural elements, or absorbed or reflected by the glass. A 70 % average transmission in a summer period can be achieved for an air-supported polyethylene structure (Canham 1967). Over 60.5 % transmission is expected for a double-skin rigid polytunnel greenhouse over a period of 2 years (O'Flaherty and Grant 1984; Van den Kieboom et al. 1981). Transmission is reduced by:

- Condensation on the inner surface of the cover, which under some conditions could persist during the day, reduces transmission (Canham 1967). This reduction is more pronounced with plastics than with glass because the higher angle of contact between the water globules and the plastic, leads to greater reflection (Morris et al. 1972).
- Dirt and dirt settling on the cover, and
- Algae growing on moist surfaces.

An ingenious Venetian-blind suspended from the roof apex intercepts and reflects downwards the light which would otherwise pass through the greenhouse (Critten 1985).

11.3 Mathematical Modelling the Internal Environment in a Greenhouse

An energy balance can be performed on each individual component of the system by calculating the steady-state rates of heat transfer between them to predict the long-term hour-by-hour quasi-steady-state conditions attainable. These components shown in Fig. 11.2 include the cover/fabric, the vegetation, the soil surface, and the internal air.

This technique is similar to modelling thermophysical environments in buildings occupied by people, except that, in a greenhouse, because of the high rates of water exchange between the plants and their environment, the energy content of the water vapour must be taken into consideration. The heat-transfer relationships for a greenhouse have been assembled in various dynamic simulation models (Businger 1963; Walker 1965; Takakura et al. 1969; Iwakiri 1971; Kimball 1973, 1986; Froehlich 1976; Van Bavel and Sadler 1979a; Slack and Clark 1973; Ahmadi and Glockner 1982). All of these models consist of a set of energy-balance equations one for each greenhouse component (Seginer and Levav 1971). The resulting set of simultaneous equations is solved for temperature, because this is usually the most significant parameter in determining the rate of plant growth. The thermal capacity of the ground is normally several orders of magnitude greater than that of the other components of the greenhouse, the heat stored in the ground is therefore significant. Heat stored in other components can usually be neglected, and, for transient analyses, their thermal responses may be regarded as rapid enough for them to be effective in a steady-state condition. Mass balances for the water vapour and carbon

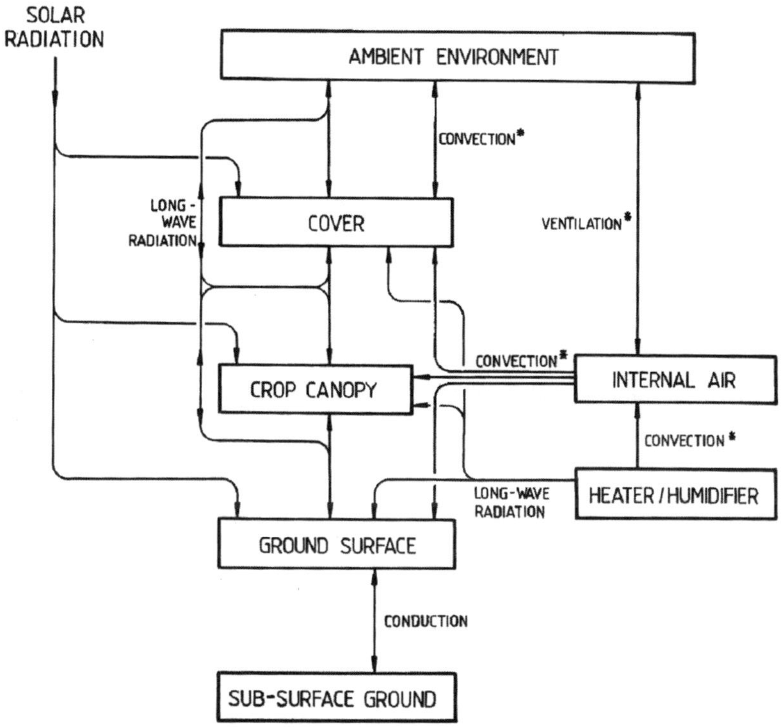

Fig. 11.2 Heat and mass transfer between the principle elements of a greenhouse

dioxide content of the internal air can be treated in a similar manner. To solve the ensuent non-linear simultaneous equations numerical-approximation methods, using iterative techniques, are employed.

An alternative approach is to determine that time-dependent response of the greenhouse's environment to external influences. As the proportionality of this response is a characteristic of the greenhouse structure as a whole no consideration is required of the complexity of the interactions between the various components of the greenhouse. Experimental data from several greenhouse types enables this approach to be used accurately when a similar greenhouse is under consideration (Takakura 1967). This approach has also been used for the adaptive control of a greenhouse environment (Udink ten Cate 1985).

For a greenhouse, a high rate of solar gain is achieved by using a cover material which transmits a high percentage of solar insolation. Most of the transmitted insolation will be absorbed by the vegetation and the soil, though it is inevitable that some will be reflected. Assuming that the sky is thermally isotropic and that the emittance of the celestial sphere is unity the net rate of radiation loss from the cover, per unit ground area can be calculated. Long-wave radiation is emitted from objects within the greenhouse, and the radiation loss from the greenhouse includes radiation allowed through the cover material. Many plastics are transparent to long-wave radiation, to which low-iron glass is opaque. The rate of heat transfer by convection is highly dependent on wind speed and, to a lesser extent, on the wind direction in relation to the greenhouse. Wind tunnel tests on bluff bodies, representative of greenhouse shapes, have shown that the steady-state convective heat-transfer coefficient h can be as high as 100 W/m^2K for an 18 $m^{-1}s$ free-stream wind velocity measured above the boundary layer (Iqbal and Kharty 1977).

Heat exchanges with the ambient environment by ventilation in conventional greenhouses are due to wind-driven infiltration, forced ventilation and buoyancy-driven ventilation. In air-supported greenhouses under normal operating conditions, only forced ventilation is present. The overall air-change rate with respect to the ambient environment is the most important parameter in determining the ventilation heat exchange. Because of the generally high rate of moisture transfer between the crop and its environment, the energy content of water vapour exchange with the external environment due to ventilation should be taken into account It is assumed, usually that the latent heat of the water vapour does not vary with temperature. Conduction losses can be accounted for simply by assuming that a constant temperature always persists at some depth below the ground's surface.

Lateral heat transfers by conduction through the ground require a specialised treatment, but in most cases where large structures are involved, can be neglected. The heat transfers between individual components within a greenhouse can be treated in a similar manner (Businger 1963; Seginer and Levav 1971; Kimball 1973, 1986; Ahmadi and Glockner 1982).

11.4 Auxiliary Heating Environmental Control and Energy Storage in Greenhouses

The most favourable orientation and optimal combination of thermal insulation to achieve desired internal collectors minimum depends to a large extent on local micro-climate weather patterns. Crop type, hanvert targets, annual and decimal variations in solar radiation intensify, diffuse fractions and prevalent winter wind directions and speeds all influence appropriate greenhouse construction and the chosen orientation. In Ireland, for example heating can account for nearly 50 % of the costs of tomato production (O'Flaherty and Maher 1978) with the leaf temperature, rather than air temperature, which governing the rate of plant growth. However, as leaf temperature is readily measured accurately, most temperature-control systems in greenhouses monitor the air temperature which is then related empirically to the rate of plant growth. The main source of heat for any greenhouse should be directly from solar radiation. However, greenhouses can require supplementary heating systems for periods during which solar heat gains are insufficient. Radiative heating systems for greenhouses directly heat the crop, without heating the intervening air. In a heating system, if radiation is to be the main form of heat transfer, then usually the heating elements must be at temperatures exceeding

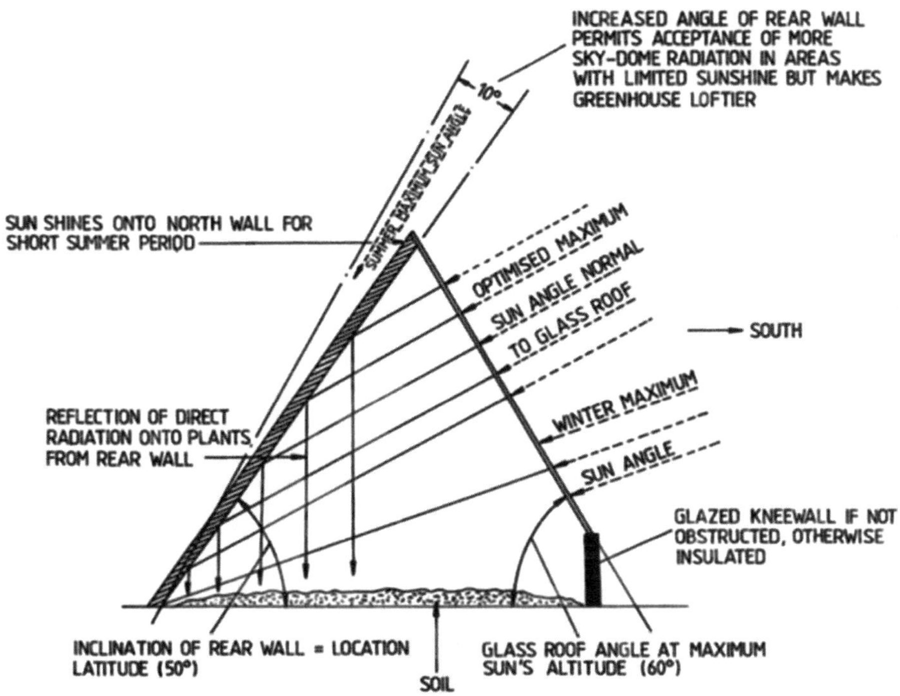

Fig. 11.3 A "solar greenhouse"

250 °C. In order to avoid scorching the plants, heating elements need to be located an appropriate distance above the crop. Radiative heat transfer to plants is significant in greenhouses utilising heating coils that usually would convey hot water at 70 °C, although convection remains the major mode of heat transfer from such systems. Heating coils which lie in contact with the ground have the advantage of providing some heat directly, by conduction, to the soil.

When convective heat transfer to the growing plants is the major mode of heating, there are clear advantages in raising the temperature of replenishing air before it enters the greenhouse. The, otherwise cold, stream of air entering the greenhouse can be warmed, with judicious use of extra heating coils distributed around the inside of the greenhouse, temperature gradients therein could be virtually eliminated. Commercial greenhouses have been heated using the relatively low-grade heat, at up to 40 °C, from the cooling thermal power-stations (Miller 1977; Pile et al. 1979; Boyd et al. 1980).

Even in temperate climates, there is frequently sufficient solar energy incident on a greenhouse during the year to satisfy the total annual heating requirements. Though, as shown in Fig. 11.3, "solar greenhouse", designs have been proposed which have high instantaneous solar gains, the need is usually to collect and store surplus heat for use later.

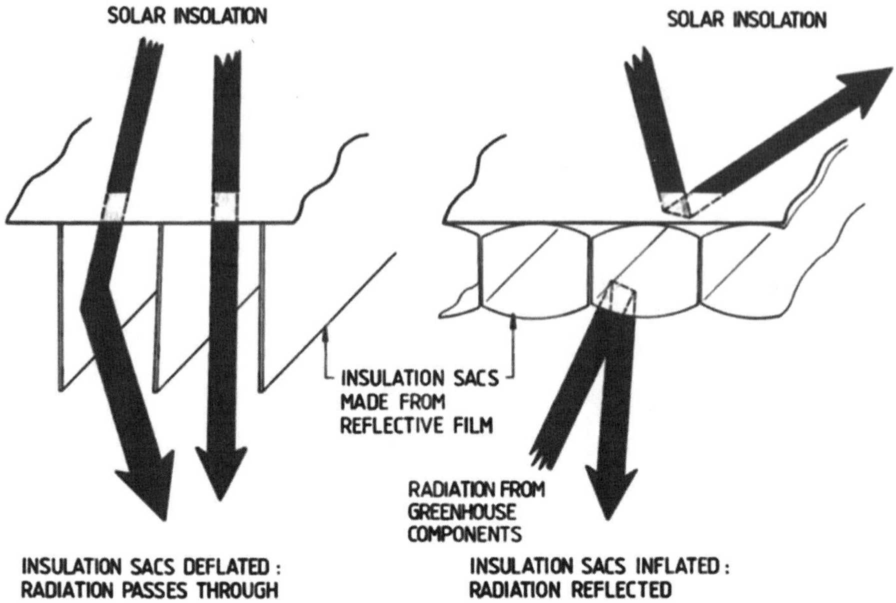

Fig. 11.4 Variable insulation and transmittance provided by inflatable sacs

Inter-seasonal storage generally requires a very large store, and is rarely justifiable commercially. Diurnal heat stores are smaller, for example, an air-heating solar collector with a rockpile as the thermal store for surplus energy (Fuller et al. 1984). Such a system could be particularly attractive where high insolation levels are common during the daytime and the nights are cold. The ground itself is a convenient medium for storage. A significant contribution can be achieved by ducting warm exhaust air from the greenhouse under the ground's surface, and an appropriate underbench sensible heat storage using water and a heat exchanger have been devised (Mears and Baird 1976). Solar ponds have also been used for greenhouse heating (Short et al. 1976).

Opaque thermal screens are often used in greenhouses, and can result in night-time energy savings of up to 60 % (Bailey 1981a, b; Meyer 1983). At night they are usually drawn across the greenhouse, above the crop canopy, in order to reduce radiation losses and to create a near-stagnant layer of air beneath the roof. However, as badly-fitting internal screens can cause cold draughts of air to circulate around the greenhouse external screens are recommended (Farrell et al. 1980). Variable thermal insulation can be provided by sleeves of opaque, but highly-reflective, plastic film hung from the roof (Laing 1967). When deflated, they form no significant barrier to radiation entering or leaving the greenhouse, but when inflated, e.g. by an auxiliary fan, they form an air-mattress, which reduces the thermal conduction loss, and become opaque to radiation as shown in Fig. 11.4. If the sleeves were made of a suitable material, then it might also be possible to use this system as a shade against excessive solar gains. When implemented in the German climate a 35 % reduction in the rates of heat-loss ensued (Stickler 1975; Jensen 1977).

The double-cladding of a greenhouse has provided reported fuel savings of up to 33 % of those for a single-skin greenhouse (O'Flaherty and Maher 1978, 1980; Short et al. 1981; O'Flaherty 1985). Insulating a greenhouse by double-cladding not only reduces heating costs, but also inhibits condensation and the consequent water dripping onto the crops below (Short et al. 1981). The reduced light transmission rate associated with double-cladding with polyethylene is usually outweighed by the benefits of the reduced rate of heat loss.

In temperate climates, greenhouse cooling is achieved usually by increasing the rate of ventilation. This is accomplished by opening ridge vents and windows to allow wind-induced and buoyancy-driven ventilation. In modern greenhouses, the area of such openings can amount to nearly 25 % of the floor area. The ventilators are usually operated automatically, the controller being activated by a sensor within an aspirated screen that measured the true temperature of the greenhouse air on hot, calm days. When this is insufficient, fan-assisted ventilation is often used. On days of very high insolation, ventilation rates of over 60 air changes per hour may be necessary.

Sun shades, of a light-coloured material, serve to reduce solar gains (Burek et al. 1989). Sun shades can be fitted externally, as well as internally. Screens which may serve both thermal screens and sunshades can be made from translucent materials. Alternatively, an inflatable insulation and shading systems, could be employed (Laing 1967; Stickler 1975).

Evaporative cooling involves either (i) directly wetting (i.e. misting) the air inside the greenhouse or (ii) wetting the ground surface or the external cover of the greenhouse. An experimental comparison of these techniques showed that misting the inside air was the most effective method, followed successively by soil wetting and wetting the outer cover of the greenhouse, although the effectiveness of all such techniques is highly sensitive to the prevailing ambient conditions (Cohen et al. 1983). Analyses of the efficacy of evaporative-cooling methods are available (Landsberg et al. 1979). Techniques for cooling greenhouses using external water cascades over the roofs are useful if either (i) low ventilation rates need to be maintained (for instance, if above-ambient levels of CO^2 are being maintained) or (ii) the water heated in the cascade is to be stored and used for later heating. Adding dyes to the water to increase the absorption of infra-red radiation can make the technique more effective (Morris et al. 1988; Canham 1962; Van Bavel and Sadler 1979a, 1979b; Chiapale et al. 1983), though toxic dyes may harm plants if the greenhouse cover leaks. An external water cascade is a more effective cooling technique on shallow-pitched roofs, because compared with a steeply-sloping surface, the liquid film is thicker providing better absorption of insolation.

To achieve their maximum growth rates different plants require different levels of ambient humidity. Water vapour concentrations inside a greenhouse depends on ventilation, evapo-transpiration from the plants themselves and evaporation from the soil. Experimental results for moisture transfers from greenhouse crops have been reported (Morris et al. 1957; Linacre et al. 1964; Hand et al. 1970).

A build-up of water vapour inside a greenhouse, is counteracted by ventilation which draws in drier air. Hot, dry conditions require some form of humidification (e.g. using evaporative coolers, wet pads or atomisers) to avoid rapid dehydration occurring within the greenhouse.

Excellent yields can be achieved in tightly-constructed greenhouses if crop management practices are adjusted to take into account their higher humidity levels (Short et al. 1981). For most plants, and especially vegetables, varieties are available to suit a wide range of growing conditions, so it is usually more economic to match a plant to the achievable environment rather than vice versa. However, high humidity levels can cause condensation on the inner surface of the cover during cold ambient conditions. Water dripping off the cover onto the crops can be harmful, especially in winter, and can cause diseases (Canham 1967, 1968). One solution is to raise nocturnal ventilation rates to avoid night-time condensation, though this increases the nocturnal heat load.

References

Ahmadi G, Glockner PG (1982) Dynamic simulation of the performance of an inflatable greenhouse in the Southern part of Alberta; part I – analysis and average winter conditions. Agric Meteorol 27:155–180

Bailey BJ (1981a) The evaluation of thermal screens in glasshouses on commercial nurseries. Acta Horti 115:663–670

Bailey BJ (1981b) The reduction of thermal radiation in glasshouses by thermal screens. J Agr Eng Res 26:215–224

Bowman GE (1962) A comparison of greenhouses covered with plastic film and with glass. In: Proceedings of 16th international horticultural conference, Brussels, pp 551–558

Boyd LL, Ashley GC, Hietala JS, Stansfield RV, Tonkinson TRC (1980) A demonstration of the beneficial uses of warm water from condensers of electricity-generating plants. Report No EPA 600/7-80 -099. Environmental Protection Agency, Washington DC

Burek SAM, Norton B, Probert SD (1989) Transmission and forward scattering of insolation through transparent and semi- transparent materials. Solar Energy 42(6):457–475

Businger JA (1963) The glasshouse (greenhouse) climate. In: Van Wijk WR (ed) Physics of the plant environment. North - Holland, Amsterdam

Canham AE (1962) Shading glasshouses with liquid films, a preliminary report. ERA report no W/T40. Electrical Research Association, Surrey

Canham AE (1967) Air-supported plastic structures – materials and design factors. In: Proceedings of the Institute of Agricultural Engineers symposium on agricultural engineering, Silsoe

Canham AE (1968) Air-supported greenhouses in England. Acta Horticult 9:105–110

Chiapale JP, van Bavel CHM, Sadler EJ (1983) Comparison of calculated and measured performances of a fluid-roof and a standard greenhouse. Energ Agr 2:75–89

Cohen Y, Stanhill G, Fuchs M (1983) An experimental comparison of evaporative cooling in a naturally-ventilated glasshouse due to wetting the outer roof and inner crop and soil surfaces. Agr Meteorol 28:239–251

Critten DL (1985) Light enhancement in conservatories and single-span greenhouses. In: Proceedings of the UK-ISES conference on greenhouses and conservatories, London, pp 83–91

Farrell JJ, Albright LD, Donohoe AJ (1980) A highly-insulative thermal curtain for greenhouses. ASAE paper no 80–4531

Froehlich DP (1976) Steady-periodic analysis of the greenhouse thermal environment. Ph.D. thesis, Cornell University, New York
Fuller RJ, Meyer CP, Sale PJM (1984) Energy analysis of a solar greenhouse. In: Institution of Engineers Australia conference on agricultural engineering, Bundaberg
Hand DW, Slack G, Machin DR (1970) Evaporation rates of capillary-watered tomatoes in an east–west glasshouse. J Hort Sci 45:3–14
Harnett RF, Sims TV, Bowman GE (1971) Comparison of glasshouse types and their orientation. Exp Hortic 31:59–66
Iqbal M, Kharty AK (1977) Wind-induced heat-transfer coefficients from glasshouses. Trans ASAE 20:157–160
Iwakiri S (1971) Prediction and control of temperature environment inside the glasshouse. Report no. 6 of the National Research Center for Disaster Prevention, Tokyo (in Japanese)
Jensen MH (1977) Energy alternatives and conservation for greenhouses. Hort Sci 12:14–24
Kimball BA (1973) Simulation of the energy balance of a greenhouse. Agri Meteorol 11:243–260
Kimball BA (1986) A modular energy balance program including subroutines for greenhouses and other latent heat devices, ARS-33. Agricultural Research Service, US Department of Agriculture, Washington
Laing N (1967) The use of solar and sky radiation for air conditioning of pneumatic structures. In: Proceedings of the IASS 1st international colloquium on pneumatic structures, Stuttgart, pp 163–177
Linacre ET, Palmer JH, Trickett ES (1964) Heat and moisture transfer from trimmed glasshouse crops. Agri Meteorol 1:165–183
Landsberg JJ, White B, Thorpe MR (1979) Computer analysis of the efficacy of evaporative cooling for glasshouses in high-energy environments. J Agric Eng Res 24:29–39
Mears DR, Baird CD (1976) Development of a low-cost solar heating system for greenhouses. In: Proceedings of the solar energy fuel-food workshop. University of Arizona, Tucson, pp 88–109
Meyer J (1983) The influence of thermal screens on energy consumption in glasshouses. Gartenbauwissenschaft 49:74–76
Miller CW (1977) The effect of a conducting wall on a stratified fluid in a cylinder. In: 12th AIAA thermophysics conference, Albuquerque
Morris LG, Neale FE, Postlethwaite JD (1957) The transpiration of glasshouse crops and its relationship to incoming solar radiation. J Agric Eng Res 2:111–122
Morris LG, Trickett EJ, Vanstone PH, Wells DA (1972) Solar radiation in greenhouses; a brief review. Int J Agric Res 22:85–97
Morris LG, Trickett EJ, Vanstone PM, Wells DA (1988) Limitation of maximum temperature in a greenhouse by use of a water film on the roof. J Agric Eng Res 3:121–130
O'Flaherty T, Maher MJ (1978) Evaluation of double-clad polythene greenhouses for energy saving in early tomato production. Acta Hortic 76:335–339
O'Flaherty T, Grant J (1984) Use of double cladding techniques to reduce greenhouse heat loss. ISHS Acta Horticulturae 148:641–650
O'Flaherty T (1985) Recent developments in greenhouse energy saving. In: Proceedings of the UK-ISES conference on greenhouses and conservatories, London, pp 64–71
O'Flaherty T, Maher MJ (1980) Fuel consumption and crop response in double-covered polythene greenhouses. Acta Hortic 107:81–85
Pile RS, Burns ER, Waddell EL, Olszewski M, Stovall TK (1979) Browns Ferry waste heat greenhouse environmental control system. Report no 7910156–1, Tennessee Valley Authority, Chattanooga
Seginer I, Levav N (1971) Models as tools in greenhouse climate design. Research no 150–085, Technion Research and Development Foundation, Haifa
Short TH, Roller WL, Badger PC (1976) A solar pond for heating greenhouses and rural residences – a preliminary report. In: Proceedings of the solar energy fuel-food workshop, University of Arizona, Tucson, pp 61–73

Short TH, Roller WL, Bauerle WL (1981) Energy research for the present and future USA greenhouse industry. Acta Hortic 115:429–436

Slack E, Clark JA (1973) The energy balance of an inflated polythene greenhouse. Acta Hortic 46:109–129

Stickler PM (1975) The use of plastic for heat insulation for greenhouses. Plasticulture 25:41–53

Takakura T (1967) Predicting temperatures in the glasshouse (I). J Meteorol Soc Jpn 45:40–52

Takakura T, Jordan KA, Boyd LL (1969) Dynamic simulation of plant growth and environment in the greenhouse. In: Proceedings of the ASAE winter meeting, Chicago, Paper 69–942

Udink ten Cate AJ (1985) Simulation models for greenhouse climate control. In: Proceedings of the Easter School on computers in agriculture, Nottingham University

Van Bavel CHM, Sadler EJ (1979a) SG79: a computer simulation program for analysing energy transformations in a solar greenhouse. Texas A & M University, College Station

Van Bavel CHM, Sadler EJ (1979b) Experimental tests on a fluid-roof greenhouse concept. In: 4th annual conference on solar energy for heating greenhouses and greenhouse/residence combinations, Rutgers University, pp 128–136

Van den Kieboom AMG, Stoffers JA, Bascetincelik A (1981) Light transmittance of alternative greenhouses. Acta Hortic 115:417–428

Van den Muijzenberg EWB (1980) A history of greenhouses. Institute for Agricultural Engineering, Wageningen

Walker JN (1965) Predicting temperatures in ventilated glasshouses. Trans ASAE 8:445–448

Whittle RM, Lawrence WJC (1959) The climatology of glasshouses: natural illumination. J Agr Eng Res 4:326–340

Wood RW (1909) Note on the theory of the greenhouse. Philosophical Magazine, 319–320

Chapter 12
Passive and Hybrid Solar Design of Buildings

Buildings, too, are children of Earth and Sun

Frank Lloyd Wright

12.1 Overview

Though there are whole or parts of buildings that exclude solar energy totally, in most buildings solar energy provides daylight, either directly or indirectly and/or heat gains. Internal temperatures within a building with this are different from external ambient temperature. Altering such internal temperatures by heating, cooling and ventilation aims to achieve acceptable comfort conditions. Buildings require energy usually solves as a consequence of being occupied, though the use of energy of energy can preceed the entry of people to the building; this can include electricity for emergency systems and lighting and thermal energy to raise or lower temperatures to comfort conditions. Where a building of high thermal mass vary intermittent occupation the energy used for the latter can be considerable. For lower thermal mass buildings, the patterns of occupancy are often the dominant determinant of space heating/cooling energy consumption. It is thus often critical, when energy consumption figures are quoted for a particular building, to know the associated occupant numbers and activities and their duration.

The environmental function of a building is to mediate between the external climate (with its seasonal and diurnal variation of temperature, illuminance and wind speed) and the more stable conditions which are normally required for human comfort. Building design has a diverse set of physical constraints (e.g. site, internal arrangement) and functional requirements (e.g. structure, use, circulation areas), climatic imperatives and regulatory requirements. Successful passive solar design is reconciled harmoniously with these. Passive solar design addresses the functional requirements in two ways, namely

B. Norton, *Harnessing Solar Heat*, Lecture Notes in Energy 18,
DOI 10.1007/978-94-007-7275-5_12,

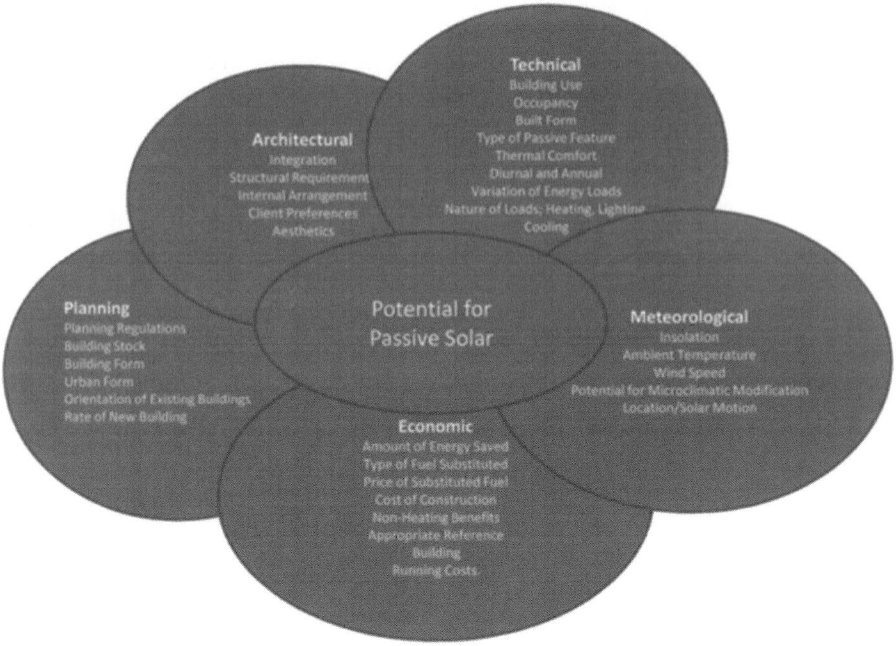

Fig. 12.1 Influences in the potential to which harnessing passive solar gain to a building can be concluded in the design process

- Reducing the auxiliary energy load (for heating, ventilating, cooling and/or lighting) and thus the running costs of the building
- Providing a pleasant, aesthetically pleasing, attractive internal environment, and possibly providing additional usable space.

By means of only the design of the form and generation of the building itself a passive solar building uses solar energy to reduce the auxiliary heating, ventilating and/or lighting load. This is distinct from an active solar building where this is accomplished by means largely of additional specialist mechanical components. A purely passive solar building uses no additional energy to collect solar heat. Climate conscious urban planning, passive solar heating, natural cooling and daylighting of buildings facilitate low energy consumption, comportable internal conditions and a more benign impact on the wider environment.

Buildings can incorporate passive features which have active elements, e.g. fans, these are termed "hybrid". However, many building features which are commonly regarded as "passive" systems transport energy to the point of use via small fans or motors. Passive solar design has been applied to domestic (Kachadorian 1997; Bainbridge 2011;), commercial (Hootman 2012; school (Norton and Hobday 1990) buildings. The potential scope for the employment of passive solar design must be cogniscent of the many factors, illustrated in Fig. 12.1.

12.2 Solar Access

If a passive solar approach is to be adopted, it is obviously essential that appropriate building facades should have access to insolation (Wright 1936). In a new urban developments where effective use of solar energy is one of the design goals, designers can have more control over the orientations of buildings, their massings, locations and adjacencies within plots of land, and over the shapes and orientations of the plots themselves. In reaching a final layout, a balance can be achieved between optimal exploitation of solar energy, urban planning requirements such as land codes, site utilization, transport access, utility services and aesthetics. Examples of such planning from the USA and UK are shown in Fig. 12.2.

The layout illustrated in Fig. 12.3 shows an area of Milton Keynes, UK planned for solar access. The plan of the houses, their orientations and locations all contribute to using winter insolation to the best advantage.

An example of the distances required between longitudinally-aligned buildings to avoid overshading, at a latitude of 52°N, required to avoid overshading is given in Fig. 12.3.

Early attempts to plan for access to solar radiation were concerned only with beam radiation. The ancient Greeks and Romans were aware of the principles of passive solar design and the Romans even enacted laws to protect a building's access to the Sun. Access to daylighting was the subject of the British Law of

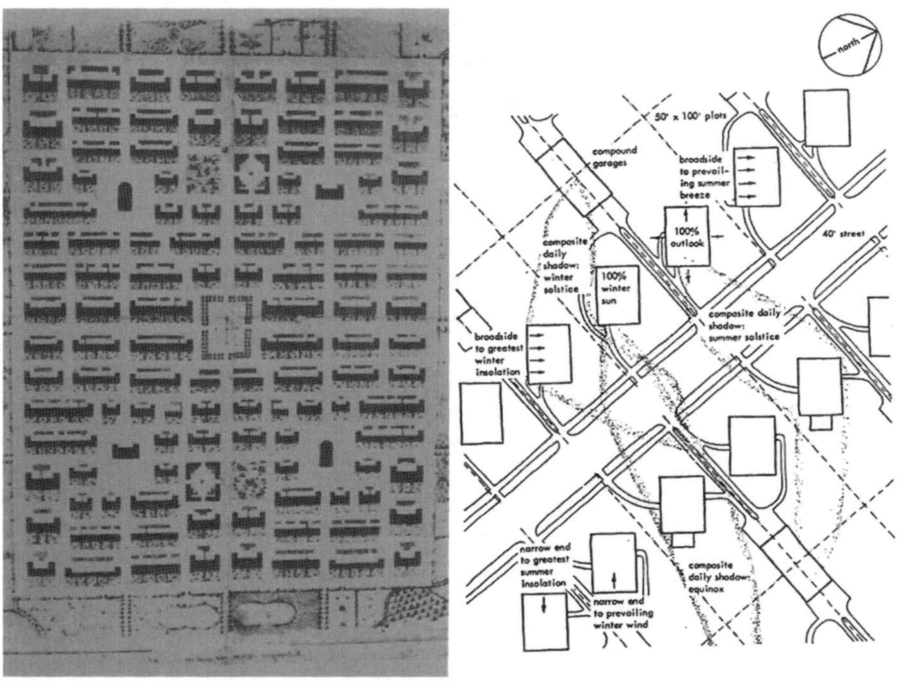

Fig. 12.2 Early examples of site planning for solar access

Fig. 12.3 Urban planning for solar access in Milton Keynes, UK, viewed from the south

Ancient Lights in embodied in statute to 1189. The Prescription Act of 1832, stated if a window in a building enjoyed uninterrupted access to daylight for a 20-year period, right to that access became permanent.

The modern equivalents of the laws used by the Romans to guarantee lighting standards are the zoning regulations developed in American cities. New York City's 1916 Zoning Ordinance was the first comprehensive municipal attempt to legally guide and control the physical form and function of a city. The form and intent of New York's zoning regulations are still the subject of debate, with the emphasis now switching to the potential for energy conservation in designing for a combination of artificial light and daylight (Bryan et al. 1981) (Fig. 12.4).

The use of a sun-path diagram enables the full annual variation of the diurnal solar motion to be taken into account. An example of such a translation of a site plan that shown in Fig. 12.5 on to a sun-path diagram is shown in Fig. 12.5.

Computer-based solar access models are available that allow a more free-ranging consideration of potential options in planning for solar access (e.g. O'Cathain 1981, 1982; Fanchiotti et al. 1983; Rowe and Willmott 1984; Dupagne and Renson 1986).

12.3 Passive Features for Temperate Climates

In temperate climates passive solar design seeks usually to provide heating in cool weather whilst avoiding overheating in warm weather. Heating involves the distribution, storage and conservation of collected solar energy, overheating prevention involves shading and ventilation. These processes are summarised in Fig. 12.7.

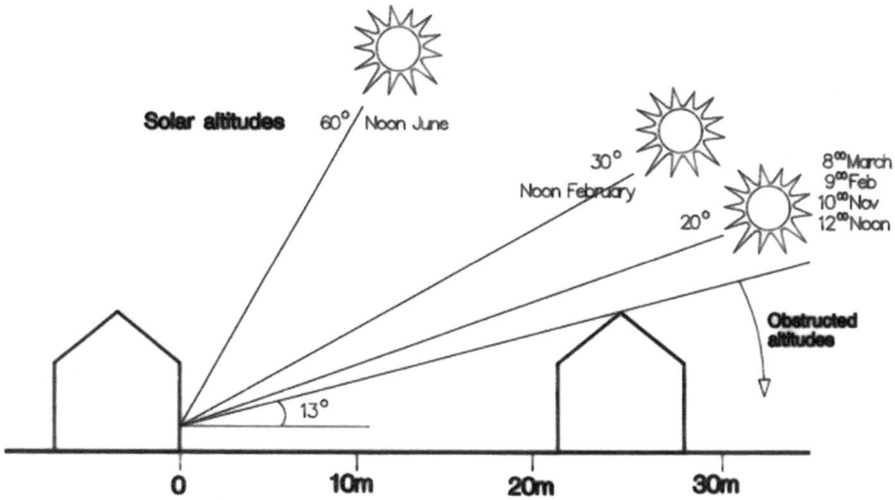

Fig. 12.4 Distances to avoid south-overshading to lower windows at a latitude of 52°N.

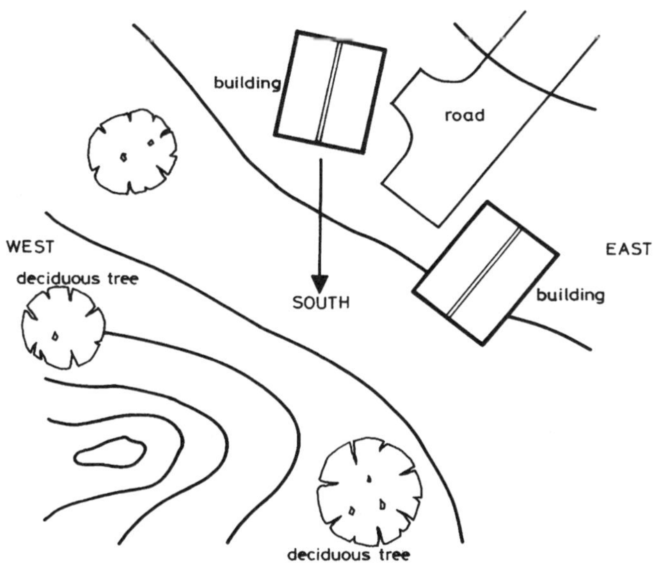

Fig. 12.5 Site plan transposed in Fig. 12.6

Generic types of passive solar features are illustrated schematically in Fig. 12.8. These generic types are often combined in a wide variety of architectural forms. Appropriate passive solar design strategies are specific to a given climate, site, building function and use and by prevailing cultural and architectural preferences. Many handbook and design guidelines for specific climates are available (Achard and Gicguel 1986; Baker 1987; Mazria 1980; Konya 1980).

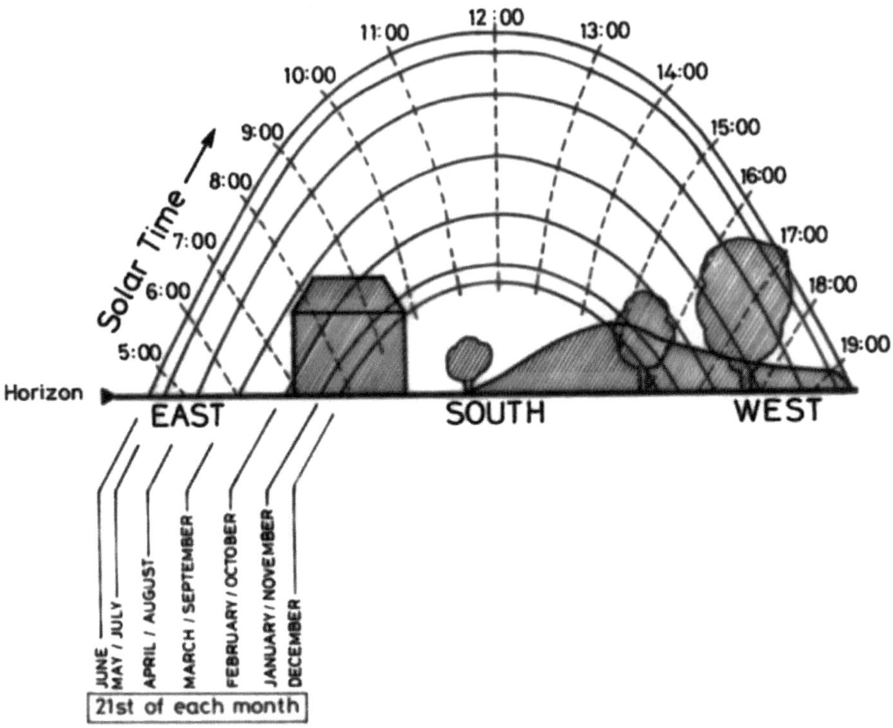

Fig. 12.6 Sun path diagram of site plan in Fig. 12.5

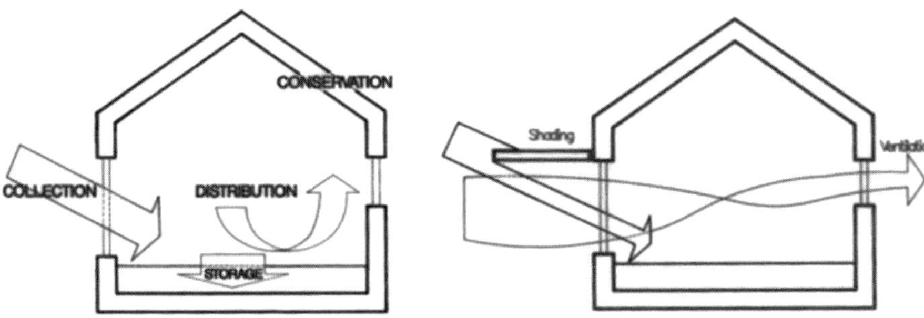

Fig. 12.7 Passive solar energy collector (on *left*) and overheating avoidance (on *right*) in temperate climates

12.3.1 Direct Gain

Direct gain in its simplest form refers to a room with a southerly-facing window. The term is often extended to include integral conservatories and atric as shown in Fig. 12.9.

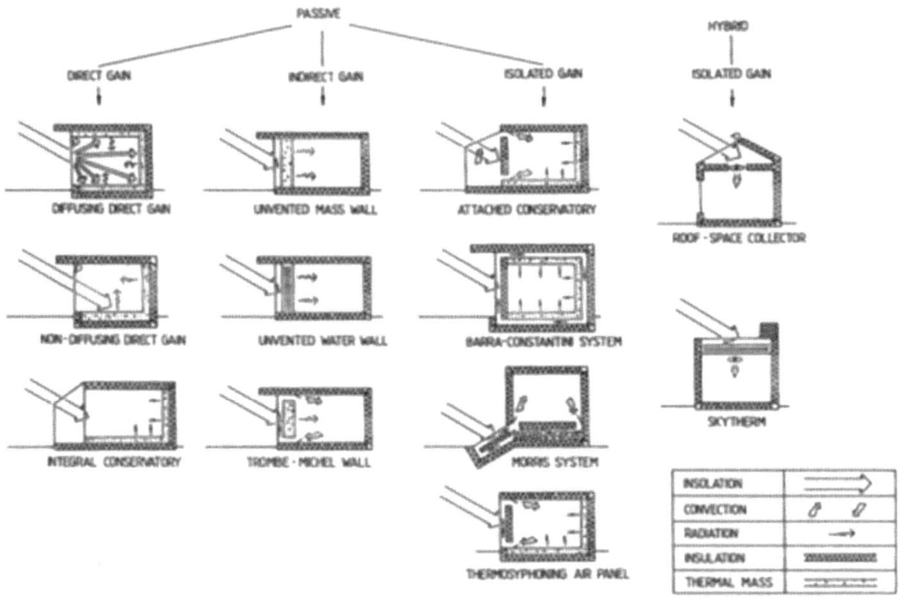

Fig. 12.8 Taxonomy of passive solar

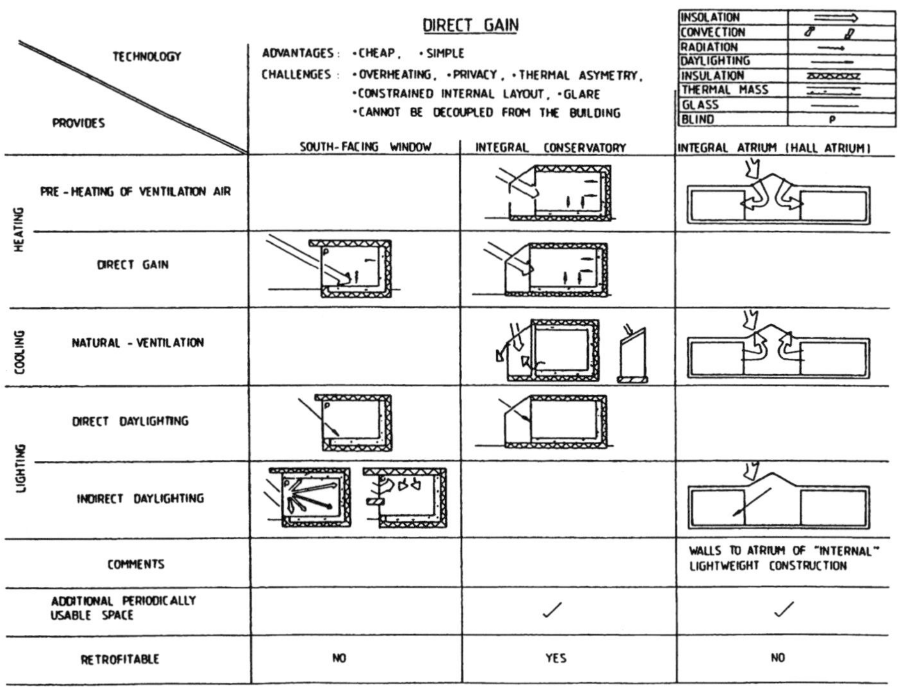

Fig. 12.9 Attributes of direct gain

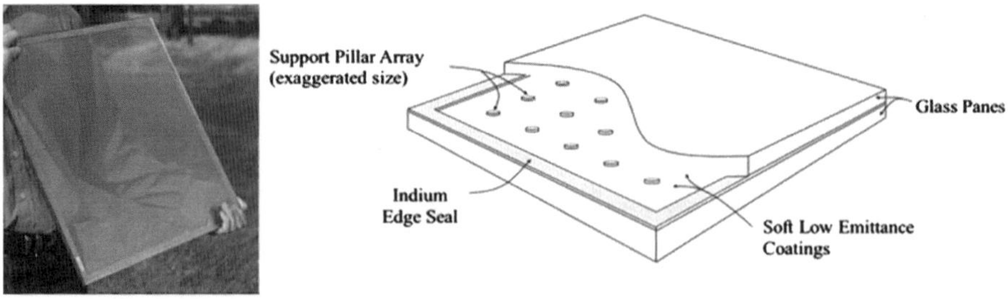

Fig. 12.10 Evacuated glazing

The particular challenges associated with direct gain include:

- Direct isolation of a large area of the room, and its occupants,
- Thermal discomfort due to radiant temperature assymetry,
- Glare problems,
- Damage from uv to fabrics and finishes,
- Maintaining privacy,
- Effectiveness likely to be reduced by occupant using net curtains,
- Summertime overheating in some instances.

Direct gain include incurs little or no extra cost and is simple with virtually self-functioning operation. The thermal behavior of single and multiple glazing units is well-understood (Elmahdy and Frank 1993; Wright and Sullivan 1995; Curcija and Goss 1994). Using various techniques, dynamic tintable or smart windows can be fabricated that can alter their solar factor and/or the transmittance to particular parts of the solar energy spectrum (Carmody et al. 2004; Baetens et al. 2010). Four modes of switchable affect can be employed;

- Electrochromic
- Gasochromic
- Photochromic
- Thermochromic

Electrochromics and gasochromics enable control of transmittance independent of both insulation or ambient temperature, thus both are particularly appropriate for use as components of windows in buildings.

Low heat loss through windows may be achieved, via using multiple panes, low long-wave omittance coatings and the inclusion between panes of inert gases, aerogels or a vacuum either singly or in combination (Muneer et al. 2000; Arasteh et al. 2006). Vacuum glazing comprises two contiguously sealed glass panes as shown in Fig. 12.10 with low emmittance films on one or both glass surfaces with the vacuum gap, separated by an array of tiny support pillars to maintain the glass separation under atmospheric pressure. The first successfully fabricated vacuum glazing used a solder glass with a melting point of 450 °C to seal the periphery of the vacuum gap (Robinson and Collins 1989). This high temperature prohibited the use of tempered glass and restricted the range of soft low emmittance coatings due to thermal degradation that resulted at this temperature. These restrictions are removed by using a low temperature edge sealing process and an indium based edge seal with a low melting point of less than 200 °C (Griffiths et al. 1998; Hyde et al. 2000).

A two dimensional model used to simulate the thermal performance of a vacuum glazing system must accurately determine the heat transfer coefficients between the two internal glass surfaces bounding the evacuated space. The vacuum space, including the pillar arrays, is usually represented in analytical models by a material whose effective thermal conductivity, the heat conduction through the pillar array and the calculated radiation heat transfer between the two internal glass surfaces within the vacuum gap. In a three-dimensional finite volume model developed to simulate the thermal performance of vacuum glazing, the support pillar array could be incorporated and modelled directly (Fang and Eames 2006). Heat flow across a full vacuum glazing is a combined function of heat conduction through the support pillars and edge seal, radiative heat flow between the two glass panes and convection and conduction in the residual gas within the vacuum space. As the vacuum pressure in a vacuum glazing is less than 0.1 Pa, so the effects of any residual gas are insignificant (Collins and Simko 1998).

12.3.2 Indirect Gain

When solar gain made to an unheated space or element is distributed to a heated space, it is termed indirect gain, the features of which are summarised in Fig. 12.11. Heat distribution is usually by (i) air convection (ii) by conduction through an adjoining wall or (iii) or a combination of the two. Non-occupyable indirect gain features include Trombe walls, mass walls and water walls.

Conservatories (or sunspaces), glazed streets, double structured facades, atria are examples of indirect gain features. A conservatory located on the south side of the building can be coupled thermally to heat a building by,

- Conduction only through separating wall
- The natural circulation of air between the indirect gain system and the heated building
- The forced circulation of air between the indirect gain system and the building
- A net flow of air from outside into the heated building via the indirect gain system.

Two or more of these operations may occur simultaneously or sequentially in a building. Appropriate sensors, controls and actuators provide an opportunity to optimise the energy collecting performance. However, the use of automatic controls incurs both a higher initial cost and, to ensure their long-term reliability, may require a significant continuing maintenance.

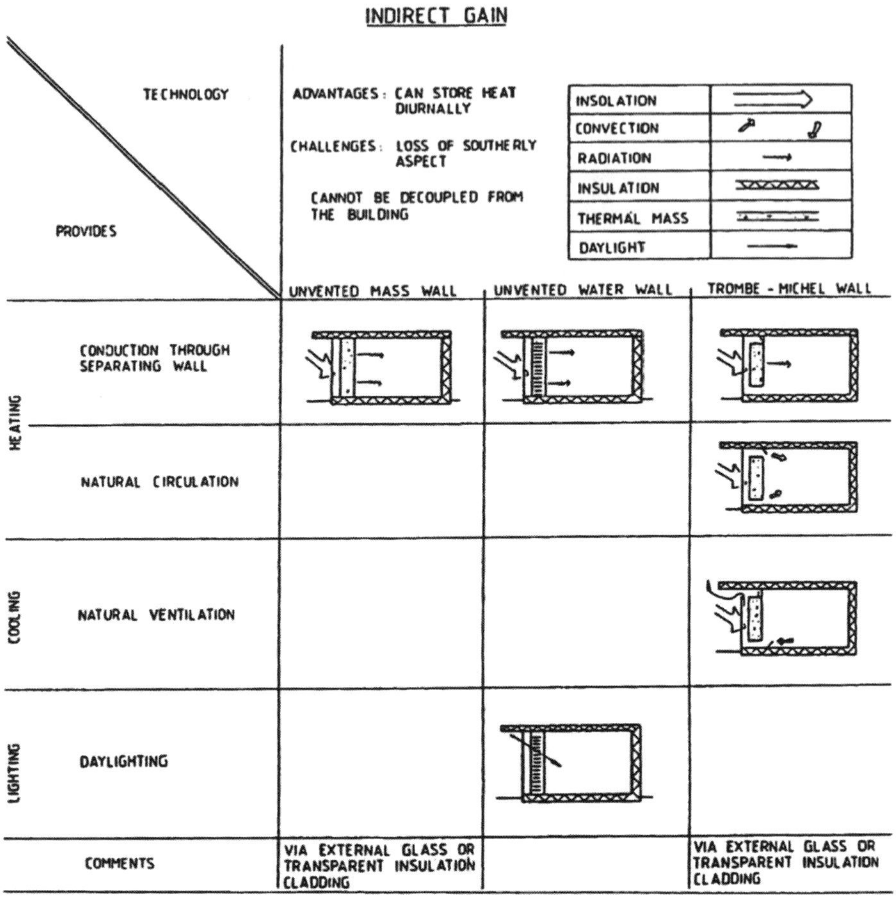

Fig. 12.11 Indirect gain

Conservatories, sunspaces and atria are occupiable, if to render occupancy possible and heating is introduced, the space will incur an increase in energy consumption disproportionately large compared to its floor area as the glazed structure will have a larger heat loss than other parts of the building fabric (Oreszczyn 1993).

Natural circulation of air between the conservatory and the heated building can occur with appropriate driving pressures when windows or doors open into the conservatory. In some cases to ensure a low resistance to air flow, ducts are provided purposely. Buoyancy-driven flow is induced by the temperature difference between the warm conservatory and the cooler adjacent room. In the northern European climates, low air flow-rates ensue for the small temperature differences encountered typically in Spring and Autumn, thus energy transport is often minimal. In Summer such action needs to be prevented in order to avoid overheating. If natural-circulation heat gain is to be encouraged then “flap valves” have to be provided to prevent nocturnal and winter reverse flow when the temperature in the conservatory is less than in the heated building. There is no heat gain from the circulation of air between the conservatory and the building, until the temperature of the air in the conservatory is greater than the heated building temperature. A contribution to a reduction of heat load however, does ensue from the reduction in the rate of thermal conduction out through the separating wall. As in well insulated buildings, with short "heating seasons", temperatures in an attached conservatory are above internal temperatures only rarely, this threshold is high. Thus the annual heating contribution is likely to be small and has to be weighed carefully, when assessing the economic feasibility, against the energy consumption of fan the forced circulation of air between the conservatory and the building and any heating that may be provided to enable extended occupancy of the conservatory (Oreszczyn 1993).

If the air entering the building via the conservatory forms the major constituent of the air required for ventilation, then the net flow of air from outside into the heated building via the conservatory provides solar ventilation pre-heating. This mode of operation of a conservatory carries the important advantage that all solar gains provide some reduction of the ventilation heat load. In addition, as buildings become better insulated generally, so the proportion of energy that is used to heat the essential requirement for ventilation air increases.

Passive solar buildings with conservatories or atria often ultimately rely upon ventilation and infiltration to provide the medium of heat transfer. Ventilation and infiltration are both dependent upon (i) the winds speed and direction, (ii) the temperature difference between the building and the ambient environment, (iii) the aerodynamic form of the building, (iv) overall building airtightness (type and position of openings) and (v) surrounding topography and obstructions. A designer may, given appropriate analytical tools, use these effects to optimise air flow.

In many countries for example France and the Netherlands, buildings ventilated naturally must be fitted generally with purpose provided ventilation stacks. These are designed to promote the effect of stake action. In contrast the requirements of the current UK building regulations assume that needs will be met either by natural infiltration or by the occupant's opening of windows.

An example of an integrated conservatory/sunspace in a school building in Tournal, Belgium is shown in Fig. 12.12.

In another indirect gain feature, the Trombe-Michel wall, the thermal coupling to the space to be heated are the same virtually as with conservatories, the relative magnitude of the conductive and convective heat transfers are different; the conductive being larger in the case of a Trombe-Michel wall. As the contribution arrives within the building after a time lag, Trombe-Michel walls delay the delivery of the solar heat. They are thus ideally suited to providing heating in the early part of a cold night after a hot sunny day, conditions encountered frequently in arid and mountainous regions. Trombe-Michel walls, unlike conservatories, provide no additional useable floor area. Thus the initial construction cost outlay must be justified solely on heating energy substitution. To better control the heat output, alternative designs of Trombe-Michel walls have been developed (Zrikem and Bilgen 1987).

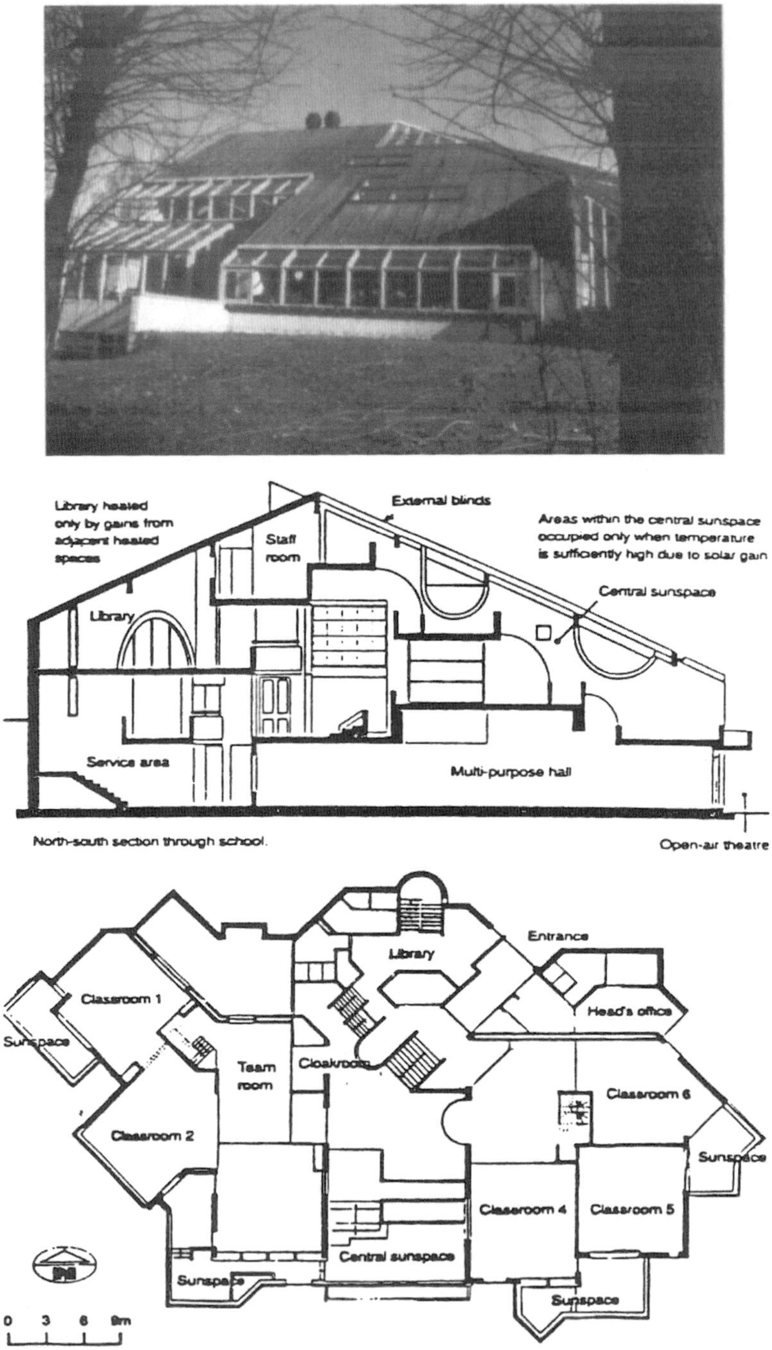

Fig. 12.12 Plan, section and photograph of a school in Belgium with extensive conservatories

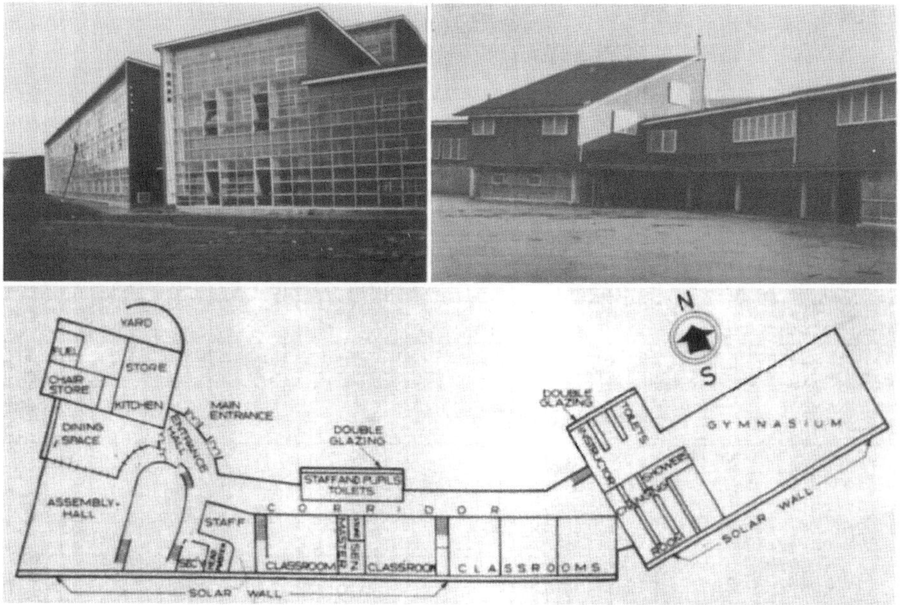

Fig. 12.13 Photographs of the front and rear elevations with the plan of Wallesey School

A school in Wallasey in north-west England constructed in 1961 included a near-south-facing double-skinned glazed wall with occasional adjustable white shutters that also comprised transparent, translucent and pinboarded elements together with minimal glazing to its north façade, as shown in Fig. 12.13. The building has been studied for over 50 years (Davies 2007). It has been found that though the building has been economical to heat, it is probably overglazed leading to some overheating in sunny conditions and increased heat losses on cloudy days. The historical significance of this building, now the "Solar Campus" os St Mary's College, has led to it being formally listed as a protected building in 1996.

12.3.3 Isolated Gain

Isolated gain features may be at specific times largely decoupled thermally from the building. Their pertinent features are summarised in Fig. 12.14.

This is accomplished via an insulating separating wall, as in thermosyphoning air panels, or by location above the building, as in roof space collectors. A more-controllable heat gain combined with – if well designed – an avoidance of summer overheating, is the primary advantage of isolated gain. With new materials becoming available it may be possible to design cladding collectors and roofspace collectors for example, at low or even zero additional cost to the basic building, and these in conjunction with ventilation pre-heating would be economically attractive.

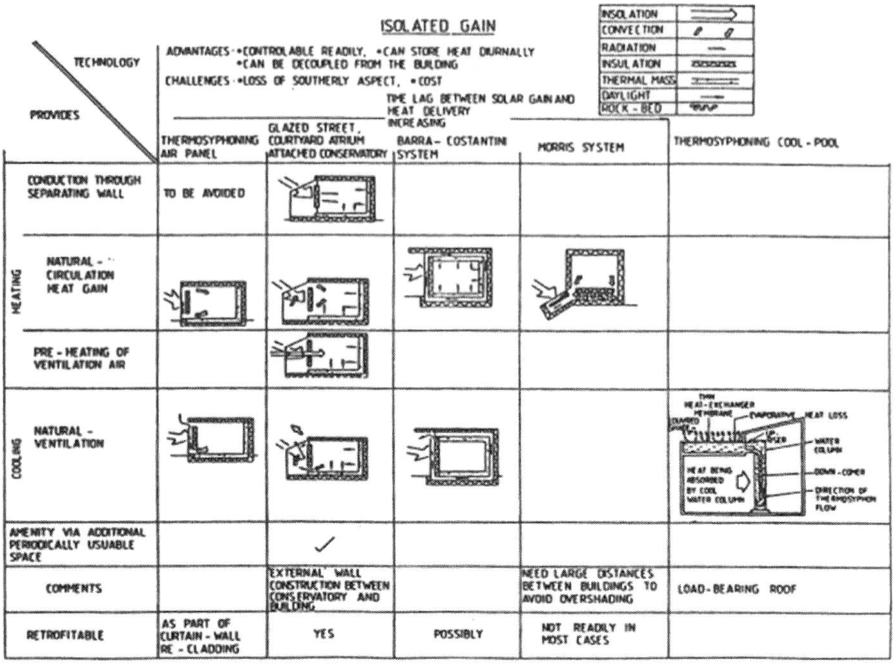

Fig. 12.14 Attributes of isolated indirect gain

For isolated systems the provision of thermal storage should concentrate upon the convective linking with the existing walls and ceiling than to provide further thermal capacity. Isolated indirect elements should not adopt storage unless it is insulated during periods of non-collection, or is remote with a controlled convective link. That is, the definition of isolated indirect features is strictly adhered to, otherwise a net heat loss will ensue.

Isolated gain collectors often form double skin facades (Poirazis 2004) on office and educational buildings. One such system is the thermosyphoning air panel (TAP), (Norton et al. 1992; Sawhney et al. 2007) overcome some of the disadvantages of indirect gain collectors by dispensing with heat storage and relying totally on convective heat gain as shown in Fig. 12.15. Heat input is almost immediate whilst heat losses during non-gain periods when the collector is isolated from the heated space, are low. This design is ideally suited to the task of providing daytime heat in cool or cold climates. A TAP operates in the same manner as the natural convection mode of a Trombe wall. However, the absorber is often made of metal, usually aluminum or steel, and the unit is insulated to prevent heat loss to, or from, the building. To control heat output, an air deflector linked to a louvre can be moved as shown in Fig. 12.16 to facilitate either heating or ventilation. Though air can be blown back into, or reverse circulate, the building, as shown in Fig. 12.18. Both reverse thermocirculation and warm air blow back are usually prevented by the inclusion of a lightweight flap-valve that only allow air to enter the panel base (Fig. 12.17).

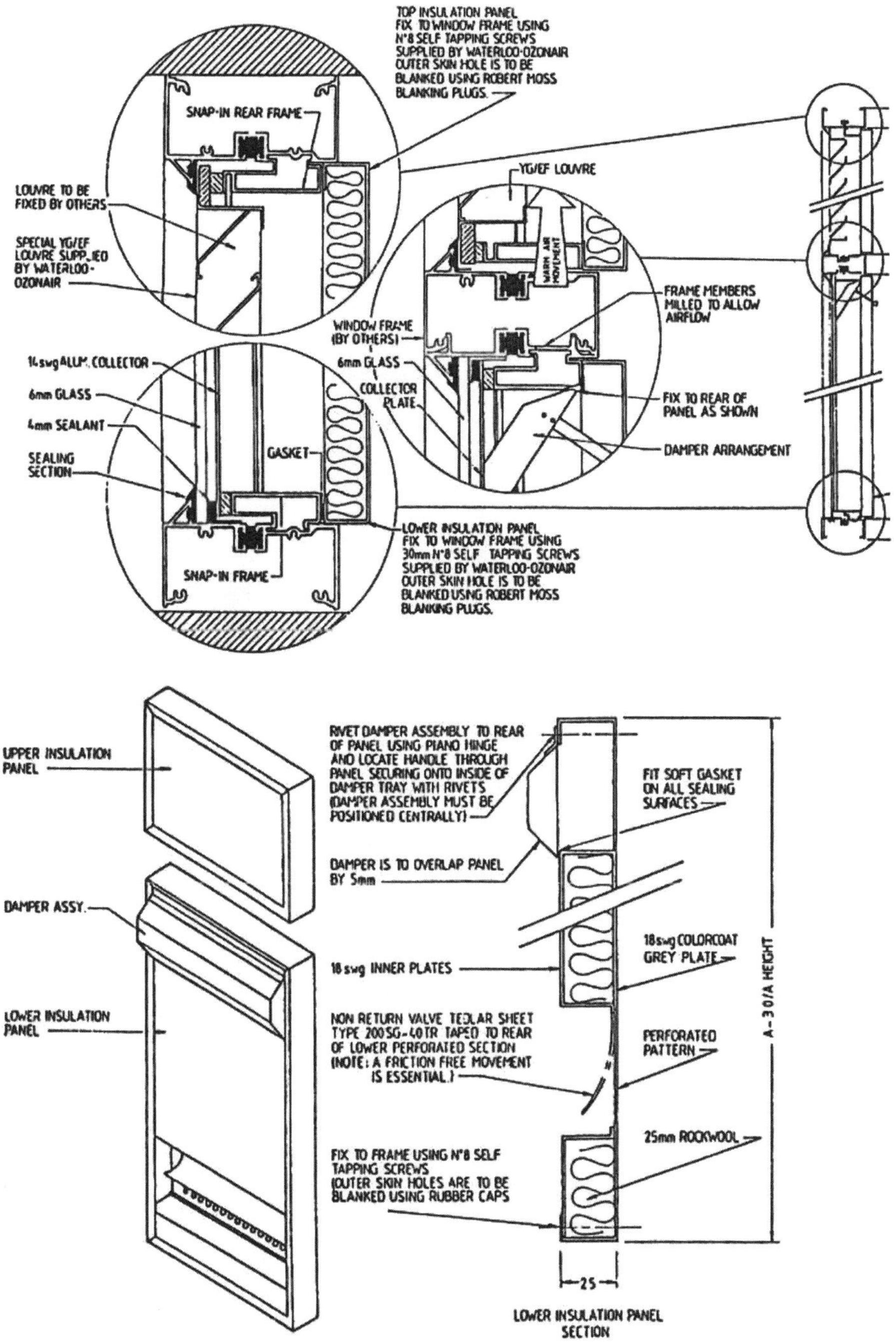

Fig. 12.15 Fabrication of TAP absorber panel

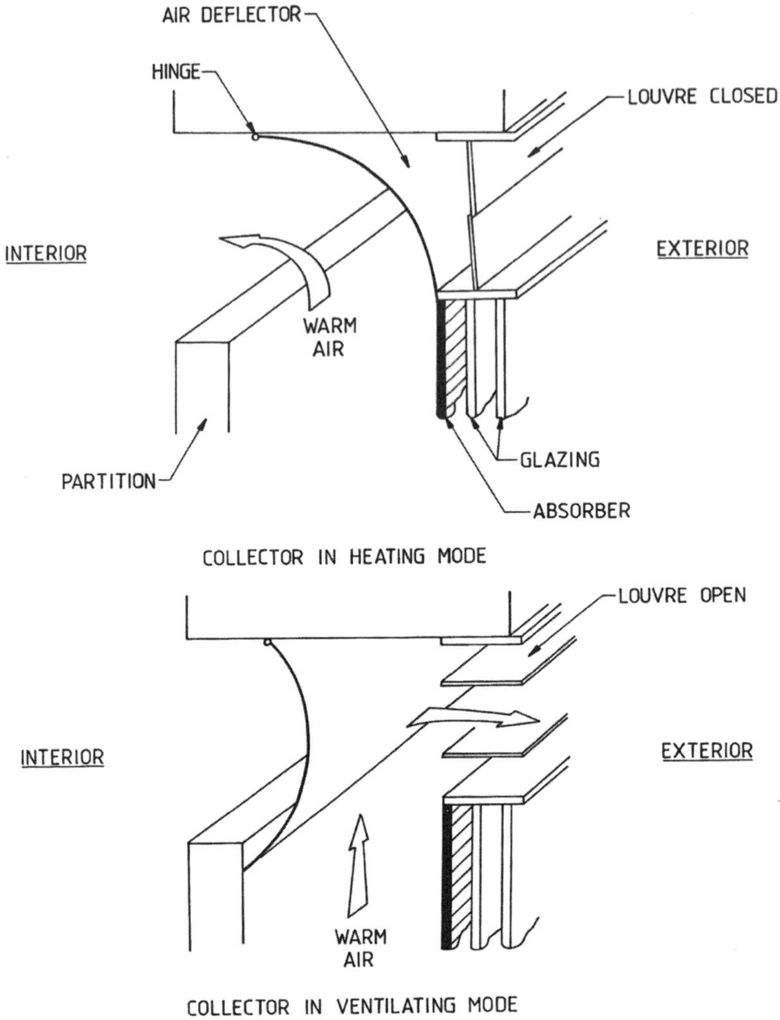

Fig. 12.16 Contained motion of an air deflector and linked louvre in a thermosyphoning air panel

Thermosyphoing air panels have been installed as part of a curtain wall replacement as shown in Fig. 12.18 at a school in Essex in south-east England (Lo et al. 1994).

Remote thermal storage can consist of a rock-store beneath the floor or in a built void between walls. The solar gain is transferred to the storage material by a specific subsystem which consists of ducts through which are is conveyed by small fans or natural convection. A similar subsystem also provides the link between the heat store and the heated space. Water is the other storage medium likely to be used, though the provision of reliable mechanisms and/or additives to prevent damage due to ice formation presents problems in the cold climates. Remote storage offers the advantage that it can be insulated, or if placed within

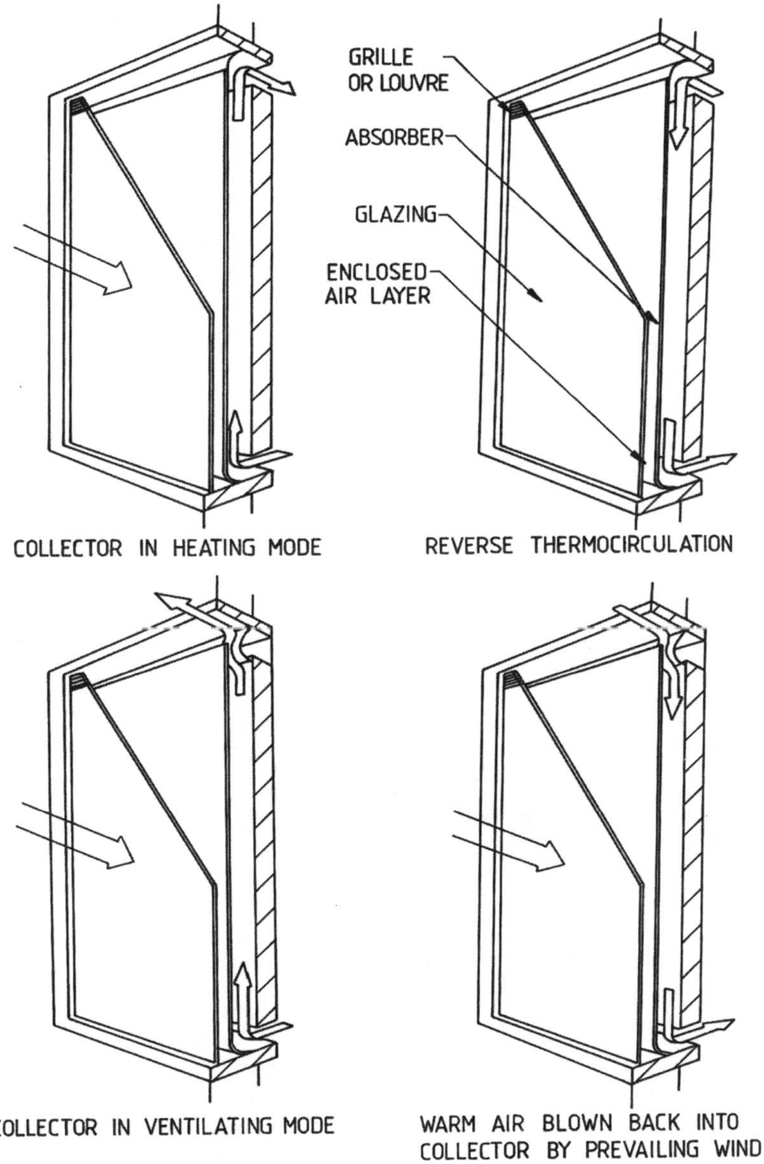

Fig. 12.17 Four possible modes of operation of a thermosyphoning air panel

the heated zone, heat losses from the store will usually be useful. However, the extra cost of remote storage and the subsystem linking the storage with the collection element and the heated space, is often high in relation to the amount of energy saved. Such systems have been installed in custom-built houses designed specifically to utilise solar air heating, located in the favourable semi-arid region of the USA below latitude 40°N (Morris 1981). The collectors are installed below

Fig. 12.18 Thermosyphoning air ponds installed as part of curtain wall refurbishment

floor-level, to facilitate natural circulation, and are coupled to rock-bed thermal storage. The thermal store which is often placed under the floor of south-facing rooms, but in the example shown in Fig. 12.19, the store has been positioned to provide radiant heat to the south-facing room and to the adjacent north-facing room.

The Barra-Costantini system (Barra et al. 1980) air heating system is a natural-convection dual-pass collector with the attributes of a Trombe-Michel wall but unlike a true Trombe-Michel wall, the storage is remote and may be decoupled from the source of solar energy collection. Figures 12.20 and 12.21 are schematic diagrams of the four modes of operation. It comprises:

(i) A low thermal capacity dual-pass absorber, (to maximise the heat transfer area between air and absorber) decoupled from the south wall of the building. The collector also acts as additional exterior insulation,
(ii) Automatic buoyancy-driven heat distribution throughout the dwelling via air channels within the ceilings, and
(iii) Additional thermal energy storage within these aforementioned ceilings and adjacent walls is radiated to the exterior environment.

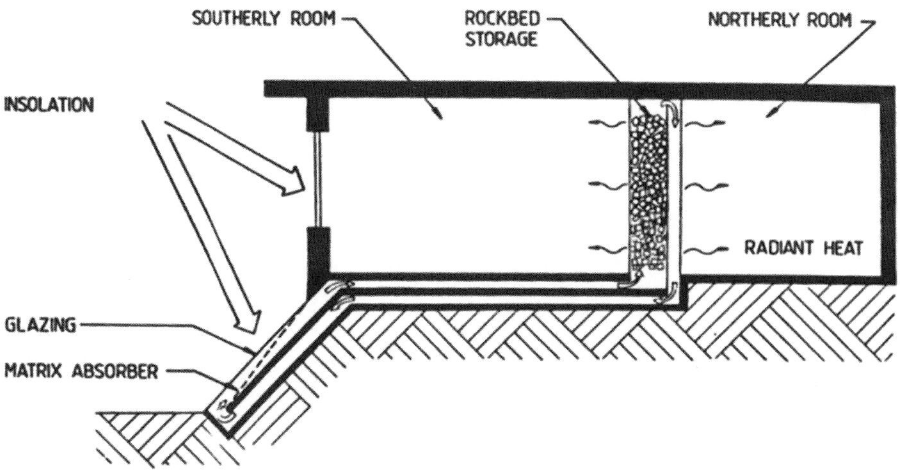

Fig. 12.19 U-tube thermosyphon solar-energy air-heating collector with rockbed storage

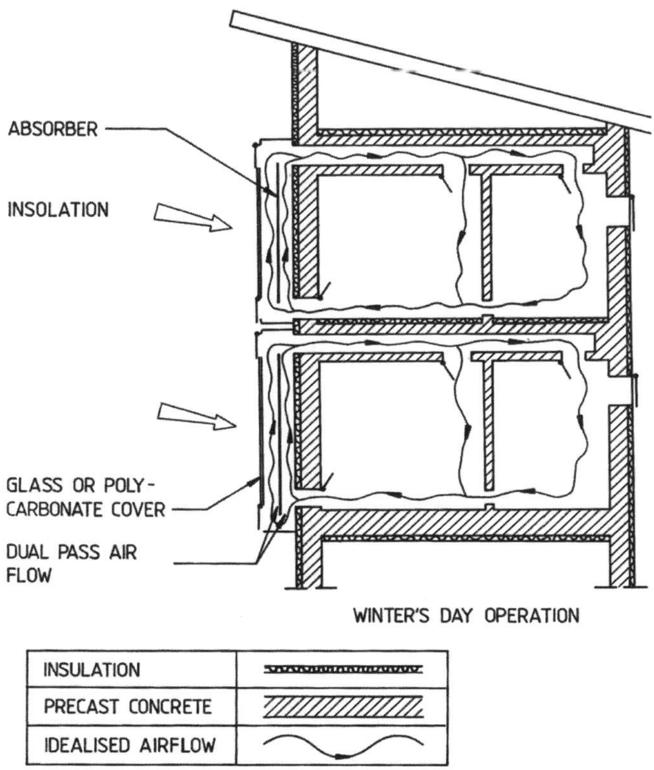

Fig. 12.20 Schematic diagram of winter's day operation of the Barra – Costantini passive solar system

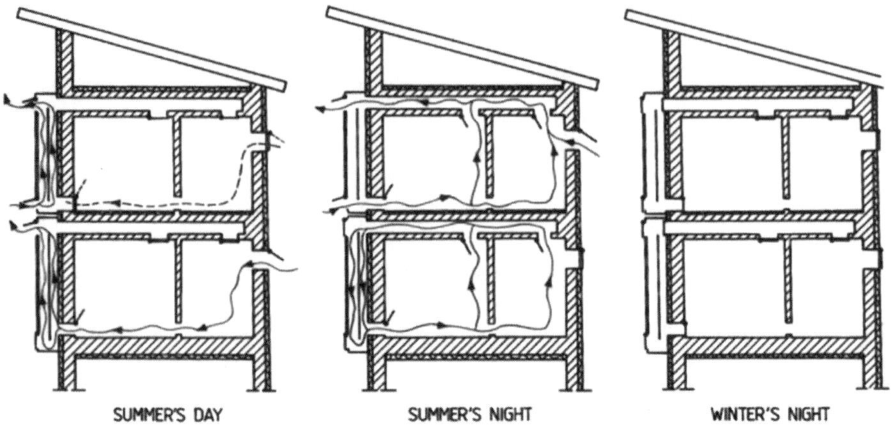

Fig. 12.21 Three modes of operation of the Barra – Costantini passive solar air system

12.3.4 Hybrid Systems

The pertinent features of hybrid systems are illustrated in Fig. 12.22.

A roof-space solar-energy collector, as shown in Fig. 12.23, is essentially a pitched-roof which is partially or fully-glazed on its southerly aspect (Norton et al. 1987). Solar-heated air from the roof space collector is conveyed by an automatically-controlled fan via a duct either directly into the living space or as a pre-heated supply to a warm-air space-heating system. The roof-space collector is replenished with air either from within the dwelling or from outside ambient environment. The fan and associated ductwork, through which the warmed air is conveyed from the roof-space collector, are existing parts of a gas-fired warm-air space-heating system. When the air from the roof-space solar-energy collector is at a lower temperature than the set level of the room thermostat, the air-stream emerging from the roof-space collector is a pre-heated supply to the gas-fired auxiliary system. A single fully-integrated wall-mounted control unit will actuate both the gas-fired auxiliary warm-air heating system and the roof-space collector.

A roof-space collector shown in Fig. 12.23 employs the passive collection and active distribution of solar heat and is thus generically a hybrid solar- energy system.

Ventilation is employed to prevent overheating in high summer. The five principal modes of operation of a roof-space collector are:

- During daytime conditions when the temperature of the roof- space collector is above that of the building, a thermostatically-controlled fan continually takes air from the building. This air is passed through the roof-space collector, is warmed and returned to the building. This mode is suitable for a dwelling that is occupied during the day.
- During the day the roof-space collector rises in temperature as solar energy is collected and no forced circulation of air occurs through the system. In the late

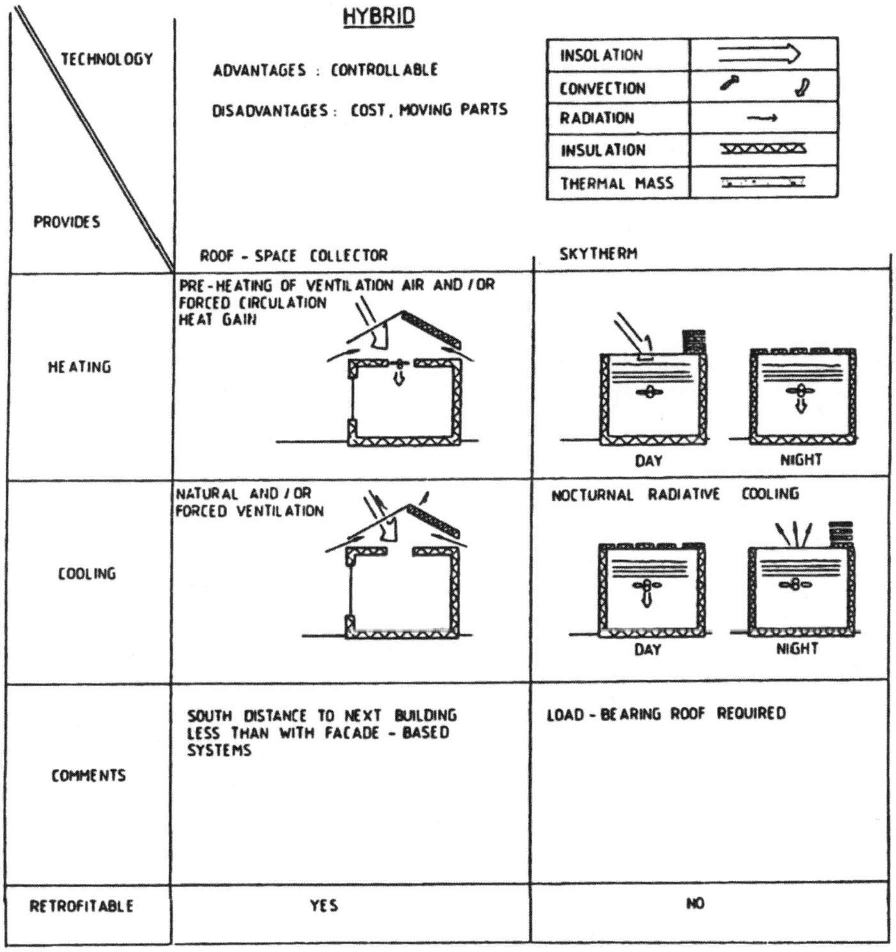

Fig. 12.22 Attributes of hybrid solar interventions in buildings

afternoon the stored heat is harnessed. Warm air is conveyed from the roof-space collector by a fan. The replenishing air to the attic may be supplied from either inside or outside the building. This mode is most appropriate when the building is unoccupied during the day.

- Air is heated as it passes, under the action of the fan, from the outside into the heated building via the roof- space collector. The volumetric flow rate through the fan in this case will be less than that used for mode (ii) operation. If this air forms a large proportion of the total ventilation, then this mode provides pre-heating of the ventilation air. Unlike modes (i) and (ii), the roof- space collector does not have to be at a temperature above that of the building to provide a beneficial effect, since any warming of the air as it passes through the attic-space will provide some reduction of the ventilation heat load. As buildings

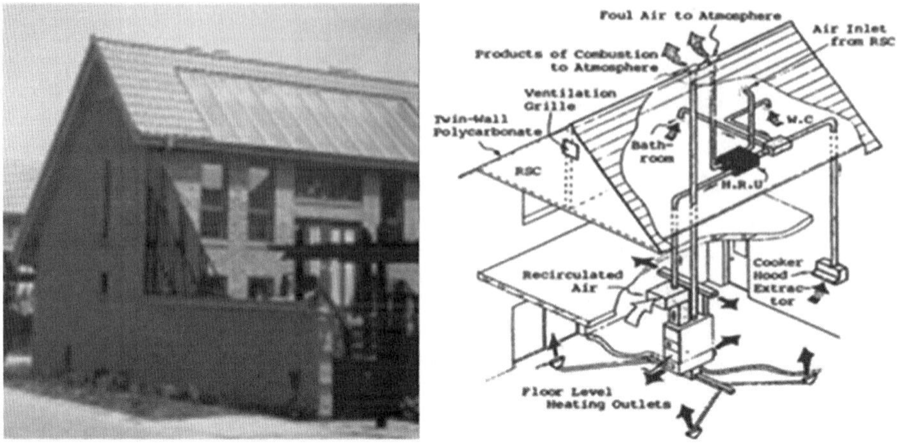

Fig. 12.23 Roof space solar energy collector

become better insulated, so the proportion of energy used to heat ventilation air will become greater. Roof-space collectors operating in this mode are thus compatible with, and complement, the use of energy conservation measures in buildings.

- Heat is passed by conduction through the floor of the roof- space collector. This will frequently be a small effect as the floor of a roof-space collector is well insulated. Good floor insulation is recommended, otherwise it is likely that the total diurnal losses would exceed gains, except perhaps if the roof-space collector were double-glazed and fitted with an insulating night blind. In this mode, the roof- space collector acts as a buffer space between the ceiling of the upper most story and the outdoor environment, thereby reducing the heat losses from the roof of the building.
- Buoyancy-driven flow occurs which draws air into the roof- space collector from the building and thence outside. The replenishing flow of air into the building is from the ambient external environment. In this mode, essentially the roof-space collector is providing ventilative cooling.

The advantages of a roof-space collector are that it has a low initial capital cost as its physical construction does not differ greatly from that of a conventional pitched roof. In addition a reduction in initial cost can arise from the employment of components (i.e. fans and controls) that would be already present in an auxiliary air heating system. A roof space collector can provide more efficient performance than a conservatory via optimal heat collection design and the practicality of attaining quite high roof-space collector temperatures; neither is possible within the occupancy constraints of a conservatory (Lo and Norton 1996). Roof-space collectors do not usually have thermal mass to provide significant energy storage. They are thus well-suited to buildings with heat loads that ensue primarily during the day. They have thus found application in school buildings with both pitched (Norton and Waterfield 1990) and domed (Williams 1989), roofs as shown in Fig. 12.24.

Fig. 12.24 Roof space collectors on Green Park School (*left*) and Perronet Thompson School (*right*) both in England

The site planning required to ensure that the roof space collectors are unshaded at all times is easier. Passive solar features on buildings are frequently at ground level. In urban locations exhibiting high housing densities, a dwelling may often experience levels of overshading at lower sun angles by neighbouring buildings such that ground-floor passive solar elements are rendered ineffective. Roof-space solar energy collectors do not cause loss of privacy as can be the case with the large glazed areas associated with direct gain systems.

12.3.5 Overheating Prevention

Since passive systems are designed to maximise solar gains, there is a high risk of overheating, not only in the summer but also towards the end of the heating season when most systems should be operating at their maximum performance. The thermal discomfort of unwanted solar gains can be avoided by:

- Preventing the initial solar gain by means of shading devices;
- Rejecting the solar gains by means of ventilation;
- Absorbing the solar gain in thermal mass in that order.

With fixed shading devices, the seasonal geometry of solar radiation permits some control of unwanted solar radiation. However, care must be given to the orientation, inclination and the geometry of fixed overhangs and fins (Givoni 1998). An important advantage of fixed shading- devices is that they are self-operating.

For a building where the solar input forms a significant proportion of the heating, a responsive control of solar gain is needed. This cannot be provided by fixing shading devices. In the temperate climates, buildings have day to day variations in heat demand within the same season. Indeed a building may go from energy surplus to deficit within a few hours. Consider the latitude 52°N, though a fixed horizontal south-facing over hand lm wide will completely shade a window about 2 m high in midsummer. Unfortunately it will also shade about 10 % in midwinter and about 50 % in the spring, a time when the performance of a passive system should be at its best. Furthermore in many climates, the annual variation in mean daily solar

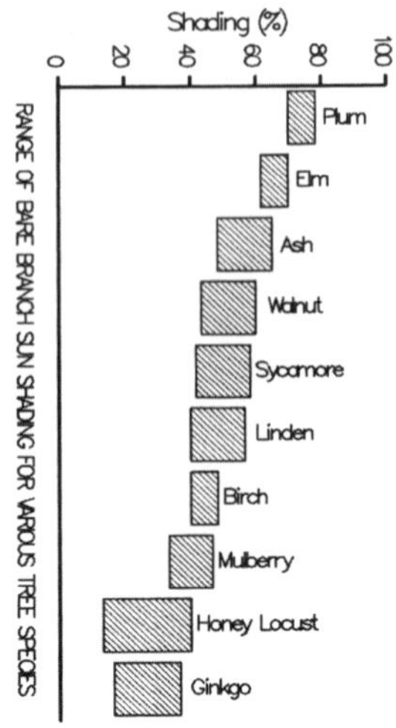

Fig. 12.25 Range of bare branch sun shading for various tree species

position and mean daily ambient temperature are not in phase. Though the solar motions are the same in September as in May, the corresponding ambient temperature are not the same. For example in England, a fixed shading device that provides the shading desired in a warm September, unfortunately will also shade during the cooler May when solar gain is useful. In appropriate climates suitably-located deciduous trees, can provide appropriate standing (Kumakara et al. 2011) however as can be seen from Fig. 12.27 for some tree species the unwanted shading by bare branches in winter can be significant (Fig. 12.25).

Shading devices also influence the view through glazing: an overhang, an opaque blind, a Venetian blind, and solar control film may all reduce the solar gain of an aperture by the same amount, but they will alter the view through that aperture very differently.

The most efficient shading is provided by external devices (e.g. awnings), as the solar energy is rejected before it enters the collector. However, external shading devices are usually expensive, since they have to be weather-proof. Weather-proofing has control implications, an awning must be withdrawn if the wind is strong even if it is sunny. The control linkage may also be difficult to install and maintain.

Indoor shading devices reflect the solar radiation which has passed through the glazing into the collecting element or zone, back out through the glazing. They are not as efficient as external shades because some of the radiation is (i) reflected and scattered by the glazing back into the collector, and (ii) is absorbed onto the surface of the shading device. An important function of all types of shading device is that they protect the occupants from direct radiation. Direct radiation elevates the effective temperature several degrees above air temperature thus lowering the threshold temperature at which thermal discomfort is reached.

Ventilation must be considered as a second line of defence against overheating since heat can only be removed when the temperature of the building has already risen above ambient. To keep the temperature elevation above ambient small by ventilation alone, very large ventilation rates would be required. This may be inconvenient or in some cases impossible to attain. However, ventilation is an important complement to shading – particularly when it convects away gains made by absorption onto internal shading surfaces. Ideally in such circumstances these gains will be removed well away from the occupants.

Ventilation is important in spaces with large areas of horizontal glazing such as atria or covered courtyards. If substantial openings at the top and inlets at ground level are provided, the ventilation can be induced by the "stack effect" even on days of zero windspeed. This vertical flow will prevent the buildup of hot air in the upper zone of the atrium.

12.4 Auxiliary Heating and Controls in Passive Solar Buildings

For solar energy systems to reduce energy consumption by displacing auxiliary heating, the control system of the latter must be able to respond to the presence of solar energy. The heating system too must respond to the control to reduce its output quickly enough.

Auxiliary heating inputs are also absorbed by thermal mass, the latter thus prolongs the warm-up period. In a building heated intermittently, this leads to a higher than required mean daily temperature which in turn leads to a greater total heat loss. In the cases of very heavyweight intermittently-heated buildings for heating dominated climates, it is doubtful if the benefit to solar utilisation is greater than the loss due to extra auxiliary heating. Thermal mass is often considered as a measure to reduce peak summertime temperatures. Whilst mass does reduce peak temperature, shading and ventilation will nearly always be more effective. The exception to this is where temperatures well below usual daytime ambient temperatures are required, such as in wine cellars and food larders, where mass must be provided.

Buildings with high levels of insulation (i.e. small overall heat loss coefficients) can attain long time constants without necessitating large quantities of thermal mass. It has been found that the best control is obtained when the response time of the control sensor, the heat emitter and the room are in ascending magnitude. Thus the thermostat must be positioned to sense the effect of the solar gains as rapidly as possible. In a building with south and north facing rooms, separate thermostats will be necessary since the south rooms will respond to the solar gains much more rapidly than the north side. The south zone thermostat should sense the direct radiation – i.e. be positioned so that the direct sunlight falls upon it, it thus senses the cause rather than the effect of the solar gain, and thus responds more quickly.

Auxiliary heating to a room should be able to respond faster than, for example, the rate of temperature declinic due to reduced solar gain to the room. With modern water based radiator systems this is possible, and with air heating systems it is easy to attain. With underfloor heating and solid fuel stoves this is more difficult to attain. Where such a slow response heating system is used it makes early sensing of solar gains by the heating control system all the more important. Heating systems which contain their own storage, such as underfloor systems, do not necessarily respond to an "off" instruction at the same speed that they will respond to an "on" instruction. Slowness in response to the "off" instruction will tend to lead to wasteful overheating, slowness to the "on" instruction will lead to underheating.

Considerations for thermal storage, usually results in many climates in passive solar buildings being moderately heavyweight – although as has already been discussed, furnishings tend usually to leave most rooms behaving as medium-weight spaces, at least when subjected to short-term temperature fluctuations. Where rooms of a passive solar building are really lightweight, then direct gains should be avoided, remote gains should be controlled with a fast response control system, and a fast response heating system should be provided.

12.5 Passive Features for Hot Climates

Shading is almost invariably used to reduce solar gain in non-air-conditioned buildings in hot climates. Natural cooling systems dissipate heat from the building to environmental heat sinks which are at lower temperatures than that of the building interior (Givoni 1994). These heat sinks may include the ambient dry bulb by ventilation (Gage et al. 2001), the ambient wet bulb by evaporative cooling (Ford et al. 2012), the night sky by radiative cooling (Dimoudi and Androutsopoulos 2006) and the ground by earth contact cooling (Al-Ajmi et al. 2006). These are summarised in Table 12.1. Passive cooling generally requires only a modest input of energy to produce significant improvements in indoor comfort in both arid and humid hot climates. Cooling fans and other sources of air motion can significantly improve comfort. The maximum temperature for comfort can be increased by several degrees. Changing relative humidity has only a modest effect; increasing the relative humidity from 20 % to 60 % is equivalent to a temperature rise of only 1 °K. Though high humidity does not reduce human comfort significantly, dehumidification below 68 % humidity may be required for the control of mould and mildew. The two methods of dehumidifying buildings are to condense the water vapour on a cooled surface or to absorb it in a desiccant. The low temperatures required for condensation cannot normally solely be achieved by passive cooling systems.

In climates where the ambient dry bulb rises significantly above the maximum temperature for comfort (i.e., climates which have a significant cooling load), the following techniques may also be employed:

- Double glazing
- Infiltration reduction, with ventilation via air to air heat exchangers
- Wall insulation
- Earth berming

Table 12.1 Classification of passive cooling systems

System	Heat sink
Ventilation	Ambient dry bulb
Earth sheltering	Ground
Direct and indirect evaporative cooling	Ambient wet bulb
Dry roof dissipation (radiative cooling)	Night sky
Wet roof dissipation (radiant and evaporative cooling)	Night sky, ambient web bulb
Wind capture	Diverted wind flow cooled by masonry heat sink

In relatively-moderate climates, the nocturnal ventilation of high mass buildings is a simple and cost-effective approach. If the night-time ambient dry bulb temperature is low enough it is possible to cool the thermal mass of a building sufficiently for it to absorb the daytime cooling load without raising the interior temperature to uncomfortable level. The night-time ventilation may be natural or forced depending on the wind speed and design of the building. In some climates, during the day it may be advantageous to restrict ventilation (to the minimum necessary for health) in order to minimise the heat gain to the building. Window placement, landscaping, and the microclimate are important design considerations if natural ventilation is to be used. The uncertainties involved in predicting the airflow patterns through buildings make quantitative prediction of system performance difficult.

Earth sheltering, i.e., burying a building in the ground either partially or completely, can reduce the cooling load in two ways:

- The earth acts as extra insulation, especially since it tends to be dry during the summer, and also tends to reduce air infiltration;
- If the temperature below ground level is lower significantly than the required internal temperature, then heat will flow out of the building into the surrounding earth.

If the cooling load is small enough and the earth is cool enough, then no further cooling will be required. However, in more extreme climates the loads are too great and the earth is too warm for significant cooling to be provided throughout the summer. The limited heat capacity and thermal conductivity of the earth restrict the amount of heat that can be absorbed from the building before the temperature of the surrounding earth becomes too high for any further heat to be absorbed.

Wind capture uses towers to divert and cool wind flows into a building. The air is cooled during its passage through the thermal mass of the masonry tower. This cooling method is a relatively common vernacular form, particularly in older market buildings in Iran. The wind capture tower on the bazaar in Kerman, Iran is shown in Fig. 12.26.

In a direct evaporative cooling, outside air is blown through a fibrous pad soaked in water. The air is adiabatically cooled and humidified and its dry bulb temperature approaches its wet bulb temperature. The air is then supplied to the building.

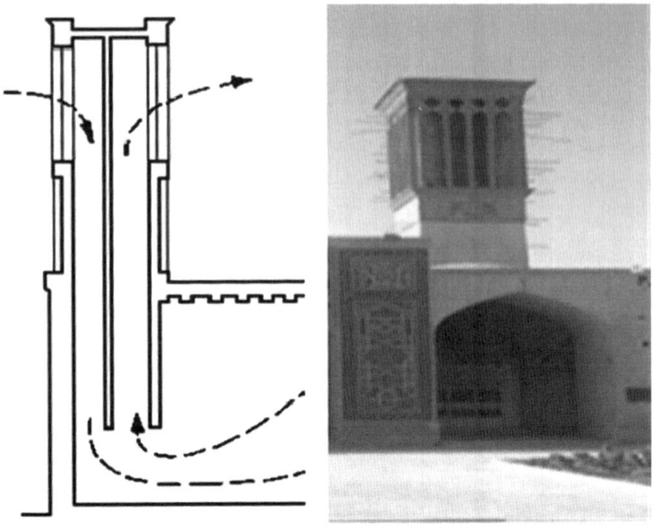

Fig. 12.26 Wind capture tower in Kerman, Iran and its principle of operation

In an indirect evaporative cooling system the evaporatively cooled air cools the air in the building via a heat exchanger rather than being supplied directly. This has the advantage that water vapour is not added to the room air, hence extending the applicability of evaporative cooling to semi-humid climates. The most effective units employ flat plate air to air counterflow heat exchangers in which one side of the heat exchange surface is kept wet, allowing evaporative cooling to take place inside the heat exchanger itself rather than in a separate unit.

In more humid climates, the wet bulb temperature is high enough to limit the amount of evaporative cooling which can be obtained without employing uneconomically large air flow rates and/or heat exchangers. In such climates roof dissipater cooling systems use the lowest temperature environmental heat sink which is the night sky. Under humid summer conditions the cloudless night sky has a radiant temperature approximately 10 °C below the ambient dry bulb temperature. However, if the sky is covered with low clouds its radiant temperature will be close to ambient. Fairly simple empirical relationships have been developed to allow the sky temperature to be predicted from the ambient dry bulb temperature and humidity and the fraction of the sky covered by clouds. A horizontal surface exposed to the night sky will dissipate 5 Wm^{-2} for every degree of temperature difference between the surface and the if the roof can be coupled to thermal storage, nocturnal cooling can be used to keep the building cool throughout the day. Such systems have been employed for the cooling of remote telecommunications repeater stations (Schmalz 1979) and of relocatable buildings in hot arid climates (Ghiraldi 1984).

The Skytherm system, (Hay and Yellot 1969) consists of approximately 15 cm of water in thermal insulating panels. At night the panels are opened to expose the water to the night sky. The water may take the form of an open pond or may be contained in

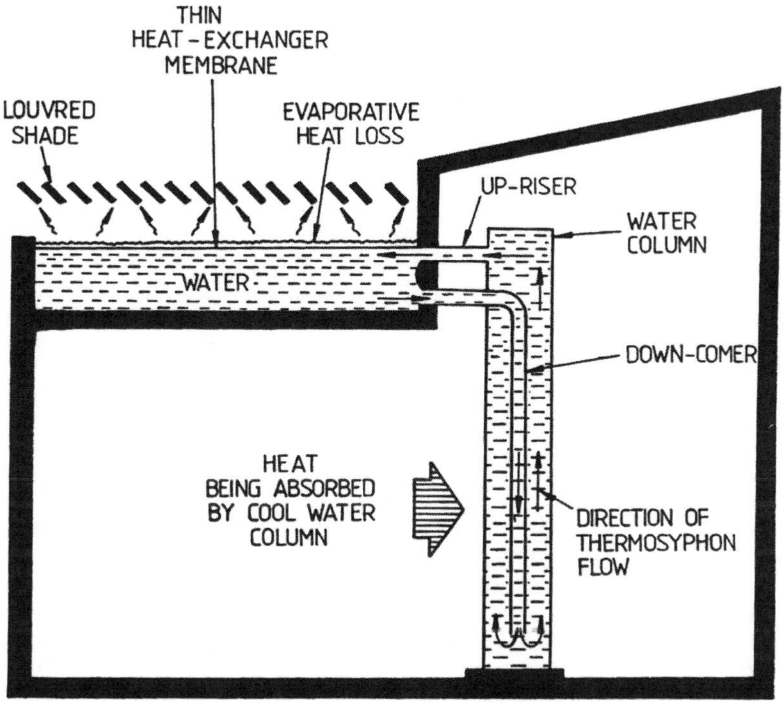

Fig. 12.27 Schematic cross-section showing the principle of operation of natural-circulation hydronic space-cooling system

plastic like a waterbed. The water cools by radiation to the night sky, and if an open pond is employed or the bags are wetted cooling by evaporation also occurs. At dawn, the panels are closed to prevent heat gain by solar radiation or by convection from the outside air. The water cools the ceiling which then cools the interior by a very effective combination of radiation and buoyancy driven natural convection. As with other passive cooling systems which produce a cool surface in the room, both convective coupling and comfort are increased by using fans to circulate the interior air. The major disadvantage of Skytherm is that it employs unconventional construction. Supporting 150 kgm^{-2} of water on the roof/ceiling whilst avoiding leaks requires considerable care in construction. Care is also needed in the construction of the moveable panels if they are to operate reliably without excessive maintenance. The system is also expensive to install although the mass production of components would presumably reduce costs and increase reliability. Skytherm can also be used to provide solar heating by opening the panels during periods of sunshine to allow the sun to heat the water. However, it is unsuited to higher latitudes and areas with appreciable snow fall. It is also unsuited to retrofitting.

A cooling system has been developed in which heat is absorbed from a room by a vertical water column, the heated water rising upwards to a shaded roof pool from which the heat is rejected to the ambient environment largely by evaporation and cooler water from the pool's bottom descends down a separate pipe to re-enter the base of the water column as shown in Fig. 12.27 (Crowther and Melzer 1979; Crowther 1979).

An advantage of this system is that, although the heat loss is achieved principally by evaporation, no extra vapour is admitted into the room. In tests conducted near Sacremento, California, U.S.A., this system provided more than the required cooling even during unusually hot weather (Hammond 1980). It has been improved to include (i) a thin heat-exchanger membrane (to separate the open evaporating pool of water from the thermosyphoning water which must be kept clean, and (ii) an effective shading device that blocks out all direct insolation but does not inhibit the evaporative cooling of the pool. To reduce the risk of disease frequently associated with stagnant pools of water in hot climates, it has been suggested that Gambusia Affinis fish that eat mosquito larvae should be housed in the evaporating pool above the heat exchanger.

References

Achard P, Gicquel R (1986) European handbook, Preliminary edition. Commission of the European Communities, Brussels

Al-Ajmi F, Loveday DL, Hanby VI (2006) The cooling potential of earth-aire heat exchangers for domestic buildings in a desert climate. Build Environ 41:235–244

Arasteh DK, Selkowitz SE, Apte JS, LaFrance M (2006) Zero energy windows. In: Summer study on energy efficiency in buildings, Pacific Grove, ACEEE

Baetens R, Jelle BP, Gustavsen A (2010) Phase change materials for building applications; a state-of-the-art review. Energy Buildings 42:1361–1368

Bainbridge DA (2011) Passive solar architecture; heating, cooling, ventilation, daylighting and more using natural flows. Chelsea Green, White River Junction

Baker NV (1987) Passive and low energy building designs for tropical Island climates. Commonwealth Secretariat, London

Barra OA, Artese G, Costantini T, Costanzo G, Franchesi L (1980) The Barra-Costantini solar passive system: experimental performances. In: Proceedings of the international congress on "building energy management", Portugal, pp 143–151

Bryan H, Moore J, Kwartler M, Jones T (1981) The utilization of daylighting as an urban design strategy. In: Proceedings of the AS/ISES annual meeting

Carmody J, Selkowitz S, Lee E, Arasteh D, Willmert T (2004) Window systems for high performance buildings. Norton, New York

Collins RE, Simko TM (1998) Current status of the sciences and technology of vacuum glazing. Solar Energy 62:180–213

Crowther K (1979) Cooling from an evaporating thermosyphoning roof pond. In: Proceedings of the passive solar conference, Kansas City, pp 409–503

Crowther K, Melzer B (1979) The thermosphoning cool pool – a natural cooling system. In: Proceedings of the third national passive solar conference, vol 3. San Jose, pp 448–451

Curcija D, Goss WP (1994) Two-dimensional finite-element model of heat transfer in complete fenestration systems. ASHRAE transactions: symposia, pp 1207–1220

Davies MG (2007) The passive solar heated school in Wallasey. V. Energy requirements for a passive school building. Int J Energ Res 11:1–20

Dimoudi A, Androutsopoulos A (2006) The cooling performance of a radiator based roof component. Solar Energy 80:1039–1047

Dupagne A, Renson M (1986) Description of a computer tool for representation, manipulation and evaluation of built forms. International seminar on urban built form and energy analysis, Cambridge

Elmahdy AH, Frank T (1993) Heat transfer at the edge of sealed insulating glass unit; comparison of hot box measurements with finite-difference model. ASHRAE transactions: symposia, pp 915–922

Fanchiotti A, Messina, G, Volpe Rinonapoli C (1983) A computer code for determining the effect of shadows on the availability of solar radiation on facades of buildings. In: Proceedings of second international PLEA conference, Crete

Fang Y, Eames PC (2006) The effect of glass coating emittance and frame rebate on the heat transfer through vacuum and electrochromic vacuum glazed windows. Solar Energy Mat Solar Cells 90:2683–2695

Ford B, Wilson R, Gillot M, Ibraheem O, Salmeron J, Sanchez FJ (2012) Passive downdraught evaporative cooling: performance in a prototype house. Build Res Inform 40:290–304

Gage SA, Hunt GR, Linden PF (2001) Top down ventilation and cooling. J Archit Plan Res 18:286–301

Ghiraldi A (1984) Brochure. Divisione Sistemi Oifesafe Progettazioni, Milan

Givoni B (1994) Passive low energy cooling of buildings. Wiley, New York

Givoni B (1998) Climate considerations in building and urban design. Wiley, New York

Griffiths PW, Leo D, Cartwright P, Eames PC, Yianoulis P, Leftheriotis G, Norton B (1998) Fabrication of evacuated glazing at low temperatures. Solar Energy 63:243–249

Hammond J (1980) Cool pool development. Quarterly technical report no 3, US Department of Energy, Washington DC

Hay HR, Yellot JI (1969) Natural air conditioning with a roof panel and movable insulation. ASHRAE Transl 75:178

Hootman T (2012) Net zero energy design: a guide for commercial architecture. Wiley, New York

Hyde TJ, Griffriths PW, Eames PC, Norton B (2000) Development of a novel low temperature edge seal for evacuated glazing. In: Proceedings of the world renewable energy congress VI, Brighton, pp 271–274

Kachadorian J (1997) The passive solar house; using solar design to heat and cool your home. Chelsea Green, White River Junction

Konya A (1980) A design primer for hot climates. Architectural Press, London

Kumakara E, Nakaokubo K, Hoyano A (2011) Numerical analysis of solar shading of a deciduous tree for seasonal change. In: Proceedings of PLEA 2011: 27th conference on passive and low energy architecture, Louvain-la-Neuve

Lo SNG, Deal CR, Norton BA (1994) School building reclad with thermosyphoning air panels. Solar Energy 5:49–58

Lo SNG, Norton B (1996) The effect of occupancy patterns on the long-term performance of roof space solar energy collectors and domestic dwellings in a northern European climate. Solar Energy 56:143–150

Mazria E (1980) Solar home book. Rodale Press, Emmaus

Morris WS (1981) A thermosyphon floor heating system: design considerations and performance data. In: Proceedings of the 6th US national passive solar conference, Portland

Muneer T, Abodahab N, Weir G, Kubie J (2000) Windows in building: thermal, acoustic, visual and solar performance. Architectural Press, Oxford

Norton B, Hobday RA (1990) Passive solar schools in the UK; features employed currently and their operation. Int J Ambient Energy 11:59–76

Norton B, Waterfield AP (1990) Green Park School-internal roof-based solar air heating, Sun at work in Europe, vol 5, pp 12–14

Norton B, Fleming PO, Ekechukwu OV (1987) Passive autarkic solar drying techniques physics and technology of solar energy. In: Proceedings of the international workshop on physics of solar energy, New Delhi, Nov–Dec 1986, vol 1. D. Reidel, pp 447–466

Norton B, Hobday RA, Lo SNG (1992) Thermosyphon air panels. Adv Solar Energy 7:495–571

O'Cathain CS (1981) A model for passive solar housing density. Int J Ambient Energy 2

O'Cathain CS (1982) Explorations with a model of passive solar housing. Energy Build 4:181–183

Oreszczyn T (1993) The energy duality of conservatory use. In: Proceedings of the 3rd European conference on architecture; solar energy in architecture and planning, Florence

Poirazis H (2004) Double slim facades for office buildings: literature review. Report EBD-R-04/3, Lund University

Robinson SJ, Collins RE (1989) Evacuated windows-theory and practice. ISES solar world congress, International Solar Energy Society, Kobe

Rowe CM, Willmott CJ (1984) Solar irradiance on flat-plate solar collectors in urban environments. Solar Energy 33:343–351

Sawhney RL, Bansal MK, Sodha MS (2007) Solar passive heating through a thermosyphon air panel. Int J Energy Res 12:217–225

Schmalz F (1979) Standard long-range operating stations in shelters with integrated thermal-syphon system. In: Proceedings of intelec '79, international telecommunications energy conference, Washington, DC

Williams A (1989) Perronet Thompson School, Building, pp 45–52

Wright MN (1936) Solar radiation as related to summer cooling and winter radiation in residences. J.B. Pierce Foundation, New York

Wright JL, Sullivan HF (1995) A two-dimensional numerical model for glazing system thermal analysis. ASHRAE Transactions: Symposia, pp 819–831

Zrikem Z, Bilgen E (1987) Theoretical study of Trombe-Michel wall solar collector system. Solar Energy 39:409–419

References

Abbot CG (1928) Quoted in Rene Bache, Harnessing the Sun. Popular Mechanics, 602

Abdel-Salam HEA, Probert SD, Norton B (1986) Predicted performance of coffered solar ponds in the U.K and Egypt. PhysioChem Hydrodynam 7(4):217–233

Anon (1965) How to make a solar cabinet dryer for agricultural produce, DIY leaflet. Brace Research Institute Canada, No L6

Anon (1977b) Methods of testing to determine the thermal performances of solar collectors. AS RAE Standard 93–77. American Society of Heating, Refrigeration and Air-conditioning Engineers, New York

Anon (1987) Proceedings of the 7th meeting of the solar collector and systems testing group, Catania

Arasteh D, Carmody J, Hescheng L, Selkowitz S Residential windows: a guide to new technologies and energy performance. W.W. Norton, New York

Arinze EA, Schoenau G, Bigsby FW (1979) Solar energy absorption properties of some agricultural products. ASAE paper, No 79–3071

Baetens R, Jelle BP, Gustavsen A (2010a) Properties, requirements and possibilities of smart windows for dynamic daylight and solar control in buildings: a state-of-the-art review. Solar Energy Mat Solar Cells 94:87–105

Baetens R, Jelle BP, Gustavsen A (2010b) Phase change materials for building applications; a state-of-the-art review. Energy Build 42:1361–1368

Bahm RJ, Nakos JC (1979) The calibration of solar radiation measuring instruments. Final report of contract EM-78-S-04-5336. United States Department of Energy

Bainbridge DA (1981a) Integral passive solar water heater performance. In: Proceedings of the 6th national passive solar conference, Portland, pp 163–167

Bainbridge DA (1981b) The integral passive solar water heater book. The Passive Solar Institute, Davis

Bala BK, Janjai S (2009) Solar drying of fruits, vegetables, spices, medicinal plants and fish: developments and potentials. In: International solar food processing conference, Dhaka, Bangladesh, pp 1–24

Bassey MW (1985) Design and performance of a hybrid crop dryer using solar energy and sawdust. In: Proceedings of the ISES Congress, INTERSOL 85, Montreal, pp 1039–1042

Bechtoft-Neilson P, Worsoe-Schmidt P (1977) Development of a solar-powered solid-absorption refrigeration system, part 1. Report No F30-77. Technical University of Denmark, Denmark

Beckett S (1958) Murphy. Routledge, London

Beckman WA, Klein SA, Duffie JA (1977) Solar heating design by the f-chart method. Wiley Interscience, New York

Best R (1979) Cassava drying, cassava information centre, CIAT Cali Columbia (Brekelbaum T, ed), Series OSEC-4

B. Norton, *Harnessing Solar Heat*, Lecture Notes in Energy 18,
DOI 10.1007/978-94-007-7275-5, © Springer Science+Business Media Dordrecht 2014

Bhadori MN (1978) Passive cooling systems in Iranian architecture. Sci Am 238:144–154
Bliss R (1959) The derivation of several plate efficiency factors, useful in the design of flat-plate solar-heat collectors. Solar Energy 3:55–62
Bolin HR, Stafford AE, Huxsoll CC (1978) Heated fruit dehydrator. Solar Energy 20:281–291
Bolin HR, Stafford AE, Huxsoll CC (1979) Solar through dryer. Solar Energy 22:455–457
Brandemuehl MJ, Beckman WA (1980) Technical note: transmission of diffuse radiation through CPC and flat -plate collector glazings. Solar Energy 24:511–513
Brenndorfer B, Kennedy L, Bateman COO, Mrema GC, Wereko-Brobby C (1985) Solar dryers – their role in post-harvest processing. Commonwealth Secretariat, London
Brace Research Institute (1980) Types of solar agricultural dryers. Sunworld 4(6):181
Casamajor AB (1979) The application of shallow solar ponds for industrial process heat: case histories 11, Sun 2. In: Proceedings of the ISES Congress, Atlanta
Chain C, Dumortier D, Fontoynont M (1999) A comprehensive model of luminance, correlated colour temperature and spectral distribution of skylight: comparison of experimental data. Solar Energy 65:259–295
Cheema LS (1978) Solar drying of cassava for alcohol production. ISES Congress, New Delhi, pp 1940–1944
Cheema LS, Ribeiro CMC (1978) Solar dryers of cashew, banana and pineapple. ISES conference, New Delhi, pp 2075–2079
Chiritescu C, Cahil DG, Ngayeu M, Johnson D, Bodapati A, Kebhiski P, Zschack P (2007) Ultralow thermal conductivity in disordered WSe_2 crystals layered. Science 315:351–353
Chow TT (2010) A review on photovoltaic/thermal hybrid solar technology. Appl Energ 87:365–379
Close DJ (1962a) The performance of solar water heaters with natural circulation. Solar Energy 6:33–40
Close DJ (1962b) CSIRO Engineering section report ED 7, Melbourne
Collares-Pereira M, Rabl A (1979a) The average distribution of solar radiation – correlations between diffuse and hemispherical and between daily and hourly insolation values. Solar Energy 22:155–164
Collares-Pereira M, Rabl A (1979b) Derivation of method for predicting long term average energy delivery of solar collectors. Solar Energy 23:223–233
Collares-Pereira M, Rabl A (1979c) Simple procedure for predicting long term average performance of non-concentrating and of concentrating solar collectors. Solar Energy 23:235–253
Diver RB, Fletcher EA (1985) Hydrogen and sulfur from H2S—III. The economics of a quench process. Energy 10:831–842
Doe PE, Ahmed M, Muslemuddin M, Sachithananthan K (1977) A polythene tent drier for improved Sun drying of fish. Food Tech Aust 29:437–441
Driver PM, Jones RW, Riddiford CL, Simpson RJ A new chrome black selective absorbing surface. Solar Energy 19: 301–306
Drude P (1904) Optische Eigenschaften und Electronen Theorie. An der Physik 14:936
Duffie JA, Beckman WA (1980) Solar engineering of thermal processes. Wiley, New York
Duffie JA, Beckman WA (1991a) Solar engineering of thermal processes, 2nd edn. Wiley Interscience, New York
Duffie JA, Beckman WA (1991b) Solar engineering of thermal processes. Wiley, New York
Eisenstadt MM, Utton AE (1976) Solar rights and their effect on solar heating and cooling. Natl Res J 16:379–388
Ekechukwu OV, Fleming PD, Norton B (1986) The potential of solar energy agricultural dryers in Nigeria. In: Proceedings of the 2nd annual conference of the Biotechnology Society of Nigeria, 16–19 May 1986
Emery M, Rogers BA (1984) On a solar collector thermal performance test method for use in variable conditions. Solar Energy 33:117–123
Ericsson J (1984) The Sun motor and the Sun's temperature. Nature 29:217–218
Excell RHB, Kornsakoo S (1978) A low cost solar rice dryer. Appropriate Tech 5(1):23–24

Exell RHB (1980a) A basic design theory for a simple solar rice dryer. J Renew Energy Res 1:1–14
Exell RHB (1980b) A simple solar rice dryer, basic design theory. Sunworld 4:186–190
Fanchiotti A, Messina G, Volpe Rinonapoli C (1983) A computer code for determining the effect of shadows on the availability of solar radiation on facades of buildings. Proceedings of second international PLEA conference, Crete
Favard GJ, Nawrocki AD (1981) Preliminary optical performance study of glazing-reflector systems in breadbox water heaters. In: Proceedings of 6th national passive solar conference, Portland, pp 188–191
Feustel JE, Hanselmann G (1985) Operational experience with three process heat plants in Europe and Arabia. In: Proceedings of the intersol '85, the 9th biennial congress of ISES, vol 2. Montreal, pp. 977–982
Fleming PD, Ekechukwu OV, Norton B, Probert SD (1986) Design, installation and preliminary testing of a natural circulation solar-energy tropical-crop dryer, solar drying in Africa. In: Proceedings of the solar drying workshop, Dakar. Published by the International Development Research Centre, Ottawa, pp 147–161
Fleming PD, Norton B, Onyegegbu SO, Probert SD, Ekechukwu OV (1986) A large-scale facility for experimental studies of natural circulation solar-energy tropical crop dryers. In: Proceedings of the 1986 international drying symposium. MIT, Cambridge, pp 685–693
Garg HP, Prakash J (2006) Solar energy for industrial process heat. In: Solar energy fundamentals and applications. Tata McGraw-Hill, New Delhi
Gawlik KM (1995) A numerical and experimental investigation of heat transfer issues in the practical utilization of unglazed transpired solar air heater. Ph.D. thesis, University of Colorado, Boulder
Gay N, Butler R (1932) The Sun has got his hat on. Sony lyrics
Gergacz JW (1980) Solar energy law; easements of access to sunlight. New Mexico Law Rev 121 (10):121–168
Gilliaert D, Tebaldi P (1987) Development of standard test procedures for the determination of the long term performance of solar domestic hot water systems, a concerted action of the commission of the European Communities, European solar collector and systems testing group. In: Proceedings of the 7th meeting, Catania
Greiga D, Siddiqui K, Panagiota K (2013) The influence of surface heating on the flow dynamics within a transpired air collector. Int J Heat Mars Transfer 56:39–402
Gueymard C (2012) Temporal variability in direct and global irradiance at various time scales as affected by aerosols. Solar Energy 86:3544–3553
Gupta CL, Garg HP (1968) System design in solar water heaters with natural circulation. Solar Energy 12:163–182
Harkness EL, Mehta ML (1978) Solar radiation control in buildings. Applied Science, London
Harris DJ, Norton B, Smith IE (1988) Design and performance of a novel Peltier effect solar powered refrigerator. In: Proceedings of the UK-ISES conference, solar electricity for development, Reading
Harrison G (1969) Here comes the Sun, Track on "Abbey Road" LP by the Beatles, Apple, London
Hastings SR, Marc D (2000) Solar air systems: a design handbook. Earth Scary, LLC, London
Hastings SR, Rostuik HN (1998) Solar air systems; product catalogue. Earth Scary, LLC, London
Hill JE, Jenkins JP, Jones DE (1977) Testing of solar collectors according to ASHRAE Standard 93–77. In: Proceedings of the annual meeting of the American section of ISES, vol 1. Orlando, pp 6–19
Hill JE, Wood BD, Reed KA (1985) Testing solar collectors. In: Boer KW, Duffie JA (eds) Advances in solar energy, 2nd edn. ASES & Plenum Press, New York
Hollands KGT, Unny TE, Raithby GD, Konicek L (1976) Free convection heat transfer across inclined air layers. ASME J Heat Transfer 98:189–193
Hottel HC, Whillier A (1958) Evaluation of flat plate collectors performance. In: Transactions of the conference, vol 2. Arizona, pp 74–104

Hull JR (1985) Comparison of heat transfer in solar collectors with heat pipe versus flow through absorbers. In: Min TC, Chiou JP (eds) Heat transfer and fluid flow in solar thermal system. American Society of Mechanical Engineers, New York, pp 1–7

Hull JR (1986) Analysis of heat transfer factors for a heat pipe absorber array connected to a common manifold. ASME J Solar Energy Eng 108:11–16

Jackson TH, Mohammed BB (1969) Sun drying of fruits and vegetables. Agricultural Services Bulletin, FAO, Rome, No 5

Jacovides CP, Timvious FS, Papaioannou DN, Theofilou CM (2006) Comparative study of various correlations in estimating hourly diffuse fraction of global solar radiation. Renew Energy 37:2492–250

Janjai S, Srisittipokakun N, Bala BK (2006) Experimental and modeling performances of a roof integrated solar drying system for drying herbs and spices. Energy 33:91–103

Kaehn HD, Geyer M, Fong O, Vignola F, McDaniels DK (1978) Experimental evaluation of the reflector-collector system. In: Proceedings of the American Section of the International Solar Energy Society, vol 2.1. Denver, p 654

Kahlen S, Wallner G, Lang RW (2010a) Aging behavior and lifetime modeling of polycarbonate. Solar Energy 84:755–762

Kahlen S, Wallner G, Lang RW, Meir M, Rekstad J (2010b) Aging behavior of polymeric solar absorber materials: aging on the component level. Solar Energy 84:459–465

Kalogirou S (2002) Parabolic trough collector for industrial process heat in Cyprus. Energy 27:813–830

Kalogirou S (2003) The potential of solar industrial process heat applications. Appl Energy 76:337–361

Kearney D, Price H (2005) Advances in parabolic trough solar power technology. In: Kreith F, Goswami DY (eds) Advances in solar energy, vol 16. ASES, Boulder

Kemp CM (1891) US patent no. 451, 384, 28 Apr 1891

Khanna ML (1973) A portable-type solar water heater. Solar Energy 15:269–272

Klein SA et al (1983) TRNSYS a transient system simulation program. Engineering Experiment Station report 38–12, University of Wisconsin, Madison

Kocher MF, Gerald R, Bodman PE, Lay M (1981) Solar grain drying and farrowing house heating with a multi-use collector. ASAE paper no 4554

Kothdiwala AF, Eames PC, Norton B (1993) The effect of variation of single inclination on the performance of low concentration ratio compound parabolic concentrating solar energy collectors. Solar Energy 55:301–309

Lampert CM (1979) Coatings for enhanced photothermal energy collection. Solar Energy Mater 1: 319–341; 2: 1–17

Landsberg JJ, White B, Thorpe MR (1979) Computer analysis of the efficacy of evaporative cooling for glasshouses in high-energy environments. J Agric Eng Res 24:29–39

Lazzarin RM (2007) Solar cooling plants: how to arrange solar collectors, absorption chillers and the load. Int J Low Carbon Technol 2:376–390

Littlefair PJ (1983) Modelling real sky daylight availability with the BRE average sky. In: Proceedings of the CIE, Amsterdam, paper D302

Lo SNG, Deal CR, Norton B (1994) A school building re-clad with thermosyphon air panels. Solar Energy 52:49–58

Ma CCY, Iqbal M (1983) Statistical comparison of models for estimating solar radiation on inclined surfaces. Solar Energy 31:313–317

Mason JJ, Brendel TA (1982) Maxorb – a new selective surface on nickel. In: Proceedings of the SPIE conference on optical coatings for energy efficiency and solar application, Los Angeles, pp 139–145

Matahar S, Alamrajubi AA (2011) An analysis of unglazed transpired solar collectors. Int J Thermodyn 13:153–163

McCullagh JC (1978) The solar greenhouse book. Rodale Press, Emmaus

McLean KA (1980) Drying and storage of combinable crops. Farm Press, Ipswich
Medved S, Azkar C, Cerne B (2003) A large-panel unglazed roof-integrated liquid solar collector-energy and economic evaluation. Solar Energy 75:455–467
Meturally MM, Abou-Zian HZ, Leathy E (1997) Performance of advanced corrugated-dust solar air collector compound with fare convectional designs. Renew Energy 10:519–537
Mills D (2004) Advances in solar thermal electricity technology. Solar Energy 76:19–31
Morris LG, Trickett EJ, Vanstone FH, Wells DA (1972) Solar radiation in greenhouses, a brief review. Isr J Agric Res 22:85–97
Morrison GL, Budihardjo I, Behnia M (2005) Measurement and simulation of flow rate in a water-in-glass evacuated tube solar water heater. Solar Energy 78:257–267
Mouchot A (1879a) Le Chaleur Solare, Paris
Mouchout A (1879b) Le chaleur solarise et les applications industrielles, 2nd edn, Paris
Mullick SC, Kandpal TC, Kumar S (1996) Testing of box-type solar cooker: second figure of most F2 and is variation with load and number of pots. Solar Energy 57:409–413
Murray J (1999) Aluminium production using high-temperature solar process heat. Solar Energy 66:133–142
Nielsen CE (1975) Salt-gradient solar ponds for solar energy utilization. Environ Conserv 2:289–292
Njomo D, Dagovemet M (2006) Sensitivity analysis of thermal performance of flat plat solar air heaters. Heat Mars Trans 42:1065–1081
Norton B, Edmonds JEJ (1991a) Aqueous propylene glycol concentrators for the freeze protection of thermosyphon solar energy water heaters. Solar Energy 47:373–382
Norton B, Edmonds JEJ (1991b) Aqueous propylene glycol concentrations for the freeze protection of thermosyphon solar energy water heaters. Solar Energy 47(5):375–382
Norton B, Prapas DE (1987) Thermal analysis of compound-parabolic concentrating solar-energy collector, physics and technology of solar energy. In: Proceedings of the international workshop on physics of solar energy, New Delhi, vol 1. D. Reidel, pp 109–135
Norton B, Probert SD, Uppal AH (1986) Thermal solar refrigeration for vaccine storage – a viable alternative to photovoltaic units in remote locations. In: Proceedings of the international workshop on solar energy for rural development, Nsukka
O'Casey S (1928) The silver tassie, purple dust and red roses for me. Abbey Theatre, Dublin
O'Flaherty T, Grant J (1984) Use of double cladding techniques to reduce greenhouse heat loss. Acta Hortic 148:641–649
Ong KS (1979) Solar drying of paddy and timber. In: Proceedings of the solar energy symposium, the utilization of solar heat in industry and agriculture, Nice, pp 339–50
Ozisik MN, Huang BK, Toksoy M (1980) Solar grain drying. Solar Energy 24:397–401
Paddick L, Grinlinton D (2009) Legal framework for solar energy. George Washington University, Washington, DC
Page JK (1981) The estimation of monthly mean values of daily total short wave radiation on vertical and inclined surfaces from sunshine records for latitudes 40N-40S. In: Proceedings of the use conference on new sources of energy, paper 598
Pasumarthi N, Sherif SA (1998a) Experimental and theoretical performance of a demonstration solar chimney model – part I, mathematical model development. Int J Energy Res 22:277–288
Pasumarthi N, Sherif SA (1998b) Experimental and theoretical performance of a demonstration solar chimney model – part II. Experimental and theoretical results and economic analysis. Int J Energy Res 22:443–461
Perez R, Stewart R, Scott JT (1985) A two-parameter description of the sky hemisphere. In: Proceedings of the fifth AMS-ASES conference on atmospheric radiation, Baltimore
Poope WI, Woomer NM (1983) Algae management practices for solar salt ponds practical approaches, progress in solar energy. American Solar Energy Society meeting, Boulder, Colorado, vol 6, pp 439–442
Prapas DE, Norton B, Probert SD (1986) Sensor system for aligning a single-axis tracker with direct solar insolation. Appl Energy 25:1–8

Prapas DE, Norton B, Probert SD (1987) Optics of parabolic trough solar-energy collectors possessing small concentration ratios. Solar Energy 39(6):541–550

Prapas DE, Norton B, Probert SD (1988a) Distributions of the heat transfer fluid for E-W and N-S alignments of CPC solar-energy collectors. Solar Technol 5:137–140

Prapas DE, Norton B, Milonidis E, Probert SD (1988b) Response function for solar-energy collectors. Solar Energy 40:371–383

Prapas DE, Norton B, Probert SD (1988c) Variations of the effective thermal capacitance in a solar-energy collector. Int J Ambient Energy 8:59–72

Pretorius JP, Kröger DG (2006a) Solar chimney power plant performance. J Solar Energy 128 (3):302–311

Pretorius JP, Kröger DG (2006b) Critical evaluation of solar chimney power plant performance. Solar Energy 80:535–544

Rao G (1974) Natural drying of cassava. Ph.D. thesis, Michigan State University, East Lansing

Rawson H (1982) Properties and applications of glass. Elsevier, Oxford

Reichmuth H, Robinson D (1982) An analytical model and associated test procedure for predicting the performance of batch type DHW systems. In: Proceedings of the 7th national passive solar conference, Knoxville, pp 963–968

Rimbaud A (1962) Soleil et Chair, Collected Poems, Olivier Bernard (ed)

Royne A, Dey CJ, Mills DR (2003) Cooling of photovoltaic cells under concentrated; illumination; a critical review. Solar Energy Mater Solar Cells 86:451–481

Sachithananthan K, Trim D, Speirs CR (1983) A solar dome dryer for drying of fish. FAO Fisheries paper, Rome, RAB/81/002/INT/LS

Santbergen R, van Zolingen RJC (2006) Modeling the thermal absorption factor of photovoltaic/thermal combi-panels. Energy Convers Manage 47:3572–3581

Schwind RG, Vliet GC (1964) Observations and interpretations of natural convection and stratification in vessels. In: Proceedings of the 1964 Heat Transfer and Fluid Mechanics Institute, Berkeley

Sherman C, Wood BD, Mason J (1979) Effect of vertical wall conductance on temperature relaxation in thermally stratified liquid thermal storage tanks. In: Proceedings of the ISES conference, Atlanta, pp 591–595

Shove GC, Barton GW, Hall MD, Peterson WH (1981) Field studies of solar grain drying. ASAE paper, no 81–4037

Shuman F (1916) New York Times, 2 July 1916

Simonson JR (1984) Computing methods in solar heating design. Macmillan, London

Singhal OP, Gupta GP (1978) Use of solar energy for parboiling and drying of paddy. ISES Congress, New Delhi, pp 1958–1963

Smyth M, Norton B (2006) Integral collector storage solar water heaters. Renew Sustain Energy Rev 10:503–538

Smyth M, Eames PC, Norton B (2001a) Evaluation of a freeze resistant Integrated Collector/Storage Solar Water Heater (ICSSWH) for Northern Europe. Appl Energy 68:265–274

Smyth M, Eames PC, Norton B (2001b) Annual performance of heat retaining Integrated Collector/Storage Solar Water Heaters in a Northern maritime climate. Solar Energy 70(5):391–401

Snail KA, O'Gallagher JJ, Winston R (1983) A stationary evacuated collector with integrated concentrator. Solar Energy 33:441–449

Sparrow EM, Ramsey JW, Mass EA (1979) Effect of finite width on heat transfer and fluid flow about and inclined rectangular plate. ASME J Heat Transfer 101:2

Steinfeld A (2002) Solar hydrogen production via a 2-step water-splitting thermochemical cycle based on Zn/ZnO redox reactions. Int J Hydro Energy 27:611–619

Tabor H (1959) Solar collector developments in Israel. Solar Energy 3:8–9

Taylor UJ, Weir AD (1985) Simulation of a solar timber drier. Solar Energy 34:249–255

Thanh NC, Muttamara S, Lohani BN (1978) Drying techniques for improvement of tapioca chips in Thailand. Tha Agr Sci 11:45–55

Thompson JF, Studer HE (1981) Sun drying of prunes continuous trays, ASAE paper, no 81–3051

Tonui JK, Tripanagnostopoulos Y (2007a) Improved PV/T solar collectors with heat extraction by forced or natural air circulation. Renew Energy 32:623–637

Tonui JK, Tripanagnostopoulos Y (2007b) Air-cooled PV/T solar collectors with low cost performance improvements. Solar Energy 81:498–511

Tschernitz JL, Simpson WT (1979) Solar heated, forced-air, lumber dryer for tropical latitudes. Solar Energy 22:563–566

Turner JS (1966) Jets and plumes with negative or reversing buoyancy. J Fluid Mech 26 (4):779–792

Umarov GG, Ikramov AI (1978) Features of the drying of fruit and grapes in solar radiation drying apparatus. Geliotekhnika 14(6):55–57

Viskanta R, Hale NW (1978) Free convection circulation in a thermal energy storage system driven by heat loss from the walls. In: Proceedings of the 14th southeastern seminar on thermal sciences, North Carolina State University

Vokas G, Christandonis N, Skittides F (2006) Hybrid photovoltaic-thermal systems for domestic heating and cooling-a theoretical approach. Solar Energy 80:607–615

Walker F (1902) US patent no 705, 167, 22 July 1902

Wang LT, Weide L, Zhai Z (1987a) Experimental study on the flow pattern inside the horizontally disposed all glass evacuated solar collector with visual technique. Acta Energiae Solaris Sinica 8:177–184

Wang XA, Xu YF, Meng XY (1987b) A filter method for transient testing of collector performance. Solar Energy 38(12):5–134

Wang LT, Weide L, Zhai Z (1989) Experimental study on the natural convective heat transfer in horizontally disposed all glass evacuated solar collector tube. Acta Energiae Solaris Sinica 10:176–185

Welford WT, Winston R (1978) The optics of non-imaging concentrators. Academic, London

Whillier A (1953a) Solar energy collection and its utilization for house heating. ScD thesis, MIT

Whillier A (1953b) The utilisation of solar energy in South Africa. J South Afr Inst Mech Eng 2:261–267

WHO (2008) Health through safe drinking water. Water sanitation and Health. World Health Organisation, Geneva

Window B Heat extraction from single ended glass absorber tubes. Solar Energy 31: 159–166

Window B, Harding GL (1983) Buoyancy effects and the maifolding of single ended absorber tubes. Solar Energy 31:153–157

Wittwer V, Platzer W, Pfluger A, Stahl W, Goetzberger A (1986) Translucent insulation materials. In: InterSol 85, Proceedings of the 9th biennial congress of ISES, vol 2. Montreal, p 1333

WMO (1971) Guide to meteorological instrumentation and observing practices, 4th edn. Secretariat of the World Meteorological Organisation, Geneva

Wood BD, Fiore PJ, Christopherson CR (1979) Application of ASHRAE Standard 93–77 for testing concentrating collectors for the purpose of predicting all day performance. In: Proceedings of the ISES conference, Melbourne, pp 487–491

Wu ST (1978) The experimental and numerical studies of liquid storage tank thermal stratification for a solar energy system. Semi-annual progress reports EG-77-S02-4479, University of Alabama

Yohanis YG, Popel O, Frid S, Norton B (2006) Geographics variation of solar water. heater performance in Europe. Proceedings of this institution of mechanical engineers, part A. J Power Energy 200: 395–407

Zepei R, Dengfu Z (1990) An investigation of three-dimensional natural convection in a large aspect ratio open cavities. J Eng Thermophy 11:187–193

Zerlaut GA (1982) Solar radiation measurements calibration and standardization efforts. Adv Solar Energy 1:1–107

Zhiqiang Y, Harding GL (1985) Comparative study of fluid-in-metal manifolds for heat extraction from single ended evacuated glass tubular collectors. Solar Energy 35:81–91

Zhiquang Y, Harding GL (1984) Water-in-glass manifolds for heat extraction from evacuated solar collector tubes. Solar Energy 32:223–230

Zondag HA, de Vries DW, van Helden WGJ, van Zulingen RJC, van Steteenhoven AA (2002) The thermal and electrical yield of a PV-thermal collector. Solar Energy 72:113–128

Zondag HA, van Helden WGJ, Bakker M, Elswijk MJ (2005) A roadmap for the development and market introduction of PVT technology. 15th symposium thermische solarmergie, Bad Stuffelstein

Index

B. Norton, *Harnessing Solar Heat*, Lecture Notes in Energy 18,
DOI 10.1007/978-94-007-7275-5,

Printed by Printforce, the Netherlands